T0135649

The Ambiguity of Morphisms in Free Monoids and its Impact on Algorithmic Properties of Pattern Languages

Vom Fachbereich Informatik
der Technischen Universität Kaiserslautern
zur Verleihung des akademischen Grades
Doktor der Naturwissenschaften (Dr. rer. nat.)
genehmigte

Dissertation

von

Daniel Reidenbach

Datum der wissenschaftlichen Aussprache: 15. Dezember 2006

Dekan des Fachbereichs: Prof. Dr. R. Gotzhein
Promotionskommission: Prof. Dr. K. Berns (Vorsitzender)
Prof. Dr. R. Wiehagen (Berichterstatter)
Prof. Dr. J. Dassow (Berichterstatter)

D 386

Bibliografische Information der Deutschen Nationalbibliothek

Die Deutsche Nationalbibliothek verzeichnet diese Publikation in der Deutschen Nationalbibliografie; detaillierte bibliografische Daten sind im Internet über http://dnb.d-nb.de abrufbar.

ISBN 978-3-8325-1449-5

Logos Verlag Berlin
Comeniushof, Gubener Str. 47,
10243 Berlin
Tel.: +49 030 42 85 10 90
Fax: +49 030 42 85 10 92
INTERNET: http://www.logos-verlag.de

Contents

Abstract

Motivated by algorithmic problems for pattern languages, the present thesis studies fundamental combinatorial questions related to the ambiguity of morphisms in free monoids. We first deal with those morphisms which map a string over an infinite alphabet Δ of *variables* onto a string over a finite alphabet Σ of *terminal symbols.* To this end, we call a string over Δ a *(terminal-free) pattern,* a string over Σ a *word,* and we say that a morphism $\sigma : \Delta^* \longrightarrow \Sigma^*$ is *unambiguous (with respect to a pattern $\alpha \in \Delta^+$)* iff there is no morphism $\tau : \Delta^* \longrightarrow \Sigma^*$ satisfying $\tau(\alpha) = \sigma(\alpha)$ and $\tau(i) \neq \sigma(i)$ for some variable $i \in \Delta$. Furthermore, we consider morphisms σ, the ambiguity of which is restricted in a particular manner (again with respect to a pattern α), i. e. every morphism τ with $\tau(\alpha) = \sigma(\alpha)$ has to conform to certain requirements, and we designate such a σ as *moderately ambiguous (with respect to α).* Finally, we introduce the term of a *morphically primitive* pattern, which covers all those $\alpha \in \Delta^+$ for which there is no $\beta \in \Delta^+$ and morphisms ϕ, ψ such that $\phi(\alpha) = \beta$, $\psi(\beta) = \alpha$ and $|\beta| < |\alpha|$; we call a pattern *morphically imprimitive* iff it is not morphically primitive. Within the scope of this setting, our research questions mainly ask for those patterns, with respect to which there *exist* unambiguous or moderately ambiguous nonerasing morphisms $\sigma : \Delta^* \longrightarrow \Sigma^*$ (with $|\Sigma| \geq 2$).

Our first crucial result states that, for every nonerasing morphism σ and for every morphically imprimitive pattern α, σ is neither moderately ambiguous nor unambiguous with respect to α. Hence, we have to restrict our search for appropriate morphisms to morphically primitive patterns. In this regard, we can demonstrate that there exist particular nonerasing morphisms $\sigma_{\text{3-seg}}$ which are moderately ambiguous with respect to *every* morphically primitive pattern. Additionally, we can give for every morphically primitive pattern α a nonerasing unambiguous morphism $\sigma_{\text{un},\alpha}$, but this morphism *must* be tailor-made, i. e. there is no nonerasing morphism which is unambiguous with respect to every morphically primitive pattern.

In accordance with the definition of pattern languages, the second main part of the thesis examines whether these comprehensive insights also hold for general patterns in $(\Delta \cup \Sigma)^+$ and for *terminal-preserving morphisms* $\sigma : (\Delta \cup \Sigma)^* \longrightarrow \Sigma^*$, i. e. those morphisms σ satisfying $\sigma(\mathtt{A}) = \mathtt{A}$ for every $\mathtt{A} \in \Sigma$. In a first step, we restrict our considerations to so-called *quasi-terminal-free* patterns, which are characterised by the fact that they do not contain at least two terminal symbols occurring in Σ. Our corresponding studies demonstrate that the results on terminal-free patterns can be largely adapted to the quasi-terminal-free patterns. In contrast to this, if we omit this restriction and consider arbitrary patterns in $(\Delta \cup \Sigma)^+$ then our results on terminal-free patterns turn out not to be extendable – hence, there exist morphically primitive patterns α (where the definition of morphic primitivity is canonically extended to terminal-preserving morphisms ϕ, ψ), with respect to which there is no moderately ambiguous or unambiguous nonerasing morphism $\sigma : (\Delta \cup \Sigma)^* \longrightarrow \Sigma^*$. This result, however, is restricted to $|\Sigma| \leq 4$, and we have to leave the case $|\Sigma| \geq 5$ open.

Due to their elementary nature, our results on the existence of moderately ambiguous and unambiguous morphisms show numerous connections to problems in other fields of discrete mathematics that are based on finite strings and morphisms. In particular, this holds for the research on equality sets and the studies on pattern languages. With regard to the latter topic, we demonstrate that our combinatorial results yield several answers to prominent open algorithmic problems for E-pattern languages. More precisely, we can conclude from our reasoning on the ambiguity of morphisms that, first, the class of terminal-free E-pattern languages over an alphabet Σ is inferrable from positive data iff $|\Sigma| \neq 2$, second, the full class of E-pattern languages (as well as that of quasi-terminal-free E-pattern languages) is not inferrable from positive data provided that $|\Sigma| \in \{3, 4\}$ and, third, Ohlebusch and Ukkonen's Conjecture on the equivalence problem for E-pattern languages is incorrect.

Acknowledgements

First of all, I am indebted to my supervisor Prof. Dr. Rolf Wiehagen. He gave significant advice whenever needed, granted freedom whenever asked for and showed elating enthusiasm about newly gained insights and newly discovered questions. His open-minded attitude towards research has always been inspiring, and so has been his admirable sense of the nature and beauty of science.

Furthermore, I wish to thank Prof. Dr. Jürgen Dassow for reviewing and backing this thesis and Prof. Dr. Karsten Berns for chairing the doctoral committee.

My cordial thanks are due to Dr. Sandra Zilles and Prof. Dr. Steffen Lange for their highly appreciated support, encouragement and numerous helpful and competent discussions. I also wish to express my gratitude to Dominik Freydenberger and Johannes Schneider for providing many valuable, intelligent and continuative ideas about the topics discussed in the present thesis. Moreover, I am grateful to my former colleagues Martin Memmel and Thorsten Michels and to the secretary Maria Pfeiffer in Prof. Wiehagen's group for providing the friendly and cooperative atmosphere I appreciate so much. Furthermore, I have benefited a lot from the esteemed advice and assistance rendered by Prof. Dr. Thomas Zeugmann and from several dialogues with Dr. Jochen Nessel. I also wish to thank the anonymous referees of my recent papers for their valuable suggestions that have indirectly improved the quality of the present thesis.

I am indebted to my family for providing continuous support in a calm, reliable and caring manner.

Finally, my most sincere thanks are due to my beloved wife Oola. She has tolerated my frequent and intense periods of absent-mindedness when thinking about intricate questions, she has always been curious about my work and she has shared my exaltation when gaining new insights. Her love and patience are the true reasons why this thesis has been finished, so that, in fact, it is hers.

Chapter 1

"*All* morphisms? That's a lot!" [1]

The perception of *strings of symbols* (or *strings*[†] for short) as a worthwhile subject of formal studies is a rather recent achievement of mathematics. It dates back to 1906, when Thue [102] published his initial work "Über unendliche Zeichenreihen" on regularities in infinite strings of symbols, which he later supplemented by some additional papers (such as [103] and [104]) on several other questions related to strings. At that time, however, his work did not lead to a wide research on the properties of strings and of *formal languages*, i. e. sets of strings. Instead of this, every now and then, a few papers were published that dealt with this subject, some of them even independently rediscovering the results of Thue (such as some work by Morse [61] and Morse and Hedlund [62]).

By the beginning of the computer era, this time of little interest and sparse progress ended abruptly because crucial aspects of computation are related to strings: computer programs, their formal specifications (and, more generally, logical formulas) and various kinds of application data are naturally represented as strings of symbols, and, in fact, computation is nothing but a manipulation of symbols. Therefore strings and formal languages rapidly turned into a focus of interest of mathematicians, and the corresponding pioneering work published around 1950 – e. g. by Post [72, 73], Markov [51], Shannon [96], Chomsky [13], Kleene [37] and Schützenberger [95] – in the numerous branches of what today is referred to as *formal language theory* became a well-established foundation of *theoretical computer science*. Consequently, (sets of) strings of symbols nowadays belong to the most popular and best explored topics in the influential field of noncommutative discrete mathematics.

From an algebraic point of view it can be immediately observed that, when referring to the *concatenation* as a binary operation on strings, the set $\mathcal{A}^*$ of all *finite* strings (including the *empty string* ε as the identity element) over some fixed set $\mathcal{A}$ of symbols actually is nothing but a *monoid* (which usually is designated as the *free* monoid *generated* by $\mathcal{A}$, because each string in $\mathcal{A}^*$ can be decomposed in a unique manner into the concatenation of symbols in $\mathcal{A}$). Therefore, the concept of a (homo-)*morphism* – one of the most basic types of functions in algebraic structures, which is characterised by the fact that it is compatible with a binary operation associated to the algebraic structure – can be directly adapted to strings. Consequently, for any sets $\mathcal{A}, \mathcal{B}$ of symbols,

[†]In contemporary literature, a "string" usually is called a "word", and this particularly holds for publications on formal languages and combinatorics on words. We, however, assign the latter term to strings over a special type of alphabet to be introduced below. Analogously, we use the common terms "letter" and "alphabet" for specific symbols and sets of symbols only.

a morphism $f : \mathcal{A}^* \longrightarrow \mathcal{B}^*$ is a function which, first, maps strings in $\mathcal{A}^*$ onto strings in $\mathcal{B}^*$ and, second, is compatible with the concatenation, i. e., for every pair of strings $s_1, s_2 \in \mathcal{A}^*$, it satisfies $f(s_1 s_2) = f(s_1) f(s_2)$. In accordance with the extensive literature on free monoids featuring an algebraic view of the subject, the morphisms are the most intensively discussed mappings on strings of symbols. In addition to this, however, there is also a remarkable combinatorial property of morphisms, which results from the special nature (and, in particular, the noncommutativity) of the concatenation: by definition, f necessarily maps a string s over $\mathcal{A}$ onto a string s' over $\mathcal{B}$ by mapping each *symbol* occurring in s onto a string over $\mathcal{B}$ and concatenating these images in correspondence with the order of the symbols in s. In other words, a morphism – that actually is a function which maps strings onto strings – can be equivalently defined by a function which merely maps symbols onto strings. While this feature seems to strongly simplify the definition of a morphism, it leads to momentous immanent relations between a string and its morphic image. These relations can be easily illustrated by the fact that, for all sets $\mathcal{A}, \mathcal{B}$ of symbols, there exist strings $s \in \mathcal{A}^*$ and $s' \in \mathcal{B}^*$ – such as $s := abba$ and $s' := aabb$ over the sets $\mathcal{A} := \mathcal{B} := \{a, b\}$ – such that *no* morphism $f : \mathcal{A}^* \longrightarrow \mathcal{B}^*$ can map s onto s', whereas, in "standard" Abelian monoids such as the set $\mathbb{N}_0$ of natural numbers (including 0) equipped with the addition $+$, we can always give a trivial homomorphism mapping any number $x \in \mathbb{N}_0$ onto any number $y \in \mathbb{N}_0$. Conversely, whenever there is a morphism mapping a string s onto a string s' then the structure of these strings has some common traits – or, more precisely and in other words, the structure of s' must follow the "pattern" determined by the structure of s. Hence, for every string s over any set $\mathcal{A}$ of symbols and for every free monoid $\mathcal{B}^*$, s can induce a partition of $\mathcal{B}^*$ by the simple question of whether, for any string $s' \in \mathcal{B}^*$, there *exists* a morphism f with $f(s) = s'$.

Thus, in addition to the common algebraic aspects related to (homo-)morphisms – which are relevant for, e. g., the foundations of coding theory as described, e. g. by Berstel and Perrin [10] – there exist rich and nontrivial combinatorial properties of the subject that have been intensively discussed in literature. A major part of the corresponding theory is concerned with a setting that Harju and Karhumäki [27] designate as *collaborating* morphisms, i. e. with the comparison of the images of one and the same string under several morphisms. The presumably most basic formal concept associated to this view is that of the *equality set* of fixed morphisms f and g, which consists of all strings s satisfying $f(s) = g(s)$. In spite of the simplicity of their definition, the expressive power of equality sets is known to cover the very foundations of computer science: for instance, the famous undecidable Post Correspondence Problem (named after Post [72]), in fact, is nothing but the emptiness problem for equality sets, and, furthermore, equality sets have been used for characterising the recursively enumerable sets (cf. Culik II [14]) and prominent complexity classes (cf. Mateescu et al. [56]). Consequently, unlike the standard algebraic setting in coding theory, where normally a *single* (injective) morphism is applied to all strings in a given monoid, the special properties of morphisms on strings imply a nontrivial theory which is based on *two* morphisms that are applied to one and the same string.

Additionally, there also exists a formal concept which is dedicated to the consideration of *all* morphic images of a single fixed string – the so-called pattern language. A *pattern* is a finite string over the union of two disjoint sets of symbols: an arbitrary *terminal* *alphabet* Σ and an infinite set X of *variables*. Henceforth, we use symbols in typewriter font (such as $\mathtt{a}, \mathtt{b}, \mathtt{c}$ and so on) as symbols (or: *letters*) in Σ, and we identify X with the set of natural numbers $\mathbb{N}$. For any pattern $\alpha \in (\mathbb{N} \cup \Sigma)^+$, the *pattern language* $L(\alpha)$ of α is the set of all *words* $w \in \Sigma^*$ such that there exists a morphism $\sigma : (\mathbb{N} \cup \Sigma)^* \longrightarrow \Sigma^*$ satisfying $\sigma(\alpha) = w$.† These morphisms σ, however, need to be *terminal-preserving*, i. e., for every letter $\mathtt{A} \in \Sigma$, necessarily $\sigma(\mathtt{A}) = \mathtt{A}$. Therefore, whenever $\alpha \in (\mathbb{N} \cup \Sigma)^+$ is not a *general* pattern, thus potentially containing variables and letters, but a *terminal-free* one , i. e. $\alpha \in X^+[:= \mathbb{N}^+]$, then the requirement that every morphism $\sigma : (\mathbb{N} \cup \Sigma)^* \longrightarrow \Sigma^*$ under consideration is expected to be terminal-preserving is evidently meaningless, so that σ simply corresponds to the *common morphism* $\sigma : \mathbb{N}^* \longrightarrow \Sigma^*$ showing no restrictions on the images of the symbols occurring in α.

Although it is a rather unorthodox approach to apply all morphisms to a single string (cf. the expertise of a distinguished mathematician [1] as expressed by the title of the present chapter), the theory of pattern languages, introduced by Angluin [3] and Shinohara [97], meanwhile is elaborate and well-established. In particular, pattern languages address the fact that there exist various natural domains, e. g. in computational biology, where crucial data consist of strings that can be interpreted as morphic images of (unknown) patterns, but – contrary to coding theory, where a single well-known and injective morphism is considered – the corresponding morphisms normally are neither known nor well-chosen. Consequently, pattern languages originally have been mainly discussed within the scope of *inductive inference* – that is a classical branch of what nowadays is referred to as *algorithmic* (or: *computational*) *learning theory* – and therefore, roughly speaking, early publications on pattern languages have mainly investigated their "decodability", i. e. the reliable computability of a pattern α from $L(\alpha)$ (or from a subset thereof).

Intuitively, it seems obvious that the problem of whether such a task can be successfully accomplished essentially depends on the existence of morphic images w which reflect the structure of their preimage α sufficiently precisely. However, the formal definition of a "structure-preserving" morphism σ in a combinatorial context is by no means obvious, because it cannot rely on the power of injectivity as easily as in an algebraic setting (such as provided by codes), where the presence of a single injective (and known) morphism unequivocally establishes an inverse morphism σ^{-1} mapping α onto w again. Contrary to this, if pattern languages are considered then, for every word w, there exist infinitely many patterns α and morphisms σ satisfying $\sigma(\alpha) = w$, and there can even exist several patterns among these α, with respect to which there exists an inverse morphism mapping w onto α.

Hence, inductive inference of pattern languages in general cannot reconstruct a unique pattern α from a single word $w \in L(\alpha)$, but it needs some (non-unary) set $W \subset L(\alpha)$; furthermore, there is some immediate evidence that the question of whether, for each word $w \in W$, there exists an injective morphism σ satisfying $w = \sigma(\alpha)$ might be unessential for inference. Instead of injectivity, a recent result presented by Reidenbach [76, 81]

†Basically, in literature, two different kinds of pattern languages are considered: the *E*-pattern language (where "E" is short for *extended* or *erasing*) and the *NE*-pattern language (where "NE" stands for *nonerasing*) of a pattern α. The former is defined as the set of all morphic images of α in the free monoid Σ^*, whereas the latter is restricted to nonerasing morphisms $\sigma : (\mathbb{N} \cup \Sigma)^+ \longrightarrow \Sigma^+$, i. e. those morphisms that do not map any symbol occurring in α onto the empty word. In the present thesis, we largely deal with E-pattern languages.

suggests that another property of a morphic image w can be crucial for inductive inference of pattern languages: the *ambiguity* of w. We call a nonempty word w *unambiguous* with respect to a pattern α provided that there is exactly one morphism σ with $\sigma(\alpha) = w$ (to this end we of course ignore all those morphisms which merely differ from σ on variables not occurring in α); correspondingly, w is said to be *ambiguous* with respect to α if there are at least two morphisms σ, τ such that, for some symbol i in α, $\sigma(i) \neq \tau(i)$, but nevertheless $\sigma(\alpha) = w = \tau(\alpha)$. Moreover, we extend this terminology from words to morphisms: if w is unambiguous with respect to α then we also say that the morphism σ satisfying $\sigma(\alpha) = w$ is unambiguous with respect to α, and if w is ambiguous with respect to α then we designate every morphism σ with $\sigma(\alpha) = w$ as ambiguous, too.

The questionable importance of injectivity and the great impact of ambiguity of words on inductive inference of pattern languages can be easily illustrated by the terminal-free example pattern

$$\alpha := 1 \cdot 2 \cdot 3 \cdot 4 \cdot 1 \cdot 4 \cdot 3 \cdot 2$$

(recall that we always define $X := \mathbb{N}$; furthermore, we separate the symbols in a pattern by a dot "$\cdot$", so as to avoid any confusion) and by the standard injective morphism $\sigma_\mathrm{c} : \mathbb{N}^* \longrightarrow \{\mathtt{a}, \mathtt{b}\}^*$ given by $\sigma_\mathrm{c}(i) := \mathtt{a}\,\mathtt{b}^i$. With little effort, we now can see that σ_c is ambiguous with respect to α since

$$\sigma_\mathrm{c}(\alpha) = \mathtt{ababbabbbabbbbababbbbabbbabb}$$

can, e. g., also be generated by the morphism τ with $\tau(1) := \mathtt{a\,b\,a\,b}^2$, $\tau(2) := \varepsilon$, $\tau(3) := \mathtt{a\,b}^3\,\mathtt{a\,b}^2$ and $\tau(4) := \mathtt{b}^2$:

$\sigma_\mathrm{c}(1)$ $\sigma_\mathrm{c}(2)$ $\sigma_\mathrm{c}(3)$ $\sigma_\mathrm{c}(4)$ $\sigma_\mathrm{c}(1)$ $\sigma_\mathrm{c}(4)$ $\sigma_\mathrm{c}(3)$ $\sigma_\mathrm{c}(2)$

a b a b b a b b b a b b b b a b a b b b b a b b b a b b

$\tau(1)$ $\tau(3)$ $\tau(4)$ $\tau(1)$ $\tau(4)$ $\tau(3)$

Thus, the variable 2 in α is not required for generating $\sigma_\mathrm{c}(\alpha)$, which, conversely, means that $\sigma_\mathrm{c}(\alpha)$ does not adequately substantiate the existence of 2 in α. Therefore we may conclude that, as a consequence of the existence of τ, $\sigma_\mathrm{c}(\alpha)$ is a misleading input for any potential inference algorithm reconstructing α from its morphic images; thus, in a combinatorial context, σ_c cannot be considered a morphism sufficiently reflecting the structure of α, and this holds in spite of its injectivity. But even if we restrict our examination to nonerasing morphisms, i. e. if we use the free semigroup $\{\mathtt{a}, \mathtt{b}\}^+$ instead of $\{\mathtt{a}, \mathtt{b}\}^*$ as the range of the morphisms (thus ignoring morphisms such as τ defined above), σ_c stays ambiguous with respect to α, as the morphism τ' given by $\tau'(1) := \mathtt{a\,b\,a}$, $\tau'(2) := \mathtt{b}^2$, $\tau'(3) := \mathtt{a\,b}^3\,\mathtt{a}$ and $\tau'(4) := \mathtt{b}^4$ satisfies $\tau'(\alpha) = \sigma_\mathrm{c}(\alpha)$:

$\sigma_\mathrm{c}(1)$ $\sigma_\mathrm{c}(2)$ $\sigma_\mathrm{c}(3)$ $\sigma_\mathrm{c}(4)$ $\sigma_\mathrm{c}(1)$ $\sigma_\mathrm{c}(4)$ $\sigma_\mathrm{c}(3)$ $\sigma_\mathrm{c}(2)$

a b a b b a b b b a b b b b a b a b b b b a b b b a b b

$\tau'(1)$ $\tau'(2)$ $\tau'(3)$ $\tau'(4)$ $\tau'(1)$ $\tau'(4)$ $\tau'(3)$ $\tau'(2)$

Consequently, we cannot significantly simplify the situation by reducing our considerations to morphisms in $\{\mathtt{a}, \mathtt{b}\}^+$ instead of $\{\mathtt{a}, \mathtt{b}\}^*$: still, the multitude of potential generating morphisms blurs the evidence of α in $\sigma_\mathrm{c}(\alpha)$ and can confuse any inference procedure.

It seems manifest that, contrary to this, certain unambiguous morphic images of a pattern might allow for drawing more reliable conclusions about their preimage, and some preliminary results by Reidenbach [76, 74] – such as a criterion on inductive inference of certain E-pattern languages which, in the present thesis, is described by Theorem 3.16 – confirm this expectation. With regard to our example pattern $\alpha = 1 \cdot 2 \cdot 3 \cdot 4 \cdot 1 \cdot 4 \cdot 3 \cdot 2$, unambiguous morphic images in $\{\mathtt{a}, \mathtt{b}\}^+$ can be found with a bit of effort – for instance, there is exactly one morphism $\sigma : \mathbb{N}^* \longrightarrow \{\mathtt{a}, \mathtt{b}\}^*$ (which, by the way, is not injective) satisfying

$$\sigma(\alpha) = \mathtt{a\,a\,b\,b\,b\,a\,a\,b\,a\,b\,a\,b},$$

namely σ given by $\sigma(1) := \mathtt{a}$, $\sigma(2) := \mathtt{a\,b}$, $\sigma(3) := \mathtt{b}$, $\sigma(4) := \mathtt{b\,a}$. In general, however, the search for unambiguous morphic images of patterns is not that easy, and, in fact, there even are simply structured patterns, such as $\beta := 1 \cdot 2$, with respect to which there is no unambiguous morphism at all.

Additional difficulties emerge from the fact that the definition of pattern languages features both common and terminal-preserving morphisms. In this regard, we can easily observe that the ambiguity of a fixed morphism can substantially change depending on the question of whether or not a pattern contains a terminal symbol. For instance, as stated above, there is no morphism that is unambiguous with respect to the terminal-free pattern $\beta = 1 \cdot 2$, but $w := \mathtt{b\,a\,b}$ and various other words are unambiguous with respect to the similarly structured pattern $\beta' := 1 \cdot \mathtt{a} \cdot 2$. Conversely, the common morphism $\sigma : \mathbb{N}^* \longrightarrow \{\mathtt{a}, \mathtt{b}\}^*$ given by $\sigma(1) := \mathtt{a\,b}$, $\sigma(2) := \mathtt{b\,a}$ is unambiguous with respect to the terminal-free pattern $\alpha := 1 \cdot 1 \cdot 2 \cdot 2$, whereas the equivalent terminal-preserving morphism $\sigma' : (\mathbb{N} \cup \{\mathtt{a}, \mathtt{b}\})^* \longrightarrow \{\mathtt{a}, \mathtt{b}\}^*$ defined by $\sigma'(i) := \sigma(i)$ is ambiguous with respect to, e.g., $\alpha' := 1 \cdot 1 \cdot \mathtt{a} \cdot 2 \cdot 2$, since there is a terminal-preserving morphism $\tau' : (\mathbb{N} \cup \{\mathtt{a}, \mathtt{b}\})^* \longrightarrow \{\mathtt{a}, \mathtt{b}\}^*$ satisfying $\tau'(\alpha') = \sigma'(\alpha')$ and $\tau'(i) \neq \sigma'(i)$ for a variable i in α':

$$\underbrace{\overbrace{\mathtt{a\,b}}^{\sigma'(1)}\,\overbrace{\mathtt{a\,b}}^{\sigma'(1)}}_{\tau'(1)}\;\underbrace{\mathtt{a}\,\overbrace{\mathtt{b\,a}}^{\sigma'(2)}\,\mathtt{b}}_{\tau'(1)}\!\!\overbrace{\phantom{\mathtt{b}}\,\mathtt{a}}^{\sigma'(2)}$$

Hence, fundamental (and, by the way, unresolved) problems in the well-established field of inductive inference of pattern languages show manifest connections to the following elementary and nontrivial combinatorial question on morphisms in free monoids:

> *Let Σ be an alphabet, and let $\alpha \in (\mathbb{N} \cup \Sigma)^+$ be an arbitrary (potentially terminal-free) pattern. Is there a (preferably injective or at least nonerasing) morphism $\sigma : (\mathbb{N} \cup \Sigma)^* \longrightarrow \Sigma^*$ such that σ is unambiguous with respect to α?*

As described above, this problem of intrinsic interest does not only arise from the research on pattern languages, but it is also closely related to the influential and powerful theory of equality sets. Nonetheless, to the author's best knowledge, it has never been systematically examined so far.[†] The present thesis is intended to fill this surprising and

[†]Note that Mateescu and Salomaa [54] study the *ambiguity of pattern languages* and, thus, a topic which seems to be related to ours. Unfortunately, however, their results and techniques cannot be applied to our questions. The reasons for this fact are briefly explained in Chapter 3.1.

appealing gap in the basic knowledge on morphisms. To this end, we formally introduce a number of types of ambiguity and of unambiguity of morphisms and formulate and discuss what we consider to be the most important questions on the subject (including a more formal version of the one given above). Thus, the main line of reasoning of our thesis can be subsumed under the field of *combinatorics on words* (cf., e. g., Lothaire [48, 49] and Choffrut and Karhumäki [12]), though it does not directly belong to the classical topics in that area.

In addition to this, we shall always keep the motivation for our studies in mind, i. e. we wish to find a morphism which optimally preserves the structure of its preimage or, in other words, which leads to a morphic image that contains as much information as possible about the preimage. Therefore we normally do not only seek for unambiguous morphisms, but rather for *unambiguous and injective* morphisms, since, in spite of our above explanations on the ambiguity of σ_c with respect to $\alpha = 1 \cdot 2 \cdot 3 \cdot 4 \cdot 1 \cdot 4 \cdot 3 \cdot 2$, the word $\sigma_c(\alpha)$ still is a more informative morphic image of α than, e. g., the word $\mathtt{a}^{20}$ (at least from an intuitive point of view). In this regard, we do not introduce a formal concept for directly quantifying to which extent the structure of a pattern is preserved by a morphism (a task which we expect to be extremely challenging), but we simply apply our combinatorial insights to several open problems for pattern languages. These endeavours indirectly confirm that the existence of an unambiguous or at least "not overly" ambiguous morphic image of a pattern α is characteristic for several properties of $L(\alpha)$ that intuitively depend on the existence of a word in $L(\alpha)$ which optimally reflects the structure of α. Consequently, our studies yield several (partial) solutions to prominent and long-term unresolved algorithmic questions on pattern languages, so that, in addition to its combinatorial nature, our work shows noticeable *algorithmic* (i. e., in particular, *learning theoretical*), *language theoretical* and even *information theoretical* traits.

The present thesis is structured as follows: In Chapter 2 we introduce the basic definitions and notations required for our formal reasoning. Additionally, this chapter lists some immediate properties of the concepts under consideration. Chapter 3 describes the current state of knowledge on combinatorial properties of morphisms, elementary properties of pattern languages and inductive inference of pattern languages. In particular, in Chapter 3.3, it contains the research questions on the inferrability of E-pattern languages that we wish to tackle in the present thesis. Furthermore, it proves several minor new or "folklore" results on pattern languages. While the specialist reader should be able to understand our thesis without studying these two chapters in detail, we consider it mandatory for every reader to take a closer look at Chapter 4, where we introduce our formal terminology on the (un-)ambiguity of morphisms and morphic images. Furthermore, the said chapter describes our six basic research questions on this main topic of our thesis, and it notes numerous simple yet important properties of the defined concepts. Subsequent to this, Chapter 5 is dedicated to the ambiguity of common morphisms $\sigma : \mathbb{N}^* \longrightarrow \Sigma^*$ (where Σ is an arbitrary alphabet with at least two distinct letters), i. e. of those morphisms which are applied to terminal-free patterns. Our studies lead to exhaustive answers to several of the abovementioned questions – in particular, we characterise those terminal-free patterns with respect to which there *exists* an unambiguous morphism. In addition to this, we use our results on the ambiguity of common morphisms for presenting fundamental positive insights into the inferrability of pattern languages generated by terminal-free patterns. In Chapter 6 we try to extend our comprehensive knowledge on common morphisms and terminal-free patterns to terminal preserving-morphisms $\sigma : (\mathbb{N} \cup \Sigma)^* \longrightarrow \Sigma^*$ and, thus, to general patterns in $(\mathbb{N} \cup \Sigma)^+$. Our corresponding results mainly reveal, first, that there exists a natural type of patterns for which this extension is largely straightforward, second, that the size of Σ gains in importance as soon as we consider general patterns and, third, that, in general, the properties related to the ambiguity of common morphisms cannot be extended to terminal-preserving morphisms. The latter insight remains restricted to alphabets with at most four distinct letters. Moreover, we show that, for arbitrary alphabets, there exists an unexpected negative relation between the injectivity of a terminal-preserving morphism σ and the question of whether σ can preserve the structure of a pattern. Again, we explain that our combinatorial statements have momentous implications on well-known problems for pattern languages: we present some strong negative results on inductive inference of general E-pattern languages over alphabet sizes 3 and 4, we disprove a well-known conjecture on the equivalence problem for E-pattern languages over these alphabets, and we demonstrate that a widely used proof technique for problems related to the inclusion of certain E-pattern languages cannot be applied to arbitrary patterns. Finally, Chapter 7 summarises the main statements of the present thesis and discusses some potential further research directions. A list of references and an index of notations and defined terms conclude this work.

Most major results of the present thesis have been previously published by the author in conference proceedings and journals. As the presentation in this thesis partially strongly differs from that in the said literature (which largely focuses on the algorithmic problems for pattern languages rather than on the combinatorial questions), the following list is meant to facilitate an easier mapping between the subsequent chapters and the corresponding articles: Chapter 5.1, its subchapter 5.1.1, an adapted version of Chapter 5.2 and its subchapter 5.2.1 have been originally presented in [78] and are contained in [74]. Furthermore, the results of Chapter 5.2 are referred to in [21] and its journal version [22], which, besides minor additional insights, extensively describe the main achievements given in Chapters 5.3 and 5.4. The crucial statements of Chapter 6.1 and its subchapter 6.1.1 can be found in [80]; the presentation in the present thesis, however, is much closer to that of the journal version, which again is contained in [74]. Finally, Chapter 6.2 and its subchapter 6.2.1 have been previously described in the conference paper [79] and the journal version [75]; in addition to this, the latter article also contains the results in Chapter 6.3 and the subchapter 6.3.1. Whenever the author cites one of his own results (thus interpreting it as a part of the literature rather than an original content of the present thesis) then this is due to the fact that it has already been presented in his diploma thesis [76] (and, additionally, possibly in [77], [81] or [74] as well). Hence, the present thesis formally is simply treated as a follow-up of [76].

Chapter 2

Basic notations and definitions

We begin the formal part of this thesis with a detailed description of basic definitions. We assume that the reader is familiar with the standard mathematical concepts and notations and, moreover, the most elementary insights in formal language theory (cf. Salomaa [87], Hopcroft and Ullman [30]) and recursion theory (cf. Odifreddi [67], Rogers [83]).

Note that, within the present chapter, we merely introduce a tentative definition on the main subject of our thesis, namely the ambiguity of morphisms and morphic images. Since this topic has not been systematically examined so far, our concepts are largely without a counterpart in previous literature. Therefore we consider it adequate to give them – together with corresponding examples and some first properties – in the separate Chapter 4.

2.1 Strings of symbols and common morphisms

Initially, we define $\mathbb{N} := \{1, 2, 3, \ldots\}$ and $\mathbb{N}_0 := \mathbb{N} \cup \{0\}$. The symbols $\subseteq$, $\subset$, $\supseteq$ and $\supset$ denote subset, proper subset, superset and proper superset, respectively.

We now introduce our general notations on strings of symbols and morphisms: An *alphabet* is an enumerable set of symbols. The *size* $|\mathcal{A}|$ of an alphabet $\mathcal{A}$ is the number of symbols in $\mathcal{A}$. A *string* s *(over an alphabet* $\mathcal{A}$*)* is a finite sequence of symbols from $\mathcal{A}$, i. e. $s = a_1 \cdot a_2 \cdot \ldots \cdot a_n$ with $n \in \mathbb{N}_0$, $a_i \in \mathcal{A}$, $1 \leq i \leq n$, and a separator symbol $\cdot \notin \mathcal{A}$. Concerning the given string s and any symbol $a \in \mathcal{A}$ with, for some i, $1 \leq i \leq n$, $a_i = a$ we say that s *contains* a and, conversely, that a *occurs* in s. Frequently we simply omit the symbol $\cdot$ in the presentation of a string provided that there is no equivocality involved in the more concise presentation. For a string $s = a_1\, a_2\, \ldots\, a_n$, $a_i \in \mathcal{A}$, $1 \leq i \leq n$, the number $n \in \mathbb{N}_0$ is called the *length* of s; moreover, we write $|s|$ for the length of s, and, thus, $|s| = n$. If $n = 0$ then we call s the *empty* string, and we write ε for this unique string of length 0.

The *concatenation* is a function which, for any alphabets $\mathcal{A}, \mathcal{B}$, maps any pair of strings $s_1 := a_1\, a_2\, \ldots\, a_m$, $m \in \mathbb{N}_0$, $a_1, a_2, \ldots, a_m \in \mathcal{A}$, and $s_2 := b_1\, b_2\, \ldots\, b_n$, $n \in \mathbb{N}_0$, $b_1, b_2, \ldots, b_n \in \mathcal{B}$, onto the string $a_1\, a_2\, \ldots\, a_m\, b_1\, b_2\, \ldots\, b_n$ over $\mathcal{A} \cup \mathcal{B}$. Similar to our presentation of a single string, which can be interpreted as the concatenation of strings of length 1, we write the concatenation of s_1 and s_2 as $s_1 \cdot s_2$ or, alternatively, simply as $s_1 s_2$. The string that results from the n-fold concatenation of a string s is occasionally denoted by s^n.

For any alphabet $\mathcal{A}$, $\mathcal{A}^*$ is the set of all (empty and nonempty) strings over $\mathcal{A}$, and $\mathcal{A}^+ := \mathcal{A}^* \setminus \{\varepsilon\}$. It can be verified effortlessly that $\mathcal{A}^*$ – when combined with the concatenation as the binary operation and with ε interpreted as the identity element – is a *monoid*, which normally is referred to as the *free monoid (generated by $\mathcal{A}$)*. While, hence and by definition, $\mathcal{A}^*$ is not only a monoid, but also a semigroup, the term *free semigroup (generated by $\mathcal{A}$)* is exclusively dedicated to $\mathcal{A}^+$ (of course, in combination with the concatenation again). Note that, within the scope of the present thesis, we use this terminology, but we do not deal with any questions that require a deeper algebraic knowledge on semigroups or monoids.

We call a string $t \in \mathcal{A}^*$ a *substring* of a string $s \in \mathcal{A}^*$ if, for some $r_1, r_2 \in \mathcal{A}^*$, $s = r_1\, t\, r_2$. In addition to this, if t is a substring of s then we say that s *contains* t and, conversely, that t *occurs* in s. A string $t \in \mathcal{A}^+$ is a *scattered substring*[†] of a string $s \in \mathcal{A}^+$ if and only if there exist $r_0, r_1, \dots, r_n \in \mathcal{A}^*$ and $t_1, t_2, \dots, t_n \in \mathcal{A}^+$, $n \in \mathbb{N}$, such that $t = t_1\, t_2\, \dots\, t_n$ and $s = r_0\, t_1\, r_1\, t_2\, r_2\, \dots\, t_n\, r_n$. If, for some $s, t_1, t_2 \in \mathcal{A}^*$, $s = t_1\, t_2$ then t_1 is a *prefix (of s)* and t_2 is a *suffix (of s)*. Additionally, we use the notations $s = \dots t \dots$ if t is a substring of s, $s = t \dots$ if t is a prefix of s, and $s = \dots t$ if t is a suffix of s. In contrast to this, if we wish to omit in our presentation some parts of a canonically given string then we henceforth use the symbol $[\dots]$, i. e., e. g., $s = a_1\, a_2\, [\dots]\, a_5$ stands for $s = a_1 a_2 a_3 a_4 a_5$. Let, for some symbol $a \in \mathcal{A}$, $t \in \{a\}^+$ and, for some strings $r_1, r_2 \in \mathcal{A}^*$, $s := r_1\, t\, r_2$; then t is called a *uniform substring (of s) (over a)* provided that $r_1 \neq \dots a$ and $r_2 \neq a \dots$. Finally, $|s|_t$ denotes the number of occurrences of a nonempty substring t in a nonempty string s – i. e., with $s = a_1 \cdot a_2 \cdot [\dots] \cdot a_n$ and $t = b_1 \cdot b_2 \cdot [\dots] \cdot b_m$, $m, n \in \mathbb{N}$, $|s|_t$ is the size of the set $\{i \mid 1 \leq i \leq n - m + 1 \text{ and } a_i = b_1, a_{i+1} = b_2, \dots, a_{i+m-1} = b_m\}$.

A *decomposition* is a function which maps a string $s \in \mathcal{A}^*$ onto an n-tuple of (sub)strings $t_1, t_2, \dots, t_n \in \mathcal{A}^*$, $n \in \mathbb{N}$, such that $s = t_1 \cdot t_2 \cdot [\dots] \cdot t_n$. Normally, we do not declare a decomposition formally (i. e. as a mapping), but we simply write s as a concatenation of the substrings $t_1, t_2, \dots, t_n$.

We say that strings $t_1, t_2 \in \mathcal{A}^+$ are *overlapping* provided that there are strings $r_1, r_2, r_3 \in \mathcal{A}^+$ such that $t_1 = r_1 r_2$ and $t_2 = r_2 r_3$ (or, alternatively, $t_1 = r_2 r_1$ and $t_2 = r_3 r_2$). In addition to this, we speak of an *overlapping occurrence of $t_1 = r_1 r_2$ and $t_2 = r_2 r_3$ in a string $s \in \mathcal{A}^+$* provided that $s = \dots r_1\, r_2\, r_3 \dots$.

The following, technically motivated notation allows to address certain substrings of a string s over an alphabet $\mathcal{A}$: If s contains n, $n \geq 1$, overlapping or non-overlapping occurrences of a substring t then for every i, $1 \leq i \leq n$, $t\langle i \rangle$ is the ith occurrence (from the left) of t in s. For that case, the substring $[s/t\langle i \rangle]$ is the prefix of s up to (but not including) the leftmost letter of $t\langle i \rangle$ and the substring $[t\langle i \rangle \backslash s]$ is the suffix of s beginning with the first letter that is to the right of $t\langle i \rangle$. Moreover, for every string s that contains at least i occurrences of a substring r, j occurrences of substring t and that satisfies $s = s_1\, r\langle i \rangle\, s_2\, t\langle j \rangle\, s_3$ with $s_1, s_2, s_3 \in \mathcal{A}^*$, we use $[r\langle i \rangle \backslash s / t\langle j \rangle]$ as an abbreviation for $[r\langle i \rangle \backslash [s/t\langle j \rangle]]$. Thus, for appropriate r, s, t, the specified substrings satisfy $s = [s/t\langle i \rangle]\, t\langle i \rangle\, [t\langle i \rangle \backslash s]$ and $s = [s/r\langle i \rangle]\, r\langle i \rangle\, [r\langle i \rangle \backslash s/t\langle j \rangle]\, t\langle j \rangle\, [t\langle j \rangle \backslash s]$; e.g., with $\{a, b, c\} \subseteq \mathcal{A}$, $s := abcabb$, $r := a$ and $t := ab$, the definition leads to $[s/r\langle 2 \rangle] = abc$, $[r\langle 2 \rangle \backslash s] = bb$, and $[r\langle 1 \rangle \backslash s/t\langle 2 \rangle] = bc$. Finally – for the sake of a more concise presentation

[†]It is a common alternative naming to choose the term "factor" instead of "substring/subword" and "substring/subword/subsequence" instead of "scattered subword". Due to our combinatorial view on the matter, however, we prefer our terminology, which has the less algebraic flavour. Analogously, we do not use the common term "factorisation", but we prefer the term "decomposition" (see below).

in the above notations $[s/t\langle i\rangle]$, $[t\langle i\rangle\backslash s]$ and $[r\langle i\rangle\backslash s/t\langle j\rangle]$ – we occassionally omit the declaration $\langle i\rangle$ (or, if applicable, $\langle j\rangle$) provided that $i = 1$ (or $j = 1$). Consequently, $[s/t] := [s/t\langle 1\rangle]$, $[t\backslash s] := [t\langle 1\rangle\backslash s]$ and $[r\backslash s/t] := [r\langle 1\rangle\backslash s/t\langle 1\rangle]$.

Since we deal with word monoids (or, alternatively, word semigroups), we consider a (homo-)*morphism* a mapping that is compatible with the concatenation, i. e. for alphabets $\mathcal{A}$, $\mathcal{B}$ and arbitrary strings $s_1, s_2 \in \mathcal{A}^*$, a mapping $f : \mathcal{A}^* \longrightarrow \mathcal{B}^*$ is a morphism if it satisfies $f(s_1 \cdot s_2) = f(s_1) \cdot f(s_2)$ and, additionally, $f(\varepsilon) = \varepsilon$. Hence, a morphism is fully explained as soon as it is declared for all symbols in $\mathcal{A}$, and the image under a morphism f of a string $s \in \mathcal{A}^*$ is composed by the concatenation of the images under f of the symbols occurring in s. Note that we restrict ourselves to total morphisms, even though we normally declare a morphism only for those variables explicitly that, in the respective context, are relevant. For any morphism f, a string s and the string $t := f(s)$, we say that f *generates* t *(when being applied to* s*)*. For the *composition* of two morphisms f, g we write $g \circ f$, i. e., for every $s \in \mathcal{A}^*$, $g \circ f(s) = g(f(s))$.

Hence, for alphabets $\mathcal{A}$, $\mathcal{B}$, let $f : \mathcal{A}^* \longrightarrow \mathcal{B}^*$ be a morphism. Then f is called *nonerasing* provided that, for every $a \in \mathcal{A}$, $f(a) \neq \varepsilon$; occasionally we refer to a nonerasing morphism by restricting its domain and range to the corresponding free semigroups, i. e. we simply write $f : \mathcal{A}^+ \longrightarrow \mathcal{B}^+$. Moreover, f is said to be *injective (on* $\mathcal{A}^*$*)* if, for any strings $s_1, s_2 \in \mathcal{A}^*$, the equality $f(s_1) = f(s_2)$ implies $s_1 = s_2$. Note that f necessarily is nonerasing provided that it is injective. Occasionally, we call an injective morphism a *code*.

Let $f : \mathcal{A}^* \longrightarrow \mathcal{A}^*$ be an (endo-)morphism, and let $s \in \mathcal{A}^+$. We say that f is *nontrivial (for* s*)* provided that, for some symbol a occurring in s, $f(a) \neq a$; otherwise, we call f *trivial (for* s*)*. If f is nontrivial for s and, additionally, $f(s) = s$ then s is called a *(nontrivial) fixed point (of* f*)*.

As explained in Chapter 1, the main subject of this thesis is a property of strings related to morphisms, namely the (un-)ambiguity. We extensively introduce and discuss our terminology on this topic in Chapter 4; therefore, within the present chapter, we are content with a preliminary, semi-formal definition (which nevertheless does not conflict with our detailed notations in Chapter 4): For any alphabets $\mathcal{A}, \mathcal{B}$, a morphism $f : \mathcal{A}^* \longrightarrow \mathcal{B}^*$ is said to be *unambiguous (with respect to a string* s*)* provided that there is no morphism g with $g(s) = f(s)$ and, for some symbol a in s, $g(a) \neq f(a)$. Additionally, a string t is said to be unambiguous (with respect to s) if there is exactly one morphism f such that $f(s) = t$. Finally, the morphism f is called *ambiguous (with respect to* s*)* if it is not unambiguous, and t is ambiguous (with respect to s) if there are morphisms f, g with $g(s) = t = f(s)$ and, for some symbol a in s, $g(a) \neq f(a)$.

We briefly discuss the current state of knowledge on general properties of strings and morphisms in Chapter 3.1.

2.2 Words, patterns and terminal-preserving morphisms

We proceed with the more specific terminology on strings of symbols depending on the particular alphabets to be considered in this thesis: We regard two types of alphabets – an infinite alphabet which is represented by $\mathbb{N}$ and an alphabet Σ. Unless stated otherwise, we assume Σ to be finite, but not unary. We postulate $\Sigma \cap \mathbb{N} = \emptyset$, and we indicate this by choosing lower case letters from the beginning of the Latin alphabet in typewriter font as symbols in Σ, i.e. $\Sigma \subseteq \{\mathtt{a}, \mathtt{b}, \mathtt{c}, \dots\}$. We strictly posit that these symbols are pairwise different. If we wish to refer to symbols in Σ that may be identical then we use upper case letters from the beginning of the Latin alphabet in typewriter font such as $\mathtt{A}, \mathtt{B}$ or indexed lower case letters such as $\mathtt{a}_1, \mathtt{a}_2$. With regard to the infinite alphabet, our choice implies that we use $\mathbb{N}$ in two different ways within this thesis: first, as a set of symbols generating the (non-commutative) free monoid $\mathbb{N}^*$ (or, alternatively, the free semigroup $\mathbb{N}^+$) – thus considering the concatenation $\cdot$ as the required operation and, if applicable, ε as the neutral element – and, second, as the totally ordered set $\mathbb{N}_0$ of natural numbers forming a commutative semiring with the addition $+$, the multiplication $\times$ (normally, we simply omit the latter operation symbol), 0 as the neutral element with respect to the addition, and 1 as the neutral element with respect to the multiplication. For instance, the expression $s = i \cdot j^{2k+1} \cdot i$, $i, j, k \in \mathbb{N}$, refers to any string which consists of an initial symbol i, followed by the string of length $2k + 1$ over the alphabet $\{j\}$ and concluded by the symbol i again; thus, for $k = 1$, $s = i \cdot j \cdot j \cdot j \cdot i \in \mathbb{N}^*$, for $k = 2$, $s = i \cdot j \cdot j \cdot j \cdot j \cdot j \cdot i \in \mathbb{N}^*$, and so on. Though we expect that this twofold use of $\mathbb{N}$ does not confuse the reader, we shall indicate the intended interpretation whenever we feel that there could be a chance of a misreading of our statements.

Subject to these particular alphabets, we replace the terms "symbol" and "string" by a more specific terminology. We designate Σ as the *terminal alphabet* and the symbols in Σ as *letters* or as *terminal symbols* (or as *terminals* for short). Additionally, henceforth, the term *alphabet* exclusively refers to Σ. Whenever we do not consider $\mathbb{N}$ as the set of natural numbers, but as the set of symbols generating the free monoid $\mathbb{N}^*$, we designate the symbols in $\mathbb{N}$ as *variables*.

For any alphabet Σ, we deal with strings in $(\mathbb{N} \cup \Sigma)^*$. We designate any string in $(\mathbb{N} \cup \Sigma)^+$ as a *pattern (over Σ)*. Among the set of all patterns, a string in $\mathbb{N}^+$ is a *terminal-free*† pattern, and a string in Σ^+ is a *(terminal) word*. Additionally, we adapt the term "substring" so that we can speak of *subpatterns* and *subwords*. We name patterns in $(\mathbb{N} \cup \Sigma)^+$ with lower case letters from the beginning of the Greek alphabet such as α, β, γ. However, if we wish to emphasise that a string is a terminal word then we normally choose u, v or w when referring to this string.

For any given pattern α, $\mathrm{var}(\alpha) \subset \mathbb{N}$ is the set of variables occurring in α and $\mathrm{term}(\alpha) \subset \Sigma$ is the set of terminals occurring in α. The maximum scattered subword over Σ in a pattern α is called the *residual* of α; we denote the residual of α by $\mathrm{res}(\alpha)$. The maximum scattered terminal-free subpattern in α is denoted by $\mathrm{tf}(\alpha)$. Consequently, for instance, $\mathrm{res}(1 \cdot \mathtt{a} \cdot 2 \cdot \mathtt{b} \cdot 3) = \mathtt{a}\,\mathtt{b}$ and $\mathrm{tf}(1 \cdot \mathtt{a} \cdot 2 \cdot \mathtt{b} \cdot 3) = 1 \cdot 2 \cdot 3$. Note that res and tf can be interpreted as morphisms by defining, for every $\mathtt{A} \in \Sigma$ and $j \in \mathbb{N}$, $\mathrm{res}(\mathtt{A}) := \mathtt{A}$, $\mathrm{res}(j) := \varepsilon$ and $\mathrm{tf}(\mathtt{A}) := \varepsilon$, $\mathrm{tf}(j) := j$.

We say that a pattern α is *in canonical form* if its variables occur "in the natural order", i.e., for some $n \in \mathbb{N}$, $\mathrm{var}(\alpha)$ equals the set $\{1, 2, \dots, n\}$ and, for any $i, j \in \mathrm{var}(\alpha)$ with $i < j$, there is a prefix β of α such that $i \in \mathrm{var}(\beta)$ and $j \notin \mathrm{var}(\beta)$. Thus, e.g., $1 \cdot 1 \cdot 2 \cdot 2 \cdot 3 \cdot 3$ and $1 \cdot \mathtt{a} \cdot \mathtt{b} \cdot 2 \cdot 1$ are in canonical form, whereas $1 \cdot 1 \cdot 3 \cdot 3 \cdot 2 \cdot 2$ and $1 \cdot \mathtt{a} \cdot \mathtt{b} \cdot 3 \cdot 1$ are not. An additional notation on canonical forms is introduced in the end

†Note that Filè [19] designates terminal-free patterns as *pure* patterns.

of the present Chapter 2.2.

A pattern α is said to be *regular*[†] provided that, for every $i \in \mathrm{var}(\alpha)$, $|\alpha|_i = 1$. This name, introduced by Shinohara [97], is derived from a property of these patterns explained in Proposition 3.3 (cf. Chapter 3.2).

We designate two patterns as *similar* if their subpatterns over Σ are identical and occur in the same order in the patterns. More formally, the patterns α, β are similar provided that – for some $m \in \mathbb{N}$, $\alpha_i, \beta_i \in \mathbb{N}^+$ with $1 \leq i < m$, $\alpha_0, \beta_0, \alpha_m, \beta_m \in \mathbb{N}^*$ and $u_i \in \Sigma^+$ with $i \leq m$ – it is $\alpha = \alpha_0 u_1 \alpha_1 u_2 \ldots \alpha_{m-1} u_m \alpha_m$ and $\beta = \beta_0 u_1 \beta_1 u_2 \ldots \beta_{m-1} u_m \beta_m$. Thus, with the patterns $\alpha := 1 \cdot \mathtt{a} \cdot 2 \cdot \mathtt{b} \cdot 3$, $\beta := 1 \cdot \mathtt{a} \cdot 2 \cdot 2 \cdot \mathtt{b}$, $\gamma := 1 \cdot \mathtt{a} \cdot \mathtt{b} \cdot 3$, α and β are similar, but neither α and γ nor β and γ are similar.

As explained in Chapter 1, we are mainly interested in the properties of those morphisms which map a string over a large alphabet onto a string over a smaller alphabet. Consequently, as suggested by the designation of the symbols in the infinite set $\mathbb{N}$ as "variables" and of the symbols in the finite alphabet Σ as "terminals", we focus on morphisms mapping strings in $\mathbb{N}^+$ onto terminal words. Additionally, however, we allow the preimages of the morphisms to be augmented with terminal symbols, which is reflected by the definition of a pattern. Therefore, we largely deal with morphisms $\sigma : (\mathbb{N} \cup \Sigma)^* \longrightarrow \Sigma^*$. If we recall our general statements on morphisms $f : \mathcal{A}^* \longrightarrow \mathcal{B}^*$ as introduced in Chapter 2.1, we thus have $\mathcal{A} = \mathbb{N} \cup \Sigma$ and $\mathcal{B} = \Sigma$ such that we allow $\mathcal{A} \cap \mathcal{B} \neq \emptyset$. Since, anyway, $\mathcal{A} \supseteq \mathbb{N}$ is an infinite alphabet, such a decision of course is meant to have some impact on the morphic images to be considered for the terminal symbols in $\mathcal{A} = \mathbb{N} \cup \Sigma$. Consequently, with regard to all symbols in Σ, we assume any morphism to leave them unchanged (which again explains the name "terminal" and which is crucial for our considerations on pattern languages, cf. Chapter 2.3). In other words, apart from very few, technically motivated and explicitly marked exceptions (such as the morphism $\mathrm{tf} : (\mathbb{N} \cup \Sigma)^+ \longrightarrow \mathbb{N}^*$ introduced above), we demand any morphism $\sigma : (\mathbb{N} \cup \Sigma)^* \longrightarrow \Sigma^*$ – and, more general, any (endo-)morphism $\phi : (\mathbb{N} \cup \Sigma)^* \longrightarrow (\mathbb{N} \cup \Sigma)^*$ – to be *terminal-preserving*, i. e., for every $\mathtt{A} \in \Sigma$, $\phi(\mathtt{A}) = \mathtt{A}$. Among the terminal-preserving morphisms, in turn, there are two crucial subtypes, the first mapping patterns onto patterns and the second (and most important) mapping patterns onto words. More precisely, we designate a terminal-preserving morphism $\phi : (\mathbb{N} \cup \Sigma)^* \longrightarrow (\mathbb{N} \cup \Sigma)^*$ as *residual-preserving*[‡] provided that, for all $i \in \mathbb{N}$, $\phi(i) \in \mathbb{N}^*$, and a terminal-preserving morphism $\sigma : (\mathbb{N} \cup \Sigma)^* \longrightarrow \Sigma^*$ is called a *substitution*. Any morphism $\bar{\sigma} : \Sigma^* \longrightarrow \mathbb{N}^*$, which thus is necessarily not terminal-preserving, is called an *inverse substitution*. We designate any injective residual-preserving morphism ϕ with, for every $i \in \mathbb{N}$, $|\phi(i)| = 1$ as a *renaming of variables*.

Obviously, if a terminal-preserving morphism ϕ is applied to a terminal-free pattern then it is simply a common morphism. Therefore, henceforth, we normally restrict the use of the terms "terminal-preserving morphism", "residual-preserving morphism" and "substitution" to formal statements and, in informal remarks, to those cases where we wish to emphasise that preimages of ϕ in $(\mathbb{N} \cup \Sigma)^* \setminus \mathbb{N}^*$ are considered.

[†]Frequently, in the recent literature, the probably more appropriate term *linear* is chosen instead of "regular". We, however, prefer the more traditional terminology.

[‡]In [79] the rather misleading term "similarity-preserving" is used instead of "residual-preserving". For a proof of the lack of precision involved in the naming chosen in [79], see the terms "similar" and "residual" introduced in the present chapter.

We conclude this chapter with a number of names for properties of patterns and variables that are related to both common and terminal-preserving morphisms: We say that patterns α, β are *(morphically) coincident* if there exist (if applicable, residual-preserving) morphisms ϕ and ψ such that $\phi(\alpha) = \beta$ and $\psi(\beta) = \alpha$; we call them *(morphically) semi-coincident* if there is either such a ϕ or such a ψ, and, for the case that there is neither such a ϕ nor such a ψ, they are designated as *(morphically) incoincident.*

A pattern α is *morphically imprimitive* if there is a pattern β such that $|\beta| < |\alpha|$ and α and β are morphically coincident. Correspondingly, a pattern α is called *morphically primitive* if it is not morphically imprimitive.

A variable j is said to be *redundant (in a pattern α)* if, for the (residual-preserving) morphism ϕ with $\phi(j) := \varepsilon$ and $\phi(k) := k$, $k \neq j$, there exists a (residual-preserving) morphism ψ such that $\alpha = \psi(\phi(\alpha))$, i.e. α and $\phi(\alpha)$ are coincident.

The relation between morphically imprimitive patterns and the redundancy of variables is by no means surprising:

Proposition 2.1. *Let α be a pattern. Then α is morphically primitive if and only if there is no variable $j \in \mathrm{var}(\alpha)$ such that j is redundant.*

Proof. Let α be morphically primitive. Then there do not exist a pattern β and morphisms ϕ', ψ' such that $|\beta| < |\alpha|$, $\phi'(\alpha) = \beta$ and $\psi'(\beta) = \alpha$. Thus, in particular, there is no $i \in \mathrm{var}(\alpha)$ such that, for the morphism ϕ given by $\phi(i) := \varepsilon$ and $\phi(j) := j$, $j \neq i$, the patterns α and $\phi(\alpha)$ are coincident. Thus, there is no redundant variable in α.

Conversely, let α be morphically imprimitive. Then there exist a pattern β and morphisms ϕ', ψ' such that $|\beta| < |\alpha|$, $\phi'(\alpha) = \beta$ and $\psi'(\beta) = \alpha$. Consequently, there exists at least one variable $i \in \mathrm{var}(\alpha)$ with $\phi'(i) = \varepsilon$. Let now the morphism ϕ be given by $\phi(i) := \varepsilon$ and $\phi(j) := j$, $j \neq i$, and the morphism ψ by $\psi(k) := \psi'(\phi'(k))$, $k \in \mathrm{var}(\alpha)$. Then, since $\phi'(i) = \varepsilon$, we have $\psi(\phi(\alpha)) = \alpha$, and therefore α and $\phi(\alpha)$ are coincident. Consequently, i is redundant in α. □

Finally, a pattern α is said to be the *canonical form of* a pattern β provided that α is in canonical form and there exists a morphism ϕ such that ϕ is a renaming of variables and $\phi(\beta) = \alpha$. Note that, for any such pattern β, α is unique, and so is ϕ when the view is restricted to the variables in α.

2.3 Pattern languages

In the present chapter, we start interpreting patterns (so far merely described as particular strings of symbols, see Chapter 2.2) as generators of languages, thus leading to the concept of pattern languages introduced by Angluin [3] and Shinohara [97].

For any alphabet $\mathcal{A}$, a *language L (over $\mathcal{A}$)* is a set of strings over $\mathcal{A}$, i. e. $L \subseteq \mathcal{A}^*$. A language L is *empty* if $L = \emptyset$; otherwise, it is *nonempty.* A *class $\mathcal{L}$ of languages (over $\mathcal{A}$)* is a set of languages over $\mathcal{A}$, i. e. $\mathcal{L} \subseteq \mathcal{P}(\mathcal{A}^*)$, where $\mathcal{P}$ refers to the power set. Some additional notations on particular classes of languages are given below and in Chapter 2.4.

Let Σ be an alphabet. If we wish to refer to the set of all patterns over Σ as generators of languages (a concept to be explained below) then we use the notation Pat_Σ, i. e. $\mathrm{Pat}_\Sigma := (\mathbb{N} \cup \Sigma)^+$. We omit the subscript Σ if there is no need to emphasise the concrete underlying alphabet. The set of terminal-free patterns (again considered as generators of languages) is denoted $\mathrm{Pat}_{\mathrm{tf}}$, i. e. $\mathrm{Pat}_{\mathrm{tf}} := \mathbb{N}^+$. We establish these additional notations due to mere presentational reasons: we use the more elementary terms such as $(\mathbb{N} \cup \Sigma)^+$ and $\mathbb{N}^+$ if we discuss the combinatorial properties of patterns and morphisms, and we prefer the terminology introduced in the present chapter – namely Pat_Σ, $\mathrm{Pat}_{\mathrm{tf}}$ and similar terms to be established in the subsequent chapters – if we deal with pattern languages. This allows for a more immediate perception of the respective topic under consideration, and it follows the prevalent terminology on pattern languages in literature.

Basically, for any $\alpha \in \mathrm{Pat}_\Sigma$, the pattern language $L_\Sigma(\alpha)$ of α is the set of all images of α in Σ^* under arbitrary (if applicable, terminal-preserving) morphisms; again, we restrict the use of the subscript Σ to those cases where we consider it necessary to explicitly refer to the alphabet under consideration. More precisely, we distinguish between two types of pattern languages: Following the definition by Angluin [3], the *NE-pattern language* $L_{\mathrm{NE},\Sigma}(\alpha)$ *(of α)* is given by $L_{\mathrm{NE},\Sigma}(\alpha) := \{w \in \Sigma^+ \mid w = \sigma(\alpha)$ for a substitution $\sigma : (\mathbb{N} \cup \Sigma)^+ \longrightarrow \Sigma^+\}$. Thus, the definition of NE-pattern languages is restricted to a consideration of nonerasing morphisms (which, by the way, explains the notation "NE"). Contrary to this, the definition of Shinohara [97] allows arbitrary morphisms; consequently, the *E-pattern language* $L_{\mathrm{E},\Sigma}(\alpha)$ *(of α)* is given by $L_{\mathrm{E},\Sigma}(\alpha) := \{w \in \Sigma^* \mid w = \sigma(\alpha)$ for a substitution $\sigma : (\mathbb{N} \cup \Sigma)^* \longrightarrow \Sigma^*\}$. In this case, of course, the notation "E" refers to "erasing" or – since, for every pattern α, $L_{\mathrm{E}}(\alpha) \supseteq L_{\mathrm{NE}}(\alpha)$ and, for every nontrivial pattern α (i. e. $\alpha \in (\mathbb{N} \cup \Sigma)^+ \setminus \Sigma^+$), even $L_{\mathrm{E}}(\alpha) \supset L_{\mathrm{NE}}(\alpha)$ – to the term "extended". In the present thesis we mainly deal with E-pattern languages, and therefore the short notation $L(\alpha)$ refers to the E-pattern language of α; thus, henceforth, $L(\alpha) := L_{\mathrm{E}}(\alpha)$. Furthermore, a pattern language $L_{\mathrm{E}}(\alpha)$ (or $L_{\mathrm{NE}}(\alpha)$) is called *terminal-free* provided that α is terminal-free, i. e. $\alpha \in \mathrm{Pat}_{\mathrm{tf}}$. If we simply speak of a *pattern language (of a pattern α)* then we mean *any* of the definitions introduced above. Finally, we canonically extend a term introduced with respect to morphisms: We say that a pattern α *generates* a language L if L equals the pattern language of α.

Obviously, two patterns α, β generate the same language if there exists a renaming of variables mapping α onto β. Thus, without loss of generality, we can restrict most considerations on generators of pattern languages to patterns in canonical form (introduced in the previous chapter):

Proposition 2.2. *Let Σ be an alphabet, and let $\alpha, \beta \in \mathrm{Pat}_\Sigma$. Then $L_{\mathrm{NE},\Sigma}(\alpha) = L_{\mathrm{NE},\Sigma}(\beta)$ and $L_{\mathrm{E},\Sigma}(\alpha) = L_{\mathrm{E},\Sigma}(\beta)$ if the canonical form of α equals the canonical form of β.*

As to be noted in the subsequent chapter, this criterion is even characteristic for NE-pattern languages (cf. Theorem 3.8), whereas it is not for E-pattern languages. The latter statement immediately follows from examples such as the patterns $\alpha := 1$ and $\beta := 1 \cdot 2$: In this case, for every morphism $\sigma : \mathbb{N}^* \longrightarrow \Sigma^*$, we can define the morphisms τ_1 and τ_2 by $\tau_1(1) = \sigma(1 \cdot 2)$ and $\tau_2(1) = \sigma(1)$, $\tau_2(2) = \varepsilon$. Then $\sigma(\alpha) = \tau_2(\beta)$ and $\sigma(\beta) = \tau_1(\alpha)$, and therefore $L_{\mathrm{E}}(\alpha) = L_{\mathrm{E}}(\beta)$.

This example suggests a partition of the set of all patterns into those that are a shortest generator of their respective E-pattern language and those that are not. In this regard, we call a pattern α *succinct (on an alphabet Σ)* if, for every pattern β with $L_{\mathrm{E},\Sigma}(\alpha) = L_{\mathrm{E},\Sigma}(\beta)$, $|\alpha| \leq |\beta|$, and we call α *prolix (on Σ)* provided that it is not succinct on Σ.

A variable $j \in \text{var}(\alpha)$ is said to be *superfluous (in a pattern α) (with respect to an alphabet Σ)* if, for the (residual-preserving) morphism $\phi : (\mathbb{N} \cup \Sigma)^* \longrightarrow (\mathbb{N} \cup \Sigma)^*$ given by $\phi(j) := \varepsilon$ and $\phi(k) := k$, $k \in \mathbb{N} \setminus \{j\}$, $L_{\text{E},\Sigma}(\alpha) = L_{\text{E},\Sigma}(\phi(\alpha))$.

A first insight into the relation between superfluous and redundant (cf. Chapter 2.2) variables follows with little effort or, alternatively, a brief consultation of Theorem 3.5 presented in Chapter 3.2.1:

Proposition 2.3. *Let α be a pattern. Then a variable $j \in \text{var}(\alpha)$ is superfluous in α with respect to any alphabet Σ if it is redundant in α.*

In Chapter 6.2.1, we discuss the problem of whether the converse of Proposition 2.3 holds as well.

After these definitions and first observations related to single pattern languages, we now introduce our notations on selected *classes* of pattern languages: The *class* ePAT_Σ *(of E-pattern languages)* is defined as $\text{ePAT}_\Sigma := \{L_{\text{E},\Sigma}(\alpha) \mid \alpha \in \text{Pat}_\Sigma\}$ and the *class* nePAT_Σ *(of NE-pattern languages)* as $\text{nePAT}_\Sigma := \{L_{\text{NE},\Sigma}(\alpha) \mid \alpha \in \text{Pat}_\Sigma\}$. Accordingly, the *class* $\text{ePAT}_{\text{tf},\Sigma}$ *(of terminal-free E-pattern languages)* is given by $\text{ePAT}_{\text{tf},\Sigma} := \{L_{\text{E},\Sigma}(\alpha) \mid \alpha \in \text{Pat}_{\text{tf}}\}$ and the *class* $\text{nePAT}_{\text{tf},\Sigma}$ *(of terminal-free NE-pattern languages)* by $\text{nePAT}_{\text{tf},\Sigma} := \{L_{\text{NE},\Sigma}(\alpha) \mid \alpha \in \text{Pat}_{\text{tf}}\}$. Once more, we abandon the subscript Σ whenever we expect this omission not to cause any confusion. If we simply speak of the *class of pattern languages* then we mean *any* of the classes introduced above.

An extensive description of some elementary properties of (classes of) pattern languages is provided in Chapter 3.2.

2.4 Indexable classes of languages

We proceed with a brief description of a particular type of classes of languages that significantly eases the application of our combinatorial results to inductive inference of pattern languages.

Hence, let $\mathcal{L}$ be a class of languages over some alphabet $\mathcal{A}$. Then $\mathcal{L}$ is said to be *indexable* provided that there exists an indexing $(L_i)_{i\in\mathbb{N}}$ of languages L_i such that, first, $\mathcal{L} = \{L_i \mid i \in \mathbb{N}\}$ and, second, there exists a total computable function χ which uniformly decides the membership problem for $(L_i)_{i\in\mathbb{N}}$ – i. e., for every $w \in \mathcal{A}^*$ and for every $i \in \mathbb{N}$, $\chi(w, i) = 1$ if and only if $w \in L_i$. In this case, we call $\mathcal{L} = (L_i)_{i\in\mathbb{N}}$ an *indexed family (of recursive languages)*. Of course, in this notation for an indexed family (which conforms with the use in the literature) the equality symbol "=" does not refer to an equality in the usual sense, but is merely a symbol indicating that $\mathcal{L}$ contains all languages in $(L_i)_{i\in\mathbb{N}}$ and vice versa.

Due to the decidability of the membership problem for ePAT and nePAT (cf. Chapter 3.2.1) and the fact that we shall not consider any sets of patterns which are not recursively enumerable, we can apply the concept of indexable classes to each class of pattern languages to be considered in our thesis:

Proposition 2.4. *Let Σ be an alphabet. Let $\text{Pat}_{\star,\Sigma}$ be a recursively enumerable set of patterns over Σ, and let $\text{PAT}_{\star,\Sigma} := \{L_{\text{E},\Sigma}(\alpha) \mid \alpha \in \text{Pat}_{\star,\Sigma}\}$ (or $\text{PAT}_{\star,\Sigma} := \{L_{\text{NE},\Sigma}(\alpha) \mid \alpha \in \text{Pat}_{\star,\Sigma}\}$) be the corresponding class of E-pattern languages (or NE-pattern languages, respectively). Then $\text{PAT}_{\star,\Sigma}$ is indexable.*

Note that we do not discuss possible recursive enumerations for Pat_Σ (which w. l. o. g. can normally be replaced by a recursive enumeration of all patterns in canonical form) or any other set of patterns that we examine. Due to the simplicity of this topic, we do not expect this omission to confuse the reader. Nevertheless, we consider it worth mentioning that, for any $n \in \mathbb{N}$, the set of terminal-free patterns in canonical form of length n corresponds to the set of all partitions of a set with exactly n elements into nonempty subsets. Thus, the number of these patterns equals the nth *Bell number* (for additional information on set partitions, see Rota [85]):

Proposition 2.5. *For each $n \in \mathbb{N}$, the number of terminal-free patterns in canonical form of length n equals the number of all partitions of a set S with $|S| = n$ into nonempty subsets.*

From this insight, standard combinatorial considerations lead to the number of general patterns in canonical form over some alphabet. Furthermore, efficient recursive enumerations of all patterns in canonical form can be easily derived from the literature on combinatorial algorithms dealing with partitions of sets (see, e. g., Nijenhuis and Wilf [66]).

2.5 Inductive inference

As stated in Chapter 1, the ambiguity of morphisms is a problem which manifestly emerges in inductive inference of pattern languages. Therefore, in the present chapter, we provide an appropriate background on this topic, which moreover shall allow to apply some of our results to open questions in the said field.

Inductive inference is the act of (algorithmically) eliciting an abstract principle that is common to a set of examples. Consequently, it can be interpreted as a recursion-theoretic view on learning phenomena and therefore it is normally subsumed under the paradigm of *algorithmic* (or: *computational*) *learning theory*. Among the various models of inductive inference, we concentrate on the initial and presumably most fundamental one, namely that of *identification in the limit* of formal languages as introduced by Gold [24] in 1967, which is based on preliminary considerations given by Solomonoff [101]. Informally, this model considers a class $\mathcal{L}$ of languages to be "learnable in the limit" if and only if there exists a computable "learner" which, for every language $L \in \mathcal{L}$, stepwise reads any infinite stream of example strings that, in the limit, fully enumerates L (or, alternatively, a stream of labelled *negative* and *positive* examples that enumerates both L and its complement) and which outputs a corresponding infinite stream of guesses that, after finitely many steps, stabilises on a distinct reference to an acceptor or generator of L. As a consequence of our focus on pattern languages we can regard a prominent, but restricted version of the original model (in this regard, we largely follow Angluin [4]) which assumes that the objects to be learned are nonempty languages forming an indexed family. Moreover, we limit the input of the learner to positive example strings for the language to be inferred.

Before we formally declare the learning model, we explain the said input of our inference algorithm: For any alphabet $\mathcal{A}$ and any nonempty language $L \subseteq \mathcal{A}^*$ we call a total function $t : \mathbb{N} \longrightarrow \mathcal{A}^*$ a *text* of L if and only if it satisfies $\{t(i) \mid i \in \mathbb{N}\} = L$. Moreover, for every text t and every $n \in \mathbb{N}$, t^n codes the first n values of t in a single string, i.e. $t^n := t(1) \triangledown t(2) \triangledown t(3) \triangledown [\dots] \triangledown t(n)$ with $\triangledown \notin \mathcal{A}$. Finally, $\text{text}(L)$ denotes the set of all (computable and non-computable, repetitive and non-repetitive) texts of a language L.

Now our definition on language identification in the limit can be given as follows: Let $\mathcal{L} = (L_i)_{i\in\mathbb{N}}$ be an indexed family of nonempty recursive languages over an alphabet $\mathcal{A}$. Then $\mathcal{L} = (L_i)_{i\in\mathbb{N}}$ is *inferrable (from positive data)* (or *learnable* for short) if and only if there exists a computable function $S : (\mathcal{A} \cup \{\triangledown\})^* \longrightarrow \mathbb{N}$ such that, for every $L \in \mathcal{L}$ and for every $t \in \text{text}(L)$,

1. $S(t^n)$ is defined for every $n \in \mathbb{N}$ and
2. there is a $j \in \mathbb{N}$ with $L_j = L$ and there is an $m \in \mathbb{N}$ with $S(t^n) = j$ for every $n \geq m$.

Then we call S a *(learning) strategy* and, for every $n \in \mathbb{N}$, $S(t^n)$ a *hypothesis* of S. Any indexable class $\mathcal{L}$ of languages is said to be inferrable (from positive data) (or learnable for short) if and only if there is an indexing $(L_i)_{i\in\mathbb{N}}$ of $\mathcal{L}$ such that $\mathcal{L} = (L_i)_{i\in\mathbb{N}}$ is inferrable from positive data. We refer to this model as the *LIM model*†.

Thus, a learning strategy is obliged to *converge* to a correct description of a potentially infinite language after having read a finite number of examples, but it does not stop the procedure, i. e. it does not need to decide on the question of whether or not it has reached its final hypothesis. This behaviour explains the term of learning "in the limit".

It is a well-known fact that the learnability of an indexable class $\mathcal{L}$ is independent from the choice of the indexing – i. e., if $(L_i)_{i\in\mathbb{N}}$ is a corresponding indexing such that $\mathcal{L} = (L_i)_{i\in\mathbb{N}}$ is inferrable from positive data then, for *every* indexing $(H_i)_{i\in\mathbb{N}}$ *comprising* $\mathcal{L}$, there is a learning strategy which, for every $L \in \mathcal{L}$ and for every text $t \in \text{text}(L)$, converges to an index j such that $H_j = L$ (cf. Lange [40]). Thus, the LIM model as introduced above can be applied to any indexable class $\text{PAT}_\star \subseteq \text{ePAT}$ (or, alternatively, $\text{PAT}_\star \subseteq \text{nePAT}$) of pattern languages in the following simplified manner: If we ask for the inferrability of $\text{PAT}_\star$ then we can simply investigate the existence of a computable learning strategy which, for every language $L \in \text{PAT}_\star$ and for every text $t \in \text{text}(L)$, converges to a *pattern* α such that $L = L_{\text{E}}(\alpha)$ (or $L = L_{\text{NE}}(\alpha)$, respectively). In fact, however, we do not argue on the learnability of pattern languages in this algorithmic way, but we use a "topological" characterisation of inferrable indexed families that is due to Angluin [5]. This proof method is explained in Chapter 3.3.

We shall not conceal that, among the wide range of literature on inductive inference, there exist various interpretations of the paradigm of learning in the limit which in parts significantly differ from our model (actually, as mentioned above, our definition on the LIM model is a mere special version of the view formalised by Gold [24], who does not restrict the considerations to indexable classes). In particular, this holds for the following aspects of the model:

- the input for the learning strategy (e. g. positive and negative examples for the languages to be identified),
- the objects to be learned (e. g. general classes of languages instead of indexed families), and, possibly as a consequence thereof,

†In literature, this model is also referred to as *EX*-learning (short for: *explanatory*).

- the output of the strategy (general hypotheses, such as arbitrary representations of language generators or acceptors, instead of indices of language decision procedures) and

- the learning goal (e. g. replacing the *syntactical convergence* of hypotheses by a *semantical convergence*, i. e. using in the above definition the condition "there is an $m \in \mathbb{N}$ with $L_{S(t^n)} = L$ for every $n \geq m$" instead of condition 2).

The latter modification leads to an approach that typically is referred to as *BC-learning* (short for: *behaviourally correct*), see, e. g., Baliga et al. [7]. Note that – despite of the more general interpretation of algorithmic learning reflected by its definition – the learning power of BC-learning does not exceed that of the LIM model as long as it is applied to indexed families with recursive indexings as hypothesis spaces (an insight that, according to Lange and Zilles [45], is folklore). Thus, all of our results to be given in Chapters 5.2.1 and 6.1.1 hold in both models (and, additionally, numerous other models of inductive inference and, more general, algorithmic learning theory not to be mentioned in the present thesis; see, e. g., Lange and Zilles [45] again); hence, our insights are not as restricted as it might be suggested by our decision to choose the LIM model.

Finally, we wish to mention that the view on learning associated to the LIM model can also be applied to classes of recursive functions instead of languages. Fundamental results and further references on this broad and well-established field are provided, e. g. by Barzdin and Freivald [8], Klette and Wiehagen [38], Angluin and Smith [5] and Odifreddi [68].

Chapter 3.3 is dedicated to several results on inductive inference of languages to be found in literature. In this regard, we focus on those insights dealing with the learnability of pattern languages.

Chapter 3

The status quo

In the present chapter, we give a brief survey on the current state of knowledge with respect to the concepts under consideration, occasionally augmented by minor new insights and corresponding proofs. Where appropriate, we complement these descriptions by compact lists of references facilitating further reading on selected topics.

Additional information on the foundations of the fields of discrete mathematics to be discussed in the subsequent chapters are provided by, e. g., Rozenberg and Salomaa [86] (w. r. t. various topics in the theory of formal languages), Lothaire [48, 49] and Choffrut and Karhumäki [12] (w. r. t. combinatorics on words), Harju and Karhumäki [27] (w. r. t. insights into morphisms), Berstel and Perrin [10] (w. r. t. the theory of codes), Mateescu and Salomaa [55] and Salomaa [90] (w. r. t. pattern languages), and finally Klette and Wiehagen [38], Angluin and Smith [5], Zeugmann and Lange [108], Odifreddi [68], Jain et al. [31] and Lange [40] (w. r. t. inductive inference). Note that we do not go into any aspects related to the time or space complexity of computable functions (cf. Papadimitriou [70]).

3.1 Codes, endomorphisms, collaborating and conflicting morphisms

In many algebraic structures, the homomorphisms, i. e. the mappings that are compatible with a binary operation of the structure, form an intensively studied type of functions. This particularly holds for word monoids, where morphisms (as we, in accordance with the major part of the related literature, prefer to call homomorphisms in this context) can be considered the probably most natural functions, as their definition immediately implies that they map a string α onto a string w by simply concatenating the images of the *symbols* occurring in α (of course in accordance with the number and order of these letters in α), so that some structural properties of α may potentially be reflected by w (especially if the corresponding morphism is injective). Consequently, there exist virtually countless studies dealing with the properties of morphisms, and therefore we do not give an exhaustive survey on the current state of knowledge and the literature related to this subject. Nevertheless, we wish to provide a brief list of references to a number of example areas, which is meant to integrate our studies on the ambiguity of morphisms in the overall research on morphisms and to discuss the question of whether previous literature contains substantial insights into this main subject of our thesis.

First of all, morphisms are essential for the field of *coding theory* (cf. Berstel and Perrin [10] and, for a different presentation, Jürgensen and Konstantinidis [34]). Coding theory explicitly addresses a problem that we tackle implicitly, namely the construction of a string over an alphabet $\mathcal{B}$ that contains as much information as possible about another string over an alphabet $\mathcal{A}$ (frequently satisfying $\mathcal{B} \subset \mathcal{A}$). To this end, however, coding theory fixes an *injective* morphism mapping the strings in $\mathcal{A}^*$ onto selected strings in $\mathcal{B}^*$, so that it largely deals with isomorphic (sub-)semigroups. In such a context, it is of course not necessary to consider the ambiguity of the morphic images in $\mathcal{B}^*$, since the respective single isomorphism under considerations establishes a reliable, traceable and invertible connection between a string in $\mathcal{A}^*$ and its counterpart in $\mathcal{B}^*$. As a consequence of its subject, coding theory largely (but not exclusively) deals with morphisms that, similarly to those considered in the present thesis, map a string over a larger alphabet (e. g. of infinite size) onto a string over some smaller alphabet (that, e. g., might merely consist of two distinct letters). Contrary to this, several branches of formal language theory and combinatorics on words are rather concerned with endomorphisms, which only play a marginal role in our thesis (e. g. as residual-preserving morphisms, cf. Chapter 2.2). It is surely not surprising that, as soon as such endomorphisms are considered, the idea of iterating these morphisms gains in importance. Among these areas, we regard it as mandatory to mention the widespread research on *morphic sequences* of various kinds (cf. Choffrut and Karhumäki [12] and Allouche and Shallit [2]), such as the famous Fibonacci word (which results from the iteration of the morphism σ_{f} given by $\sigma_{\mathrm{f}}(\mathtt{a}) := \mathtt{a}\,\mathtt{b}$ and $\sigma_{\mathrm{f}}(\mathtt{b}) := \mathtt{a}$ when starting by $\mathtt{a}$) and the Thue-Morse word (which is generated by the iteration of the morphism σ_{tm} given by $\sigma_{\mathrm{tm}}(\mathtt{a}) := \mathtt{a}\,\mathtt{b}$ and $\sigma_{\mathrm{tm}}(\mathtt{b}) := \mathtt{b}\,\mathtt{a}$). The extensively studied Lindenmayer systems or *L systems* (cf. Kari et al. [36]) constitute a second topic which in parts is based on iterated endomorphisms. Although this way of applying morphisms does not at all conform with our concept, many of our subsequent results are based on fixed points of morphisms (cf. Chapter 5.1), the importance of which for L systems is evident. Furthermore, the definition of *word equations* (cf. Choffrut and Karhumäki [12]) essentially rests upon morphisms, as a word equation is a pair of strings α, β over some alphabet, with respect to which one examines the existence of an (if applicable, terminal-preserving) morphism σ such that $\sigma(\alpha) = \sigma(\beta)$. In the subsequent Chapter 3.1.1, we briefly discuss a famous problem for word equations and give some corresponding references. Finally, numerous aspects in the *algebraic theory of automata* (cf. Pin [71]) nontrivially depend on various types of morphisms.

In all of these fields, the usage of morphisms has a common trait, namely the fact that normally a single fixed morphism is considered which is applied to each string in some set of strings. In contrast to this, we are rather interested in a setting where several morphisms are applied to one and the same string. In the light of an algebraic view on word monoids – which, such as in coding theory, largely deals with isomorphic free monoids and related topics – this idea of *collaborating* morphisms (as it is designated by Harju and Karhumäki [27]) is surely rather unorthodox; nevertheless, it is well-established, though probably not as extensively studied as the above areas. The perhaps most basic approach to this view on morphisms is formalised by the concept of the *equality set* E of just two distinct morphisms σ, τ, which is defined by $E(\sigma, \tau) := \{\alpha \mid \sigma(\alpha) = \tau(\alpha)\}$ (cf. Harju and Karhumäki [27]). Equality sets and, hence, the combinatorics on morphisms are deeply embedded into the theory of computation; for instance, by equality sets, the recursively enumerable sets can be characterised (cf. Culik II [14]) as well as major complexity classes (cf. Mateescu et al. [56]). Furthermore, the famous undecidable Post Correspondence Problem (PCP) can be immediately interpreted as a problem on equality sets. We discuss the latter subject in the subsequent Chapter 3.1.1. Besides this computational aspect involved in equality sets, additional combinatorial questions of intrinsic interest arise from the related theory. With regard to the focus of the present thesis on the ambiguity of morphisms, the examination on equality sets as conducted by Mateescu and Salomaa [52, 53] and Lipponen and Păun [47] is of particular interests since these authors study the shape of shortest strings in equality sets. Consequently, in a sense, an important part of our studies complements their research as we, among other topics, deal with the problem of finding patterns for which there exists an unambiguous morphism – or, in terms of equality sets, the problem of finding patterns α and morphisms σ such that, for any other morphism τ, $\alpha \notin E(\sigma, \tau)$. As to be explained below, referring to the terminology of the Post Correspondence Problem, this means that [47, 53, 52] seek for shortest solutions to some instance of the PCP, whereas we enquire for the non-solutions to the PCP.

While combinatorics on equality sets, thus, is based on the comparison of *two* (fixed) morphisms, the ambiguity of morphisms – which is tentatively defined in Chapter 2.1 and extensively and formally introduced in Chapter 4 – initially considers *all* (if applicable, terminal-preserving) morphisms mapping any (fixed) string in an infinitely generated free semigroup onto strings in a finitely generated free monoid. Therefore the challenging aspect of our research is associated to the understanding of the variety of potentially "equivalent" morphisms (with respect to a fixed pattern) rather than of the multitude of finite strings over infinite alphabets that can be contained in the same equality set. As a consequence thereof, in the light of our approach to combinatorics on morphisms, it seems as if whenever we face an ambiguous word w (with respect to any pattern α) then, ironically, any morphisms σ, τ satisfying $\sigma(\alpha) = w = \tau(\alpha)$ actually are not collaborating, but *conflicting* since they suggest contradictory interpretations of the structure of α (cf. our motivating explanations given in Chapter 1). Nevertheless, as briefly mentioned above, some of our subsequent results can be immediately interpreted as statements on the PCP; this is explicitly demonstrated by Corollary 5.13. Conversely, however, we do not know any method to transform known facts on equality sets into profound results answering basic questions on the ambiguity of morphisms.

Subject to the large and prominent research fields listed above, our combinatorial view on morphisms, hence, shows major immediate connections to equality sets only. Still, there exists a minor yet well-established branch of algorithmic learning theory and formal language theory dealing with a related topic, namely the *pattern languages* as defined in Chapter 2.3 and to be further discussed in Chapters 3.2 and 3.3. By definition, and according to our motivating explanations in Chapter 1, the concept of a pattern language as the set of all morphic images of a fixed pattern exactly corresponds to our combinatorial setting, and therefore one might expect that some properties of pattern languages depend on the existence of particular morphisms that are (un-)ambiguous with respect to a generating pattern of the language under consideration. The theorem by Reidenbach [76, 81] on inductive inference of E-pattern languages mentioned in Chapter 1 and formally described by Theorem 3.16 strengthens this assumed connection, and our various results on the ambiguity of morphisms to be presented in the subsequent chapters

– which, in Chapters 5.2.1, 6.1.1, 6.2.1 and 6.3.1, are used for presenting (partial or exhaustive) answers to several and well-known open problems for pattern languages – further substantiate this relation. Conversely, however, the current state of knowledge on pattern languages does not contain any noteworthy insights into the ambiguity of morphisms. At first glance, this seems a bit surprising as there exists a profound work by Mateescu and Salomaa [54] on the so-called *ambiguity of pattern languages* which, with respect to any fixed pattern α, studies the existence of a bound $n \in \mathbb{N}$ such that, for all words $w \in L(\alpha)$, there are at most n distinct morphisms σ satisfying $\sigma(\alpha) = w$. The main result presented in [54] states that there are patterns for which this bound n exists and is strictly greater than 1. In other words, Mateescu and Salomaa enquire for the "maximum ambiguity" of the words in a pattern language (with respect to any generating pattern of that language), and they demonstrate that, for particular patterns, this value is nontrivially bounded. Hence, they canonically apply the concept of "ambiguity" as normally understood in formal language theory to pattern languages. Contrary to this, as explained in Chapter 1 and to be formally declared in Chapter 4, we largely discuss the question of whether, with respect to any pattern, we can find an unambiguous morphism – or at least a morphism, the ambiguity of which is restricted in a particular way. Thus, we rather deal with the "minimum ambiguity" in a pattern language. In addition to the problem that the goal of Mateescu and Salomaa [54] differs from that of our thesis, the technique introduced in [54] is connected to challenging problems on word equations, whereas we have to develop our own technical concepts which (apart from a particular subject that utilises some knowledge on fixed points of morphisms, cf. Chapter 5.1) are not based on a well-established methodology.

Summarising the present chapter, we thus have to state that previous literature on morphisms is not very profitable for studying the ambiguity of morphisms, as it chooses different settings or poses different questions: Coding theory basically deals with mappings of strings over large alphabets onto strings over smaller alphabets and it shares our vague motivation of finding morphisms that preserve the structure of the preimage (cf. Chapter 1), but it normally considers a single fixed morphism which is applied to all strings in a given set. The research on morphic sequences focusses on iterated endomorphisms, and so does the relevant part of L system; therefore it strongly differs from our studies. Word equations are based on two patterns which are mapped by a single morphism onto the same word. Consequently, they can only potentially serve for answering particular technical questions on the ambiguity of morphisms (and we do not make use of this option). Contrary to this, at first glance, the concept of equality sets shows deeper connections to our work, because it studies two morphisms which, when applied to a pattern, lead to same image. However, it considers the morphisms to be fixed, whereas we regard a fixed pattern α (and, if appropriate, a given morphism σ) and we ask for the existence and, if applicable, the shape of the morphisms τ satisfying $\sigma(\alpha) = \tau(\alpha)$. Finally, the definition of pattern languages deals with the set of all morphic images (in some fixed free monoid) of a given pattern, which thus exactly matches with our view, but the previous approaches to pattern languages either do not consider the ambiguity of these words at all, or they yield insights into the maximum number of morphisms which can map the pattern onto the same word, whereas we ask for the existence of a word with exactly one generating morphism, or of a word such that all of its generating morphisms satisfy particular structural requirements.

Consequently, our subject shows only few immediate or well-known relations to prominent research fields dealing with morphisms, and all these connections so far are rather shallow. In particular, to our best knowledge, there is not a single paper directly or indirectly addressing our concrete research questions to be posed in Chapter 4. Our considerations in the subsequent Chapters 5 and 6, however, shall reveal that the ambiguity of morphisms, due to its fundamental character, in fact is deeply interwoven with numerous important properties of other domains based on morphisms.

3.1.1 Decision problems

The present chapter is dedicated to two prominent general decision problems for strings and morphisms derived from our explanations in the above Chapter 3.1. Note again that we assume the reader to be familiar with basic insights in the theory of computation and, more precisely, the theory of decidable sets (cf. Odifreddi [67]), and therefore we do not provide any general explanations on this topic.

We begin with the probably most prominent problem in this field, namely the *Post Correspondence Problem (PCP)* (cf. Post [72]), dealing with a general setting that is similar to the problem of ambiguity of morphisms: in a terminology that is adapted to our subject, the PCP asks for the existence of a total computable function χ such that, for every pair of alphabets $\mathcal{A}, \mathcal{B}$ and for every pair of morphisms $\sigma, \tau : \mathcal{A}^* \longrightarrow \mathcal{B}^*$, $\chi(\sigma, \tau) = 1$ if and only if there exists at least one pattern $\alpha \in \mathcal{A}^+$ satisfying $\sigma(\alpha) = \tau(\alpha)$;[†] in this case, α is said to be a *solution* to the PCP for σ and τ. Thus – in terms of the equality set $E(\sigma, \tau)$ – the PCP is the *emptiness problem* for $E(\sigma, \tau)$.

As shown by Post, the function χ in question does not exist:

Theorem 3.1 (Post [72]). *The Post Correspondence Problem is undecidable.*

Since the PCP is one of the most influential decision problems – which results from its seeming simplicity and the fact that it has been among the first undecidable problems to be known which do not immediately originate from a recursion-theoretic context – there exists a large number of publications on the subject, many of them discussing decidable subproblems. An extensive respective survey is provided by Harju and Karhumäki [27].

The second well-known decision problem on morphisms to be introduced in the present chapter is the *solvability problem* for word equations. This problem asks for the existence of a total computable function χ which, for every alphabet $\mathcal{A}$ and every pair of strings $\alpha, \beta \in \mathcal{A}^+$, decides on whether or not there exists an alphabet $\mathcal{B}$ and an (if necessary, terminal-preserving) morphism $\sigma : \mathcal{A}^* \longrightarrow \mathcal{B}^*$ such that $\sigma(\alpha) = \sigma(\beta)$. This famous problem, constituting sort of a noncommutative counterpart of the problem of the solvability of Diophantine equations (cf. Matiyasevich [57]), is known to be decidable:

Theorem 3.2 (Makanin [50]). *The solvability problem for word equations is decidable.*

For further information on the problem and Makanin's algorithm see, e. g., Schulz [94], Gutiérrez [25] and Lothaire [49].

Note that the setting of the solvability problem can also be adapted such that it matches with problems related to the ambiguity of morphisms. This option is mentioned

[†] Post [72] as well as parts of the subsequent literature use a different terminology which, nevertheless, is fully equivalent to our description.

by Mateescu and Salomaa [54], but does not serve our purposes, and therefore we do not discuss it in detail.

3.2 Properties of pattern languages

We proceed with a description of elementary properties of the class of pattern languages (as introduced in Chapter 2.3). Although numerous basic properties of pattern languages are not completely resolved yet, the related theory meanwhile is well-established and discussed in a large number of papers. Therefore – and due to the fact that the present thesis is primarily dedicated to combinatorics on morphisms – we do not give a comprehensive survey on the subject, but rather focus on some aspects that are crucial for demonstrating the applicability of our combinatorial results (and, hence, for substantiating the relevance of our approach). For additional information on pattern languages, the reader can consult, e. g., the surveys presented by Salomaa [88, 89], Mateescu and Salomaa [55] and Salomaa [90], the papers by Angluin [3], Filè [19], Jiang et al. [32, 33] and Ohlebusch and Ukkonen [69] dealing with basic properties of pattern languages and related decision problems, the abovementioned paper by Mateescu and Salomaa [54] on the ambiguity of pattern languages, the numerous considerations on modifications and enhancements of the original model – e. g. by Dassow et al. [16], Kari et al. [35], Mitrana et al. [60] and Georgescu [23] – and, finally, the pertinent references in all these publications. With regard to the large number of *learning theoretical* papers on pattern languages, we present and briefly discuss a selection of important references in Chapter 3.3.

Among the properties of pattern languages to be presented within the scope of this thesis, we first of all wish to mention that the class of pattern languages is known to be reasonably (and for many applications sufficiently) rich:

Proposition 3.1 (Angluin [3]). *Let Σ be a finite alphabet, $|\Sigma| \geq 2$. Then the class of pattern languages over Σ is incomparable to the class of regular languages over Σ, and it is incomparable to the class of context-free languages over Σ.*

Additionally, the relation of pattern languages and context-sensitive languages is also understood. The corresponding result is sort of folklore and can be proven using the characterisation of context-sensitive languages as the class that is exactly accepted by linear bounded automata:

Proposition 3.2. *Let Σ be a finite alphabet. Then the class of pattern languages over Σ is a proper subclass of the class of context-sensitive languages over Σ.*

Despite of the general insights presented in Proposition 3.1 and Proposition 3.2, many details on the relation between pattern languages and the above languages of the Chomsky hierarchy are still open. In particular, this holds for the existence of context-free pattern languages that are not regular:

Open Problem 3.1. *Let Σ be a finite alphabet, $|\Sigma| \geq 2$. Are the class of pattern languages over Σ and the class of nonregular context-free languages over Σ disjoint?*

While a general characterisation of patterns generating regular languages is still open, some sufficient conditions are known, the first of which is evident (and reflected in the name of regular patterns as introduced in Chapter 2.2):

Proposition 3.3. *Let α be a regular pattern. Then $L_{\text{NE}}(\alpha)$ is regular and $L_{\text{E}}(\alpha)$ is regular.*

We do not provide a proof for Proposition 3.3 since, for every regular pattern α, it can be straightforward verified by constructing an appropriate automaton (or Type 3 grammar) accepting (or generating, respectively) $L_{\text{NE}}(\alpha)$ or $L_{\text{E}}(\alpha)$.

The next condition on regular pattern languages[†] shows close connections to the concept of *unavoidability* of patterns as examined in combinatorics on words (cf., e. g., Thue [102], Bean et al. [9], Lothaire [49]). In this context, a terminal-free pattern α is said to be *unavoidable* (on an alphabet Σ and an $m \in \mathbb{N}$) provided that *every* word $w \in \Sigma^*$ with $|w| \geq m$ contains a subword v that "follows" α, i. e. there exists a morphism σ such that $\sigma(\alpha) = v$. Otherwise, i. e. if there is no such bound m, α is called *avoidable* (on Σ). The most elementary nontrivial unavoidable pattern is the *square* (i. e. $\alpha := 1 \cdot 1$), which is unavoidable on binary alphabets and words of length at least 4, but avoidable on alphabets with more than two letters (shown by Thue [102]). Note that this connection of unavoidable patterns and pattern languages also affects the inclusion problem for NE-pattern languages (cf. Chapter 3.2.1).

Due to manifest focus of these considerations concerning unavoidability on terminal-free patterns – after all, a pattern $\alpha \in \text{Pat} \setminus \text{Pat}_{\text{tf}}$ is always avoidable on non-unary alphabets – we restrict our second condition on regular pattern languages to those generated by terminal-free patterns:

Proposition 3.4. *Let Σ be an alphabet. Let $\alpha, \beta \in \text{Pat}_{\text{tf}}$ be patterns such that $\beta = i_1 \cdot \alpha \cdot i_2$ with $i_1, i_2 \notin \text{var}(\alpha)$ and $i_1 \neq i_2$. Then $L_{\text{NE},\Sigma}(\beta)$ is regular if there exists an $m \in \mathbb{N}$ such that α is unavoidable on Σ and m.*

Proof. This proof for Proposition 3.4 extends an argumentation given by Lange [41] on the pattern $1 \cdot 2 \cdot 2 \cdot 3$.

If α is unavoidable on Σ and m then, for every word w with $|w| \geq m+2$, $w \in L_{\text{NE},\Sigma}(\beta)$. Thus, $L_{\text{NE},\Sigma}(\beta)$ is a cofinite language (i. e. the complement of a finite language). Since every finite language is regular and the class of regular languages is closed under complement, every cofinite language is regular as well. Consequently, $L_{\text{NE},\Sigma}(\beta)$ is regular. □

In Proposition 3.4, it is mandatory to include the variables i_1 and i_2 in the pattern β since an unavoidable pattern α merely covers a *sub*word of a sufficiently long word w. Consequently, in general, we may observe that $L(\alpha)$ is not regular, a fact which is beautifully illustrated by one of the most basic and prominent examples for a non-context-free language, namely $L_{\text{NE},\Sigma}(1 \cdot 1)$, when compared to the regular language $L_{\text{NE},\Sigma}(1 \cdot 2 \cdot 2 \cdot 3)$, $|\Sigma| = 2$. Furthermore, note that it does not lead to a more meaningful statement if we

[†] We do not use the term "regular pattern language" in the way it is normally understood in literature: Frequently, and due to Proposition 3.3, this term is merely meant to denote a language L for which there is a regular pattern α with $L = L(\alpha)$. Contrary to this, we call any language a *regular pattern language* provided that it is both regular and a pattern language. If we wish to refer to the restricted former type then – in a possibly slightly cumbersome, but surely unequivocal manner – we speak of a *language generated by a regular pattern*.

extend Proposition 3.4 to E-pattern languages – though the proof is verbatim the same in that case. This results from the fact that, for every pattern β satisfying the condition of the said proposition, $L_{\mathrm{E},\Sigma}(\beta)$ due to the single occurrences of the variables i_1 and i_2 necessarily equals Σ^*, which of course is a trivial regular language. Moreover, we wish to mention that the restriction of Proposition 3.4 to terminal-free patterns can be weakened easily by extending it to patterns such as $\beta = i_1 \cdot \mathtt{a}_1 \cdot i_2 \cdot \alpha \cdot i_3 \cdot \mathtt{a}_2 \cdot i_4$ for some letters $\mathtt{a}_1, \mathtt{a}_2 \in \Sigma$. Any formal statement on such an extension, however, requires a slightly more sophisticated reasoning as the resulting language is not cofinite (if $|\Sigma| \geq 2$).

Finally we may state that, for a unary alphabet, even the full class of pattern languages is a proper subclass of that of regular languages:

Proposition 3.5. *Let Σ be an alphabet, $|\Sigma| = 1$. Then, for every pattern $\alpha \in \mathrm{Pat}_\Sigma$, $L_{\mathrm{NE},\Sigma}(\alpha)$ is regular and $L_{\mathrm{E},\Sigma}(\alpha)$ is regular.*

Proof. We do not provide any general explantions on regular languages and automata – which are required in order to understand this formal proof. Instead of that, we refer the reader to Yu [107].

To begin with, note that there is hardly any standard proof technique for showing that a language is regular – such as the construction of an appropriate grammar or the characterisation of regular languages given in the so-called Myhill-Nerode Theorem (cf. Myhill [63], Nerode [64]) – which does not immediately lead to the stated result. Therefore we only give a proof on one of the two cases covered by Proposition 3.5, namely on E-pattern languages. Moreover, this proof does not use any of the techniques mentioned above, but explains how to construct a *nondeterministic finite automaton (NFA)* exactly accepting the language of a given pattern α.

Hence, let $\Sigma := \{\mathtt{a}\}$ be an alphabet, and let $\alpha \in \mathrm{Pat}_\Sigma$. If $\mathrm{var}(\alpha) = \emptyset$ then Proposition 3.5 trivially holds true. Thus, we assume that α contains at least one variable. Let F be the set of all frequencies of variables in α, i. e. $F := \{m \in \mathbb{N} \mid |\alpha|_j = m \text{ for some } j \in \mathrm{var}(\alpha)\}$. Obviously, F is nonempty and finite; consequently, for some $n \leq |\mathrm{var}(\alpha)|$, we can write $F = \{m_1, m_2, \ldots, m_n\}$. Finally. let $p := |\mathrm{res}(\alpha)|$.

We now can observe that Σ being unary implies that the language of α does not depend on the order of its variables, but solely on p and F. Therefore, the following NFA A_α exactly accepts $L_{\mathrm{E},\Sigma}(\alpha)$: A_α contains $1+p+(m_1-1)+(m_2-1)+\ldots+(m_n-1)$ different states, designated as s_0 (the initial state), $s_1, s_2, \ldots, s_p$ (where s_p is the single accepting state of A_α) and $s_{1,1}, s_{1,2}, \ldots, s_{1,m_1-1}$, $s_{2,1}, s_{2,2}, \ldots, s_{2,m_2-1}$, and so on up to $s_{n,1}, s_{n,2}, \ldots, s_{n,m_n-1}$. The transition relation of A_α leads to a single path of length p connecting s_0 and s_p via $s_1, s_2, \ldots, s_{p-1}$ (note that s_0 equals s_p if and only if $\alpha \in \mathrm{Pat}_{\mathrm{tf}}$) and n different loops of length m_i, $1 \leq i \leq n$ starting and ending at s_p. Thus, more formally,

- for every i' with $0 \leq i' \leq p-1$, there is a transition which connects $s_{i'}$ to $s_{i'+1}$,
- for every i with $1 \leq i \leq n$, there is a transition which connects s_p to $s_{i,1}$ and a transition which connects s_{i,m_i-1} to s_p again,
- for every i with $1 \leq i \leq n$ and for every i' with $1 \leq i' \leq m_i-2$, there is a transition which connects $s_{i,i'}$ to $s_{i,i'+1}$ and

there are no other transitions in the NFA. Consequently, A_α looks as follows:

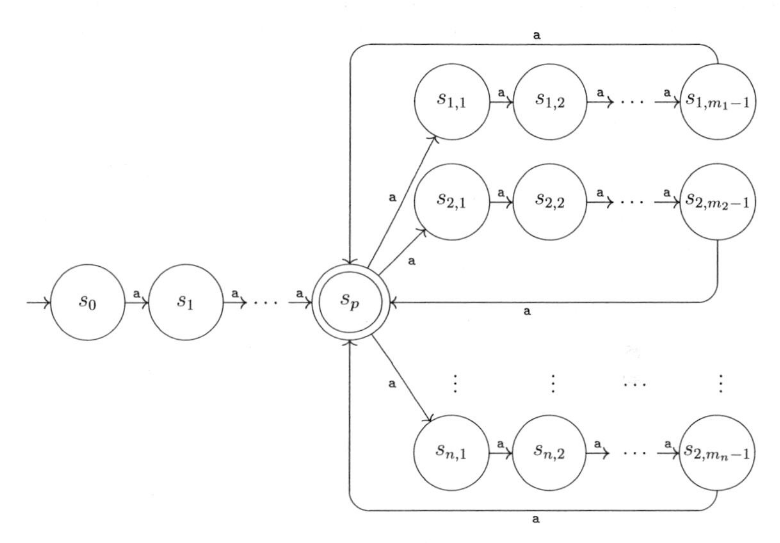

Now it can easily be verified that the language accepted by A_α equals $L_{\mathrm{E}}(\alpha)$: For any morphism σ and for every variable $j \in \mathrm{var}(\alpha)$, there is a $c_j \in \mathbb{N}_0$ such that $\sigma(j) = \mathtt{a}^{c_j}$. Thus, given $\sigma(\alpha)$ as an input, A_α reaches the accepting state s_p when choosing, for every $j \in \mathrm{var}(\alpha)$, c_j times the loop of length $|\alpha|_j$ (which exists due to the definition of F). Conversely, for any word $w \in \Sigma^*$ accepted by A_α we know that, for some $d_i \in \mathbb{N}_0$, $1 \leq i \leq n$, the procedure passes d_i times the loop of length m_i. If we now, for every i, choose any variable j with $|\alpha|_j = m_i$ (which exists due to the definition of F) and define a morphism σ by $\sigma(j) := \mathtt{a}^{d_i}$ and, for any variable k which is never chosen in this procedure, $\sigma(k) := \varepsilon$ then $\sigma(\alpha) = w$.

With regard to NE-pattern languages, we have to define $p := |\alpha|$ instead of $p = |\mathrm{res}(\alpha)|$. Then, with marginal modifications, we can apply the same construction and argumentation as given on E-pattern languages. This proves Proposition 3.5. □

We illustrate our explanations on pattern languages over unary alphabets being regular with an example for the construction of an NFA – which, by the way, does not need to be the minimum one accepting the corresponding pattern language – as introduced in the proof of Proposition 3.5:

Example 3.1. Let $\Sigma := \{\mathtt{a}\}$ and $\alpha := \mathtt{a} \cdot 1 \cdot 2 \cdot \mathtt{a} \cdot \mathtt{a} \cdot 2 \cdot 3^4 \cdot 1$. Consequently, $F = \{2, 4\}$, $p = 3$ and, thus, A_α looks as follows:

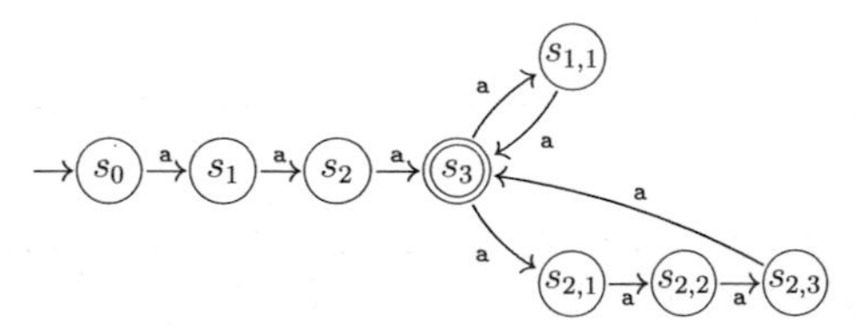

Evidently, the language accepted by A_α equals $L_{\mathrm{E},\Sigma}(\alpha)$. A minimum NFA with the same property is obtained from A_α by deleting the states $s_{2,1}, s_{2,2}, s_{2,3}$ and the corresponding transitions – a modification which, moreover, leads to a DFA. ◇

On the one hand, it is evident that pattern languages over *unary* alphabets strongly differ from other pattern languages since they do not contain any word that can reflect the order of variables in the generating pattern (as shown by Proposition 3.5 and its proof). On the other hand, pattern languages over *infinite* alphabets always contain a word w that, in a sense, is simply a "renaming" of the generating pattern α, i. e. w is generated by a morphism σ such that, for all $i \in \mathrm{var}(\alpha)$, it is $|\sigma(i)| = 1$ and $\sigma(i) \notin \mathrm{term}(\alpha)$ and, for every $i' \in \mathrm{var}(\alpha)$ with $i \neq i'$, $\sigma(i) \neq \sigma(i')$ (see, e. g., the proof for Proposition 3.12). Thus, roughly speaking, morphisms over unary alphabets show very weak and morphisms over infinite alphabets show very strong capabilities to preserve the structure of a preimage – by the way, this also explains the marginal role of suchlike alphabets in the research on codes and many other fields related to morphisms. As a consequence of this, for the said two alphabet sizes, many properties of corresponding pattern languages can be understood much easier than for finite alphabets with at least two letters; this holds, e. g., for decision problems (cf. Chapter 3.2.1) and for many questions related to inductive inference of pattern languages (cf., e. g., Chapter 3.3 and Chapter 6.1.1). Therefore, in the present thesis, we shall largely deal with the "nontrivial" alphabets of at least binary, but still finite size.

We conclude this chapter with some basic closure properties of the class of pattern languages:

Proposition 3.6 (Angluin [3]). *The class of pattern languages is closed under concatenation and reversal. The class of pattern languages is not closed under union, complement and intersection.*

Note that Angluin [3] proves Proposition 3.1 and Proposition 3.6 for nePAT only, but the corresponding argumentation can be immediately extended to ePAT. With regard to many different properties, however, E-pattern languages and NE-pattern languages show a different behaviour. For instance, this holds for several decision problems. We examine some of these problems in the subsequent chapter.

3.2.1 Decision problems

We now complete our description of selected properties of pattern languages by extensive notes on some fundamental decision problems; for the sake of a more concise presentation, we do not introduce these problems in the general terminology for arbitrary classes of formal languages, but we tailor them for our needs. Again, we refer the reader to Odifreddi [67] for a description of the necessary recursion-theoretic background of our explanations.

A first decidability result on pattern languages can immediately be concluded from the decidability of the solvability problem for word equations (cf. Chapter 3.1.1). Hence, let Σ be an alphabet. Let $\mathrm{PAT}_{\star,\Sigma} \subseteq \mathrm{ePAT}_\Sigma$ (or $\mathrm{PAT}_{\star,\Sigma} \subseteq \mathrm{nePAT}_\Sigma$) be a class of pattern languages over Σ. Then we say that the *disjointness* problem is *decidable* for $\mathrm{PAT}_{\star,\Sigma}$ if and only if there exists a total computable function χ such that, for every pair of patterns $\alpha, \beta \in \mathrm{Pat}_\Sigma$ with $L_{\mathrm{E},\Sigma}(\alpha), L_{\mathrm{E},\Sigma}(\beta) \in \mathrm{PAT}_{\star,\Sigma}$ (or, alternatively, $L_{\mathrm{NE},\Sigma}(\alpha), L_{\mathrm{NE},\Sigma}(\beta) \in$

$\mathrm{PAT}_{\star,\Sigma}$), χ decides on whether or not $L_{\mathrm{E},\Sigma}(\alpha) \cap L_{\mathrm{E},\Sigma}(\beta) = \emptyset$ (or, alternatively, $L_{\mathrm{NE},\Sigma}(\alpha) \cap L_{\mathrm{NE},\Sigma}(\beta) = \emptyset$).

The connection to the solvability problem now reads as follows: We regard any morphism ϕ that is a renaming of variables and that leads to $\mathrm{var}(\alpha) \cap \mathrm{var}(\phi(\beta)) = \emptyset$. Then the language of α and the language of $\phi(\beta)$ (which of course is equivalent to the language of β) are disjoint if and only if the word equation of α and $\phi(\beta)$ has no solution. Hence, the disjointness problem is just a special case of the solvability problem and therefore we may conclude:

Corollary 3.1. *Let Σ be an alphabet. Then the disjointness problem for* nePAT_Σ *is decidable, and the disjointness problem for* ePAT_Σ *is decidable.*

While we present this simple insight into the disjointness problem merely because, first, it follows immediately from Theorem 3.2 and, second, it has not been mentioned in literature so far (although it is known to have some learning theoretical impact, see Lange and Wiehagen [43]), the subsequent decision problems are crucial for the application of our results on the ambiguity of morphisms to a number of problems for pattern languages.

Let Σ be an alphabet. Let $\mathrm{PAT}_{\star,\Sigma} \subseteq \mathrm{ePAT}_\Sigma$ (or $\mathrm{PAT}_{\star,\Sigma} \subseteq \mathrm{nePAT}_\Sigma$) be a class of pattern languages over Σ. Then we say that the *membership* problem is *decidable* for $\mathrm{PAT}_{\star,\Sigma}$ if and only if there exists a total computable function χ such that, for every pair of a pattern $\alpha \in \mathrm{Pat}_\Sigma$ with $L_{\mathrm{E},\Sigma}(\alpha) \in \mathrm{PAT}_{\star,\Sigma}$ (or, alternatively, $L_{\mathrm{NE},\Sigma}(\alpha) \in \mathrm{PAT}_{\star,\Sigma}$) and a word $w \in \Sigma^*$, χ decides on whether or not $w \in L_{\mathrm{E},\Sigma}(\alpha)$ (or, alternatively, $w \in L_{\mathrm{NE},\Sigma}(\alpha)$).

This problem is decidable for the class of NE-pattern languages:

Proposition 3.7. *Let Σ be an alphabet. Then the membership problem is decidable for* nePAT_Σ.

The correctness of Proposition 3.7 can be seen immediately since, for any pattern α, any word w and the question of whether or not there exists a morphism σ with $\sigma(\alpha) = w$, the number of morphisms to be tested is bounded by the length of w and the number of letters occurring in w. According to Angluin [3], the membership problem for the full class of NE-pattern languages is NP-complete, a result, that due to Ehrenfeucht and Rozenberg [17] even holds for the subclass of terminal-free NE-pattern languages.

Moreover, the analogous result holds for E-pattern languages:

Proposition 3.8. *Let Σ be an alphabet. Then the membership problem is decidable for* ePAT_Σ.

Obviously, we can apply the above proof idea on Proposition 3.8 as well. Furthermore, adapting the corresponding proof by Angluin [3], Jiang et al. [32] show that the membership problem for the class of E-pattern languages is NP-complete, too.

Concerning the central topic of our thesis, namely the *ambiguity of morphisms* and morphic images, we can directly conclude from Propositions 3.7 and 3.8 and our related proof notes that this problem at least is decidable:

Proposition 3.9. *There exists a total computable function χ such that, for every alphabet Σ, for every $\alpha \in \mathrm{Pat}_\Sigma$ and for every $w \in \Sigma^*$, χ decides on whether or not w is ambiguous with respect to α.*

Thus, it is worth mentioning that in a constellation very similar to the Post Correspondence Problem (see Chapter 3.1.1) – we consider two different morphisms that, when applied to the same pattern, generate the same word – there is a different result on the decidability: we have a decidable problem if one morphism and the pattern are given, whereas we face an undecidable problem if both morphisms are given and the pattern is sought for.

We proceed with another elementary decision problem for pattern languages: For any alphabet Σ and any class $\mathrm{PAT}_{\star,\Sigma} \subseteq \mathrm{ePAT}_\Sigma$ (or $\mathrm{PAT}_{\star,\Sigma} \subseteq \mathrm{nePAT}_\Sigma$) of pattern languages over Σ, the *inclusion* problem is said to be *decidable* if and only if there exists a total computable function χ such that, for every pair of a patterns $\alpha, \beta \in \mathrm{Pat}_\Sigma$ with $L_{\mathrm{E},\Sigma}(\alpha), L_{\mathrm{E},\Sigma}(\beta) \in \mathrm{PAT}_{\star,\Sigma}$ (or, alternatively, $L_{\mathrm{NE},\Sigma}(\alpha), L_{\mathrm{NE},\Sigma}(\beta) \in \mathrm{PAT}_{\star,\Sigma}$), χ decides on whether or not $L_{\mathrm{E},\Sigma}(\alpha) \subseteq L_{\mathrm{E},\Sigma}(\beta)$ (or, alternatively, $L_{\mathrm{NE},\Sigma}(\alpha) \subseteq L_{\mathrm{NE},\Sigma}(\beta)$).

Because of the importance of the inclusion problem for our considerations on inductive inference (cf. Chapter 2.5 and, in particular, Chapter 3.3), we discuss the corresponding phenomena in a bit more detail. First of all, Jiang et al. [33] show that there is not a single procedure uniformly deciding the inclusion problem for all alphabet sizes, and this holds for both the NE and the E case. Thus, the *general inclusion problem* for pattern languages is undecidable:

Theorem 3.3 (Jiang et al. [33]). *There is no total computable function χ_{NE} which, for every alphabet Σ and for every pair of patterns $\alpha, \beta \in \mathrm{Pat}_\Sigma$, decides on whether or not $L_{\mathrm{NE},\Sigma}(\alpha) \subseteq L_{\mathrm{NE},\Sigma}(\beta)$.*

There is no total computable function χ_{E} which, for every alphabet Σ and for every pair of patterns $\alpha, \beta \in \mathrm{Pat}_\Sigma$, decides on whether or not $L_{\mathrm{E},\Sigma}(\alpha) \subseteq L_{\mathrm{E},\Sigma}(\beta)$.

In literature, Theorem 3.3 and its sophisticated proof occasionally seem to be interpreted in such a manner as if the inclusion problem for the full class of pattern languages over some/any *fixed* alphabet (just as defined in the present chapter) is shown to be undecidable, too. However, Theorem 3.3 as presented in [33] does *not* imply any suchlike statement, and we wish to emphasise that, according to Salomaa [91], there is no straightforward option to extend its proof such that it covers our more specific definition, which attaches the inclusion problem to any class of pattern languages over a distinct alphabet.

In fact, if we understand the inclusion problem in the way formally introduced above, its decidability is still unresolved for all standard alphabets:

Open Problem 3.2. *Let Σ be any finite alphabet, $|\Sigma| \geq 2$. Is the inclusion problem decidable for* nePAT_Σ*? Is the inclusion problem decidable for* ePAT_Σ*?*

Contrary to this, when considering unary or infinite terminal alphabets, appropriate answers can be given easily. With regard to NE-pattern languages over unary alphabets we can directly conclude that the decidability of the inclusion problem follows from the well-known decidability of the same problem for the class of regular languages and the fact that we can algorithmically transform a pattern into an equivalent nondeterministic finite automaton (as demonstrated by Proposition 3.5 and its proof):

Proposition 3.10. *Let Σ be an alphabet, $|\Sigma| = 1$. Then the inclusion problem is decidable for* nePAT_Σ.

Similar to the situation concerning the membership problem, the outcome for E-pattern languages equals that for NE-pattern languages:

Proposition 3.11. *Let* Σ *be an alphabet,* $|\Sigma| = 1$. *Then the inclusion problem is decidable for* ePAT_Σ.

Our argumentation on infinite alphabets is based on a simple (and decidable) sufficient condition, which holds for every alphabet size and which directly follows from the definitions of morphisms and pattern languages:

Theorem 3.4 (Angluin [3]). *Let* Σ *be an alphabet, and let* $\alpha, \beta \in \text{Pat}_\Sigma$. *Then* $L_{\text{NE},\Sigma}(\beta) \subseteq L_{\text{NE},\Sigma}(\alpha)$ *if there exists a terminal-preserving morphism* $\phi : (\mathbb{N} \cup \Sigma)^+ \longrightarrow (\mathbb{N} \cup \Sigma)^+$ *such that* $\phi(\alpha) = \beta$.

Additionally, with a small and evident modification, the above criterion can be adapted to E-pattern languages:

Theorem 3.5 (Jiang et al. [32]). *Let* Σ *be an alphabet, and let* $\alpha, \beta \in \text{Pat}_\Sigma$. *Then* $L_{\text{E},\Sigma}(\beta) \subseteq L_{\text{E},\Sigma}(\alpha)$ *if there exists a terminal-preserving morphism* $\phi : (\mathbb{N} \cup \Sigma)^* \longrightarrow (\mathbb{N} \cup \Sigma)^*$ *such that* $\phi(\alpha) = \beta$.

As shown by the following example, with regard to finite alphabets, both Theorem 3.4 and Theorem 3.5 do not establish a necessary condition for the inclusion of their corresponding type of pattern languages:

Example 3.2. Let $\Sigma := \{\mathtt{a}_1, \mathtt{a}_2, \dots, \mathtt{a}_n\}$, $n \geq 2$, be an alphabet of pairwise distinct letters. Let the patterns $\alpha, \beta \in \text{Pat}_\Sigma$ be given by $\alpha := 1 \cdot 2 \cdot 2 \cdot 3$ and $\beta := 1 \cdot \mathtt{a}_1 \cdot 1 \cdot \mathtt{a}_2 \cdot 1 \cdot [\dots] \cdot 1 \cdot \mathtt{a}_n \cdot 1$. Then $L_{\text{NE},\Sigma}(\alpha) \supset L_{\text{NE},\Sigma}(\beta)$ (since each word in $L_{\text{NE},\Sigma}(\beta)$ contains an inner square), but there is no terminal-preserving nonerasing morphism ϕ with $\phi(\alpha) = \beta$ (since β itself does not contain any square).

We do not know an analogous general example for E-pattern languages. However, for small terminal alphabets such as $\Sigma := \{\mathtt{a}, \mathtt{b}\}$ we can give tailor-made pairs of patterns α, β, namely, e. g., $\alpha := 1 \cdot \mathtt{a} \cdot \mathtt{b} \cdot 2$ and $\beta := 1 \cdot \mathtt{a} \cdot 2 \cdot \mathtt{b} \cdot 3$. Then $L_{\text{E},\Sigma}(\alpha) = L_{\text{E},\Sigma}(\beta)$, but there is no terminal-preserving morphism ϕ with $\phi(\alpha) = \beta$. Concerning selected other alphabet sizes, corresponding examples can be found in Chapter 6.2.1. ◇

Returning to the inclusion of NE-pattern languages over infinite alphabets, we can prove with little effort that the condition in Theorem 3.4 is characteristic, and therefore the inclusion problem is decidable in that case:

Proposition 3.12. *Let* Σ *be an alphabet,* $|\Sigma| = \infty$. *Then the inclusion problem is decidable for* nePAT_Σ.

Proof. It is sufficient to show the following statement: For every $\alpha, \beta \in \text{Pat}_\Sigma$, $L_{\text{NE},\Sigma}(\beta) \subseteq L_{\text{NE},\Sigma}(\alpha)$ if and only if there exists a terminal-preserving morphism $\phi : (\mathbb{N} \cup \Sigma)^+ \longrightarrow (\mathbb{N} \cup \Sigma)^+$ such that $\phi(\alpha) = \beta$. From this insight we then can directly conclude that Proposition 3.12 is correct, since the existence of ϕ is decidable (according to our notes on the membership problem, cf. Proposition 3.7 and the subsequent remarks). Moreover, the *if* direction of this statement evidently holds due to Theorem 3.4. Thus, we only have to prove the *only if* direction.

Hence, let $\alpha, \beta \in \text{Pat}_\Sigma$ with $L_{\text{NE},\Sigma}(\beta) \subseteq L_{\text{NE},\Sigma}(\alpha)$. Let $\sigma_{\text{iso}} : \text{Pat}_\Sigma \longrightarrow \Sigma^+$ be any substitution such that, for every $i \in \text{var}(\beta)$, $|\sigma_{\text{iso}}(i)| = 1$ and $\sigma_{\text{iso}}(i) \notin \text{term}(\alpha) \cup \text{term}(\beta)$ and, for every $j \in \text{var}(\alpha)$ with $i \neq j$, $\sigma_{\text{iso}}(i) \neq \sigma_{\text{iso}}(j)$. The existence of σ_{iso} is granted by the condition $|\Sigma| = \infty$. Since $w := \sigma_{\text{iso}}(\beta)$ is actually nothing but a "renaming" of β, it is obvious that there exists an inverse morphism $\sigma_{\text{iso}}^{-1} : \Sigma^+ \longrightarrow \text{Pat}_\Sigma$ such that $\sigma_{\text{iso}}^{-1}(w) = \beta$. Furthermore, the definition of σ_{iso} implies that, for every $\mathtt{A} \in \text{term}(w)$, we have $\sigma_{\text{iso}}^{-1}(\mathtt{A}) \in \mathbb{N}$ if and only if $\mathtt{A} \notin \text{term}(\alpha) \cup \text{term}(\beta)$.

As $L_{\text{NE},\Sigma}(\beta) \subseteq L_{\text{NE},\Sigma}(\alpha)$, there is a nonerasing substitution σ' such that $w = \sigma'(\alpha)$. We now define $\phi := \sigma_{\text{iso}}^{-1} \circ \sigma'$. Then ϕ is terminal-preserving because σ' is a substitution (which implies that it is terminal-preserving) and, for every letter $\mathtt{A}$ in w that is mapped by σ_{iso}^{-1} onto a variable, $\mathtt{A} \notin \text{term}(\alpha)$. Furthermore, ϕ is nonerasing and, evidently, $\phi(\alpha) = \beta$. This proves Proposition 3.12. □

In addition, we can omit the restriction to nonerasing morphisms in the above proof, so that it can also be used to show the analogue for E-pattern languages (based on Theorem 3.5):

Proposition 3.13. *Let Σ be an alphabet, $|\Sigma| = \infty$. Then the inclusion problem is decidable for* ePAT_Σ.

Finally, the condition in Theorem 3.5 does not only characterise the inclusion of the full class of E-pattern languages over infinite alphabets, but also that of several *sub*classes of ePAT over arbitrary finite alphabets with at least two distinct letters. We examine some of the related results in Chapter 6.1.1 and Chapter 6.2.1; for the moment, however, we can focus on the most elementary one, which covers the inclusion of terminal-free E-pattern languages:

Theorem 3.6 (Jiang et al. [33])**.** *Let Σ be an alphabet, $|\Sigma| \geq 2$, and let $\alpha, \beta \in \text{Pat}_{\text{tf}}$. Then $L_{\text{E},\Sigma}(\beta) \subseteq L_{\text{E},\Sigma}(\alpha)$ if and only if there exists a morphism $\phi : \mathbb{N}^* \longrightarrow \mathbb{N}^*$ such that $\phi(\alpha) = \beta$.*

Theorem 3.6 and, in particular, the corresponding proof technique are vital for many different examinations of pattern languages. We discuss its extensibility to a more general class of E-pattern languages in Chapter 6.3.1. These considerations are motivated by the equivalence problem for E-pattern languages (see below).

Once more, we can utilise that the existence of the morphism ϕ introduced in Theorem 3.6 is decidable:

Theorem 3.7 (Filè [19], Jiang et al. [33])**.** *Let Σ be an alphabet, $|\Sigma| \geq 2$. Then the inclusion problem is decidable for* $\text{ePAT}_{\text{tf},\Sigma}$.

It is a well-known fact that – in contrast to the relation between Theorem 3.5 and Theorem 3.6 – the criterion in Theorem 3.4 does not characterise the inclusion of terminal-free NE-pattern languages. Presumably among other reasons, this is caused by the connection of the problem to the classical topic of *unavoidable patterns* as briefly introduced in Chapter 2.3: for instance, if $|\Sigma| = 2$, $\alpha := 1 \cdot 2 \cdot 2 \cdot 3$ and $\beta := 1 \cdot 2 \cdot 3 \cdot 4 \cdot 5 \cdot 6$ then $L_{\text{NE},\Sigma}(\alpha) \supset L_{\text{NE},\Sigma}(\beta)$, but obviously there is no nonerasing morphism mapping α onto β. The fact that the language of α is a superset of the language of β results again from unavoidability of a square $v := u^2$ in each word $w \in \Sigma^+$ with $|w| \geq 4$, and each word in

$L_{\mathrm{NE},\Sigma}(\beta)$ over a binary alphabet contains such a subword v generated by the subpattern $2 \cdot 3 \cdot 4 \cdot 5$ of β. Additional information on this beautiful and momentous connection between the inclusion of terminal-free NE-pattern languages and the unavoidability of patterns is provided, e. g., by Jiang et al. [32]. As a consequence of the large number of open questions concerning the latter subject, it is not surprising that the decidability of the inclusion problem for terminal-free NE-pattern languages is still open, and, thus, it might be a major obstacle when tackling the more general inclusion problem for nePAT over finite terminal alphabets (see Open Problem 3.2).

We conclude this chapter with another important decision problem for pattern languages, which has to be examined separately whenever the inclusion problem is undecidable (or at least open) for the class of pattern languages in question: the equivalence problem. For any alphabet Σ and any class $\mathrm{PAT}_{\star,\Sigma} \subseteq \mathrm{ePAT}_{\Sigma}$ (or $\mathrm{PAT}_{\star,\Sigma} \subseteq \mathrm{nePAT}_{\Sigma}$) of pattern languages over Σ, the *equivalence* problem is said to be *decidable* if and only if there exists a total computable function χ such that, for every pair of patterns $\alpha, \beta \in \mathrm{Pat}_{\Sigma}$ with $L_{\mathrm{E},\Sigma}(\alpha), L_{\mathrm{E},\Sigma}(\beta) \in \mathrm{PAT}_{\star,\Sigma}$ (or, alternatively, $L_{\mathrm{NE},\Sigma}(\alpha), L_{\mathrm{NE},\Sigma}(\beta) \in \mathrm{PAT}_{\star,\Sigma}$), χ decides on whether or not $L_{\mathrm{E},\Sigma}(\alpha) = L_{\mathrm{E},\Sigma}(\beta)$ (or, alternatively, $L_{\mathrm{NE},\Sigma}(\alpha) = L_{\mathrm{NE},\Sigma}(\beta)$).

The decision procedure for the equivalence problem for the class of NE-pattern languages is incomplex, and, again, it is related to the existence of a particular morphism:

Theorem 3.8 (Angluin [3]). *Let Σ be a finite alphabet, $|\Sigma| \geq 2$, and let $\alpha, \beta \in \mathrm{Pat}_{\Sigma}$. Then $L_{\mathrm{NE},\Sigma}(\alpha) = L_{\mathrm{NE},\Sigma}(\beta)$ if and only if there exists a morphism $\phi : (\mathbb{N} \cup \Sigma)^+ \longrightarrow (\mathbb{N} \cup \Sigma)^+$ such that ϕ is a renaming of variables and $\phi(\alpha) = \beta$.*

From Theorem 3.8 and the observation noted in Proposition 2.2 we can immediately conclude that two patterns α, β generate the same NE-language if and only if the canonical form of α equals the canonical form of β. The latter question is decidable, and therefore we may conclude the following fact:

Corollary 3.2. *Let Σ be an alphabet. Then the equivalence problem is decidable for* nePAT_{Σ}.

With regard to unary alphabets, the correctness of Corollary 3.2 is a direct consequence of Proposition 3.10, and for infinite alphabets it follows from Proposition 3.12 (actually, Theorem 3.8 also holds for infinite alphabets, but Angluin [3] exclusively considers finite alphabets).

In contrast to the equivalence of NE-pattern languages – and unlike the situation concerning the inclusion problem for $\mathrm{nePAT}_{\mathrm{tf}}$ and $\mathrm{ePAT}_{\mathrm{tf}}$, where the NE case is the more intricate one – the understanding of the equivalence of E-pattern languages seems to be more difficult. In fact, despite of a considerable number of publications dealing with it, little is known about the equivalence problem for ePAT[†]:

Open Problem 3.3. *Let Σ be a finite alphabet, $|\Sigma| \geq 2$. Is the equivalence problem decidable for* ePAT_{Σ}*?*

In Chapter 6.2.1 and, indirectly, in Chapter 6.3.1, we apply our results concerning the ambiguity of morphisms to Open Problem 3.3.

[†]Note that the corresponding proof in [15], which claims to show the decidability of the equivalence problem, is not correct. This is already pointed out by Ohlebusch and Ukkonen [69].

Note that the equivalence problem for E-pattern languages is decidable provided that a unary or infinite alphabet is considered. With regard to the former alphabet size, this results from Proposition 3.11 and, for the latter alphabet size, from Proposition 3.13.

As we have the characteristic criterion for the inclusion of terminal-free E-pattern languages presented in Theorem 3.6, we can at least extend that insight to the equivalence problem:

Corollary 3.3. *Let Σ be an alphabet, $|\Sigma| \geq 2$, and let $\alpha, \beta \in \mathrm{Pat_{tf}}$. Then $L_{\mathrm{E},\Sigma}(\alpha) = L_{\mathrm{E},\Sigma}(\beta)$ if and only if α and β are morphically coincident.*

Again, we can state that morphical coincidence is computable (due to an argumentation that is very similar to our notes on the membership problem), and therefore we may record a positive decidability result for terminal-free E-pattern languages:

Corollary 3.4. *Let Σ be an alphabet. Then the equivalence problem is decidable for $\mathrm{ePAT}_{\mathrm{tf},\Sigma}$.*

With regard to unary alphabets, which are not covered by Corollary 3.3, the correctness of Corollary 3.4 again follows from Proposition 3.11.

We extensively discuss additional previous insights into the equivalence problem for general E-pattern languages in Chapters 6.2.1 and 6.3.1. For the moment, we are content with some necessary conditions:

Theorem 3.9 (Jiang et al. [32]). *Let Σ be an alphabet, $|\Sigma| \geq 3$, and let $\alpha, \beta \in \mathrm{Pat}_\Sigma$. If $L_{\mathrm{E},\Sigma}(\alpha) = L_{\mathrm{E},\Sigma}(\beta)$ then α and β are similar.*

If there are at least four distinct letters in the alphabet then Theorem 3.9 can be refined as follows:

Theorem 3.10 (Jiang et al. [33]). *Let Σ be an alphabet, $|\Sigma| \geq 4$, and let $\alpha, \beta \in \mathrm{Pat}_\Sigma$. If $L_{\mathrm{E},\Sigma}(\alpha) = L_{\mathrm{E},\Sigma}(\beta)$ then α and β are similar – i. e., for some $m \in \mathbb{N}$, $\alpha_i, \beta_i \in \mathbb{N}^+$ with $1 \leq i < m$, $\alpha_0, \beta_0, \alpha_m, \beta_m \in \mathbb{N}^*$ and $u_i \in \Sigma^+$ with $i \leq m$, it is $\alpha = \alpha_0\, u_1 \alpha_1\, u_2\, \ldots\, \alpha_{m-1}\, u_m \alpha_m$ and $\beta = \beta_0\, u_1 \beta_1\, u_2\, \ldots\, \beta_{m-1}\, u_m \beta_m$ – and, for every i, $0 \leq i \leq m$, $L_{\mathrm{E},\Sigma}(\alpha_i) = L_{\mathrm{E},\Sigma}(\beta_i)$.*

It is open whether Theorem 3.10 also holds for alphabet size 3. However, if so then the proof might be much more complicated than the one for $|\Sigma| = 4$. Consequently, Theorem 3.9 and Theorem 3.10 give a first hint that there are problems for pattern languages which do not allow to argue for all finite alphabets with at least two different letters in the same way. We shall mention that this phenomenon – which, in many "classical" results on pattern languages, does not occur such that most early corresponding papers do not have to bother about the alphabet size – is caused by the fact that the existence or non-existence of particular ambiguous and unambiguous morphisms may essentially depend on the size of the terminal alphabet. Consequently, suchlike discontinuous properties of pattern languages are ubiquitous in our thesis.

We conclude this chapter with a brief note concerning some decision problems on pattern languages which are given by Mateescu and Salomaa [54]. As described in Chapter 3.1, the authors in the said paper examine the (non-)existence of patterns α that have an upper bound $n \in \mathbb{N}$ such that, for every word w in the language of α there exist at most n different morphisms σ with $\sigma(\alpha) = w$. Mateescu and Salomaa show that there is a procedure which, for every pattern α and for every $k \in \mathbb{N}$, decides on whether or not $n \leq k$. Contrary to this, they explicitly leave the question for the decidability of the sheer existence of such a bound n (subject to any given pattern α) unanswered. Although these decision problems on the so-called *ambiguity of pattern languages* show manifest connections to our topic of ambiguity of morphisms, none of our subsequent results contributes nontrivial insights to the research initiated in [54], and neither do the corresponding findings by Mateescu and Salomaa provide any tools for our reasoning. Therefore we abstain from a formal description of the problems and the related results. Recall that the reasons for this non-transferability are briefly explained in Chapter 3.1.

3.3 Inductive inference of pattern languages

In the present chapter, we describe the current state of knowledge on inductive inference of pattern languages (cf. Chapter 2.5). However, before we go into this rather specific matter, we introduce a powerful general insight into the inferrability of arbitrary indexed families provided by Angluin [4].

Considering an arbitrary inference procedure within the LIM model as introduced in Chapter 2.5, it can be easily seen that we face severe difficulties whenever there is an *overgeneralisation* (also referred to as the *subset problem*), i.e. a situation where the strategy outputs a hypothesis that describes a proper superset of the actual language to be identified. Evidently, in this case, the lack of negative input data causes major challenges for the strategy to revise its incorrect guess. This problem significantly limits the power of inductive inference from positive data, as shown by one of Gold's main theorems (adapted to our limited definition of the model), which discusses the learnability of so-called *superfinite* classes of languages:

Theorem 3.11 (Gold [24]). *Let $\mathcal{L}$ be an indexable class of languages that contains all finite and at least one infinite language. Then $\mathcal{L}$ is not inferrable from positive data.*

Consequently, Theorem 3.11 particularly states that the classes of regular, context-free and context-sensitive languages are not inferrable from positive data, and this result also holds in more general versions of the limit learning approach.

In fact, finding a way to avoid or at least correct overgeneralisations even is the sine qua non for the identifiability of any indexed family; this is expressed in the following celebrated theorem by Angluin, that gives a characterisation of LIM-learnable indexed families:

Theorem 3.12 (Angluin [4]). *Let $\mathcal{L} = (L_i)_{i \in \mathbb{N}}$ be an indexed family of nonempty recursive languages. Then $\mathcal{L} = (L_i)_{i \in \mathbb{N}}$ is inferrable from positive data if and only if there exists an effective procedure which, for every $j \in \mathbb{N}$, enumerates a set T_j such that*

- $T_j \subseteq L_j$,
- *T_j is finite, and*
- *there does not exist a $j' \in \mathbb{N}$ with $T_j \subseteq L_{j'} \subset L_j$.*

If there exists a set T_j satisfying the conditions of Theorem 3.12 then it is called a *telltale* (for L_j) (with respect to $\mathcal{L} = (L_i)_{i\in\mathbb{N}}$).

Consequently, inductive inference from positive data, for every language L in some class $\mathcal{L}$, requires the existence of a (uniformly recursively enumerable) finite sublanguage, namely a telltale, which allows to distinguish L from all of its sublanguages in $\mathcal{L}$. Hence, Theorem 3.12 inherently allows to decide on the *algorithmic* problem of learnability in a purely *language theoretic* way. Our results on inductive inference of E-pattern languages, which are presented in Chapters 5.2.1 and 6.1.1, are completely obtained with the aid of Theorem 3.12. Therefore, since the concept of a telltale is closely connected to what Lange et al. [42] designate as a set of *good examples* for a language, our negative results on the subject entail the insight that there exist particular pattern languages L such that even the best examples for L are not good enough (for unequivocally identifying L among certain classes of pattern languages).

While the recursive enumerability of telltales is characteristic for arbitrary indexed families that are learnable in the LIM model as introduced in Chapter 2.5, the mere *existence* of these significant sublanguages additionally characterises the learnable classes subject to other prominent models of inductive inference. In particular, this holds for the LIM model when applied to indexed families with a decidable inclusion problem:

Theorem 3.13 (Angluin [4]). *Let $\mathcal{L} = (L_i)_{i\in\mathbb{N}}$ be an indexed family of nonempty recursive languages for which there exists an effective procedure that, given any $j, j' \in \mathbb{N}$, decides on whether or not $L_j \subseteq L_{j'}$. Then $\mathcal{L} = (L_i)_{i\in\mathbb{N}}$ is inferrable from positive data if and only if, for every $j \in \mathbb{N}$, there exists a set T_j such that*

- $T_j \subseteq L_j$,
- *T_j is finite, and*
- *there does not exist a $j' \in \mathbb{N}$ with $T_j \subseteq L_{j'} \subset L_j$.*

Furthermore, the same criterion is characteristic for BC-learnable classes (cf. Baliga et al. [7] and our brief definitional notes in Chapter 2.5) and non-computable limit learners (cf. Jain et al. [31]). Thus, the subset problem and its solution based on telltales is fundamental in inference from positive data.

With regard to the learnability of pattern languages, the current state of knowledge remarkably differs when comparing NE- and E-pattern languages. Concerning the NE case, the most elementary question could be answered in the initial paper by Angluin [3] introducing pattern languages:

Theorem 3.14 (Angluin [3]). *For every alphabet Σ, nePAT_Σ is inferrable from positive data.*

Actually, as mentioned above, Angluin [3] assumes the alphabet to have at least two distinct letters. The proof for unary alphabets, however, is simple and can be conducted, e. g., by a straightforward adaptation of the proof for the learnability of E-pattern languages over unary alphabets given by Mitchell [59] (see below).

Based on the fundamental insight presented by Theorem 3.14 – that, after the rather discouraging finding in Theorem 3.11, due to its positive result for a rich class of languages gave new life to the research on inductive inference from positive data – there exist numerous additional papers dealing with the LIM-learnability of (classes of) NE-pattern languages, many of them seeking for particular learning strategies. For instance, Shinohara [98] shows that two subclasses of NE-pattern languages are inferrable in polynomial time (where this term, because of the infinite inference procedure, is meant to cover the time for computing a single "meaningful" hypothesis), namely that of all languages generated by regular patterns and that of *non-cross* NE-pattern languages resulting from patterns where the variables are monotonic increasing. While Shinohara's approach is based on a *consistent* learning strategy – which always computes a hypothesis that can generate all words presented to the strategy so far – the approach by Lange and Wiehagen [43] proves the *full* class of NE-pattern languages to be learnable in polynomial time by a potentially *inconsistent* learner, i. e. by an algorithm that converges correctly, but that may output intermediate hypotheses not covering all previous input words. Wiehagen and Zeugmann [105] examine the power of inconsistent learning strategies in more detail, and they partly complement the insight in [43] by the result that there does *not* exist any learning strategy consistently inferring the class of NE-pattern languages in polynomial time from positive and negative data (where consistency now means that each hypothesis exactly separates the positive from the negative examples).[†] Thus, though consistency seems easier to achieve for positive data only, this result suggests that inconsistent learning strategies might not only be sufficient, but also necessary for inferring nePAT from positive data in polynomial time. Rossmanith and Zeugmann [84] use the algorithm by Lange and Wiehagen [43] for constructing an inductive learning algorithm that stops after finitely many steps correctly identifying any NE-pattern language with high probability (the so-called *stochastic finite learning*). Several authors, such as Reischuk and Zeugmann [82] and Erlebach et al. [18], deal with the class of *one-variable* NE-pattern languages generated by patterns α that satisfy $|\operatorname{var}(\alpha)| = 1$ (introduced by Angluin [3]). In addition, there exist many other papers discussing the inferrability of other classes of languages which are derived from pattern languages; this particularly holds for unions of NE-pattern languages (cf., e. g., Wright [106], Arimura et al. [6] and Shinohara and Arimura [100]), a topic which is closely connected with inductive inference of E-pattern languages (see below). Finally, the learnability of (NE- and E-)pattern languages has also been examined in other models of algorithmic learning theory such as the PAC model (cf., e. g., Schapire [92] and Mitchell et al. [58]) and the Query model (cf., e. g., Nessel and Lange [65] and Lange and Zilles [44]). A survey covering selected results up to 1995 is given by Shinohara and Arikawa [99]. Note that, in the present thesis, we do not discuss the manifest connections between our learning theoretical approach of pattern *inference* and the topics of pattern (or alternatively: string) *matching* in the sense of, e. g., Knuth et al. [39] or pattern *discovery* as understood by, e. g., Brazma et al. [11].

In contrast to the broad knowledge on the learnability of NE-pattern languages, many fundamental questions on inductive inference of E-pattern languages are still open; all results known before only hold for subclasses or particular alphabet sizes. In the ini-

[†]Note that this result is based on the assumptions that, first, $\mathrm{P} \neq \mathrm{NP}$ and, second, the learner is only allowed to output hypotheses referring to NE-pattern languages. If the latter requirement is omitted, i. e. if we allow intermediate hypothesis which describe other than pattern languages, then there *is* a consistent learning strategy which infers nePAT from positive and negative data in polynomial time. Such an approach, however, does not meet our definition of learning in the limit of classes of pattern languages.

tial work, Shinohara [97] proves the class of E-pattern languages generated by *regular* patterns to be learnable in polynomial time. To this end, Shinohara assumes that the underlying terminal alphabet consists of at least three distinct letters. From the abovementioned work by Wright [106] and an insight on the relation between E-pattern languages and finite unions of NE-pattern languages noted by Jiang et al. [32], it can additionally be derived that any class of E-pattern languages generated by a set $\mathrm{Pat}_\star$ of patterns over a *finite* set of *variables* – i.e., for any $\mathbb{N}_\star \subset \mathbb{N}$ with $|\mathbb{N}_\star| \neq \infty$ and for any alphabet Σ, $\mathrm{Pat}_\star = (\Sigma \cup \mathbb{N}_\star)^+$ – is inferrable from positive data. Mitchell [59], in turn, extends Shinohara's positive result on E-pattern languages generated by regular patterns to so-called *quasi-regular* E-pattern languages generated by the set of all patterns α where, for all $i, j \in \mathrm{var}(\alpha)$, $|\alpha|_i = |\alpha|_j$. Unlike the approach by Shinohara, Mitchell's argumentation is not directly algorithmic, but combinatorial, and therefore it does not involve any considerations on the time-complexity of potential learning strategies; on the other hand, Mitchell's result holds for all alphabet sizes. Additionally, Mitchell shows that the *full* class of E-pattern languages is learnable for *unary and infinite alphabets*:

Proposition 3.14 (Mitchell [59]). *Let Σ be an alphabet, $|\Sigma| \in \{1, \infty\}$. Then ePAT_Σ is inferrable from positive data.*

Due to the properties of E-pattern languages over these alphabets discussed after Example 3.1, this result is not surprising and can be proven easily. Finally, Reidenbach [76] claims that the class of *terminal-free non-cross* E-pattern languages is inferrable from positive data. The corresponding proof presented in [76], however, is incomplete. We complete the proof in Chapter 5.4 of the present thesis.

The main result of Reidenbach [76], presented in [77] and finally published in [81], consists of the first negative learnability result to be known on a class of E-pattern languages, namely the class of terminal-free E-pattern languages over a binary alphabet:

Theorem 3.15 (Reidenbach [76, 81]). *Let Σ be an alphabet, $|\Sigma| = 2$. Then $\mathrm{ePAT}_{\mathrm{tf},\Sigma}$ is not inferrable from positive data.*

By definition, this implies the same result for the full class:

Corollary 3.5 (Reidenbach [76, 81]). *Let Σ be an alphabet, $|\Sigma| = 2$. Then ePAT_Σ is not inferrable from positive data.*

The proof for Theorem 3.15 is based on the pattern $\alpha_{\mathsf{ab}}^{\mathrm{tf}} := 1 \cdot 1 \cdot 2 \cdot 2 \cdot 3 \cdot 3$ and on the concept of a passe-partout:

Definition 3.1 (Passe-partout). *Let Σ be an alphabet, $\alpha \in \mathrm{Pat}_\Sigma$ a pattern and $W \subset L_{\mathrm{E},\Sigma}(\alpha)$ a finite set of words. Then a pattern $\beta \in \mathrm{Pat}_\Sigma$ is said to be a* passe-partout (for α and W) *provided that $W \subseteq L_{\mathrm{E},\Sigma}(\beta) \subset L_{\mathrm{E},\Sigma}(\alpha)$.*

Obviously, the definition of a passe-partout directly aims at the concept of telltales as determined by Theorem 3.12: If, for some pattern α and set $W \subset L(\alpha)$, a pattern β is a passe-partout for α and W, then W is not a telltale for $L(\alpha)$ with respect to any class of languages containing both $L(\alpha)$ and $L(\beta)$.

Reidenbach [76, 81] shows that, for $|\Sigma| = 2$, for $\alpha_{\mathsf{ab}}^{\mathrm{tf}}$ and for every finite $W \subseteq L_{\mathrm{E},\Sigma}(\alpha_{\mathsf{ab}}^{\mathrm{tf}})$, there exists a terminal-free passe-partout β, which by Theorem 3.12 implies the correctness of Theorem 3.15. As the crucial reason for the existence of β, the publications [76, 74] identify the ambiguity of particular words in $L_{\mathrm{E},\Sigma}(\alpha^{\mathrm{tf}}_{\mathtt{ab}})$. Alternatively, this can be demonstrated by the following characterisation of telltales for terminal-free E-pattern languages:

Theorem 3.16 (Reidenbach [76, 74]). *Let Σ be an alphabet, $|\Sigma| \geq 2$, and let $\alpha \in \mathrm{Pat}_{\mathrm{tf}}$ be a succinct pattern. Let $T_\alpha := \{w_1, w_2, \ldots, w_n\} \subseteq L_{\mathrm{E},\Sigma}(\alpha)$, $n \geq 1$. Then T_α is a telltale for $L_{\mathrm{E},\Sigma}(\alpha)$ with respect to $\mathrm{ePAT}_{\mathrm{tf},\Sigma}$ if and only if, for every $j \in \mathrm{var}(\alpha)$, there exists a $w \in T_\alpha$ such that, for every morphism $\sigma : \mathrm{Pat}_{\mathrm{tf}} \longrightarrow \Sigma^*$ with $\sigma(\alpha) = w$, there is an $\mathtt{A} \in \Sigma$ with $|\sigma(j)|_{\mathtt{A}} = 1$ and $|\sigma(\alpha)|_{\mathtt{A}} = |\alpha|_j$.*

Consequently, Theorem 3.15, Corollary 3.5 and Theorem 3.16 do not only significantly strengthen the knowledge on inductive inference of E-pattern languages, but Theorem 3.16 additionally yields first insights into the close connection between the ambiguity of morphic images and elementary properties of E-pattern languages as it characterises the existence of telltales by examining, for each word in a given set, all of its generating morphisms (with respect to a fixed pattern α).

From this state of knowledge, we derive two questions on inductive inference of E-pattern languages which shall partially guide our considerations on the ambiguity of morphisms. The first asks for the completion of Theorem 3.15 with respect to the open alphabet sizes:

Question 3.1. *Let Σ be a finite alphabet, $|\Sigma| \geq 3$. Is $\mathrm{ePAT}_{\mathrm{tf},\Sigma}$ inferrable from positive data?*

We discuss this problem in Chapter 5.2.1. For an immediate answer to Question 3.1, the impatient reader might wish to take a look at Theorem 5.5.

If Question 3.1 has an answer in the affirmative then the problem of the inferrability of the full class of E-pattern languages (over alphabets with three or more letters), which since 1980 has attracted a lot of interest in learning theory and is considered to be "one of the outstanding open problems in inductive inference" (Mitchell [59]), remains open. Therefore, we additionally investigate the extensibility of our findings on Question 3.1 to ePAT (a task which is trivial provided that Question 3.1 has an answer in the negative) – or, in other words, we examine the learnability of this class with regard to any alphabet size not considered by Corollary 3.5:

Question 3.2. *Let Σ be a finite alphabet, $|\Sigma| \geq 3$. Is ePAT_Σ inferrable from positive data?*

Our findings on Question 3.2 are presented in Chapter 6.1.1. The corresponding main result is given in Corollary 6.3.

Chapter 4

The ambiguity of morphisms: paradigms, properties and questions

In the present chapter, we formally introduce the main subject of this thesis as motivated in Chapter 1, namely the (un-)ambiguity of morphisms and of morphic images of strings. For that purpose, we first give a number of definitions which allow to formally address this phenomenon. Furthermore, we note a number of basic properties which immediately result from these definitions. A list of elementary questions – which shall determine sort of a schedule for our examinations in the subsequent Chapters 5 and 6 – conclude the chapter.

4.1 Elementary types of (un-)ambiguity and their basic properties

We begin our definitions with the most basic type of unambiguity, which is extensively discussed in Chapter 1. With regard to any pattern α, it precisely describes those words $w \neq \mathrm{res}(\alpha)$ for which there is exactly one morphism σ in a free monoid – to this end, of course, we ignore all those morphisms which merely differ from σ on variables not occurring in α – such that $\sigma(\alpha) = w$:

Definition 4.1 (Unambiguity). *Let Σ be an alphabet. Let $\alpha \in (\mathbb{N} \cup \Sigma)^+$, and let $\sigma : (\mathbb{N} \cup \Sigma)^* \longrightarrow \Sigma^*$ be a (terminal-preserving) morphism with $\sigma(\alpha) \neq \mathrm{res}(\alpha)$. Then both σ and $\sigma(\alpha)$ are called* unambiguous (with respect to α) *provided that there exists no (terminal-preserving) morphism $\tau : (\mathbb{N} \cup \Sigma)^* \longrightarrow \Sigma^*$ such that $\tau(\alpha) = \sigma(\alpha)$ and, for some $i \in \mathrm{var}(\alpha)$, $\tau(i) \neq \sigma(i)$.*

We now illustrate Definition 4.1 by a number of examples:

Example 4.1. If, for a pattern α, $|\mathrm{var}(\alpha)| = 1$ then, for every (if applicable, terminal-preserving) morphism $\sigma : (\mathbb{N} \cup \Sigma)^* \longrightarrow \Sigma^*$, $\sigma(\alpha)$ is unambiguous.

Contrary to this, with regard to the pattern $\alpha := 1 \cdot 2$, there is obviously no unambiguous morphism σ: Since, by definition, $\sigma(\alpha) \neq \mathrm{res}(\alpha) = \varepsilon$, there must be a variable i in α with $\sigma(i) \neq \varepsilon$. Hence, with $\tau(i) := \varepsilon$ and $\tau(j) := \sigma(\alpha)$, $j \neq i$, it is $\tau(\alpha) = \sigma(\alpha)$ and $\tau(i) \neq \sigma(i)$.

Concerning the pattern $\alpha := 1 \cdot 2 \cdot 2$, in turn, there are unambiguous words such as $w := \mathtt{a}$; corresponding morphisms, however, have to map the variable 2 onto the empty word. ◇

We proceed with a more complex pattern, which is our initial example in Chapter 1 (and, furthermore, our favourite example throughout this thesis):

Example 4.2. Let $\alpha := 1 \cdot 2 \cdot 3 \cdot 4 \cdot 1 \cdot 4 \cdot 3 \cdot 2$.

Let the morphism σ be given by $\sigma(1) := \mathtt{a}$, $\sigma(2) := \mathtt{a\,b}$, $\sigma(3) := \mathtt{b}$, $\sigma(4) := \mathtt{b\,a}$. Then σ and, thus, the word $\sigma(\alpha) = \mathtt{a\,a\,b\,b\,b\,a\,a\,b\,a\,b\,a\,b}$ are unambiguous with respect to α.

Let the morphism σ_{c} be given by $\sigma_{\mathrm{c}}(i) := \mathtt{a\,b}^i$, $i \in \mathbb{N}$. Then, as stated in Chapter 1, σ_{c} and, thus, the word $w := \sigma_{\mathrm{c}}(\alpha) = \mathtt{a\,b\,a\,b}^2\,\mathtt{a\,b}^3\,\mathtt{a\,b}^4\,\mathtt{a\,b\,a\,b}^4\,\mathtt{a\,b}^3\,\mathtt{a\,b}^2$ are not unambiguous with respect to α as w can also be generated, e.g., by the morphism τ with $\tau(1) := \mathtt{a\,b\,a\,b}^2$, $\tau(2) := \varepsilon$, $\tau(3) := \mathtt{a\,b}^3\,\mathtt{a\,b}^2$ and $\tau(4) := \mathtt{b}^2$:

$$\overbrace{\mathtt{a\,b}}^{\sigma_c(1)}\overbrace{\mathtt{a\,b\,b}}^{\sigma_c(2)}\overbrace{\mathtt{a\,b\,b\,b}}^{\sigma_c(3)}\overbrace{\mathtt{a\,b\,b\,b\,b}}^{\sigma_c(4)}\overbrace{\mathtt{a\,b}}^{\sigma_c(1)}\overbrace{\mathtt{a\,b\,b\,b\,b}}^{\sigma_c(4)}\overbrace{\mathtt{a\,b\,b\,b}}^{\sigma_c(3)}\overbrace{\mathtt{a\,b\,b}}^{\sigma_c(2)}$$

$$\underbrace{\mathtt{a\,b\,a\,b\,b}}_{\tau(1)}\underbrace{\mathtt{a\,b\,b\,b\,a\,b\,b}}_{\tau(3)}\underbrace{\mathtt{b\,b}}_{\tau(4)}\underbrace{\mathtt{a\,b\,a\,b\,b}}_{\tau(1)}\underbrace{\mathtt{b\,b}}_{\tau(4)}\underbrace{\mathtt{a\,b\,b\,b\,a\,b\,b}}_{\tau(3)}$$

Note that, actually, there is a total of 20 different morphisms mapping α onto w, a fact which can be verified straightforward (albeit with a bit of effort, of course). ◇

Our next example is derived from Chapter 3.3:

Example 4.3. Let $\alpha_{\mathrm{ab}}^{\mathrm{tf}} := 1 \cdot 1 \cdot 2 \cdot 2 \cdot 3 \cdot 3$.

As shown by Reidenbach [81], this pattern allows to draw momentous conclusions on inductive inference of terminal-free E-pattern languages, which is caused by the ambiguity of particular morphic images of $\alpha_{\mathrm{ab}}^{\mathrm{tf}}$. Nevertheless, there are unambiguous words with respect to $\alpha_{\mathrm{ab}}^{\mathrm{tf}}$:

Let again the morphism σ_{c} be given by $\sigma_{\mathrm{c}}(i) := \mathtt{a\,b}^i$, $i \in \mathbb{N}$. Then σ_{c} and, thus, the word $\sigma_{\mathrm{c}}(\alpha_{\mathrm{ab}}^{\mathrm{tf}}) = \mathtt{a\,b\,a\,b\,a\,b}^2\,\mathtt{a\,b}^2\,\mathtt{a\,b}^3\,\mathtt{a\,b}^3$ are unambiguous with respect to $\alpha_{\mathrm{ab}}^{\mathrm{tf}}$.

Contrary to this, the morphism σ given by $\sigma(1) := \mathtt{a\,b}$, $\sigma(2) := \mathtt{b}$, $\sigma(3) := \mathtt{b}$ and, thus, the word $\sigma(\alpha_{\mathrm{ab}}^{\mathrm{tf}})$ are not unambiguous with respect to $\alpha_{\mathrm{ab}}^{\mathrm{tf}}$ since, e.g., $\sigma(\alpha_{\mathrm{ab}}^{\mathrm{tf}}) = \mathtt{a\,b\,a\,b}^5 = \tau(\alpha_{\mathrm{ab}}^{\mathrm{tf}})$ with $\tau(1) := \varepsilon$, $\tau(2) := \mathtt{a\,b}$, $\tau(3) := \mathtt{b}^2$. In addition to this, of course, τ is not unambiguous with respect to $\alpha_{\mathrm{ab}}^{\mathrm{tf}}$ either.

$w := \mathtt{a\,b\,a}^3\,\mathtt{b}$ is not unambiguous with respect to $\alpha_{\mathrm{ab}}^{\mathrm{tf}}$ as there is no morphism σ with $\sigma(\alpha_{\mathrm{ab}}^{\mathrm{tf}}) = w$. ◇

If we regard patterns in $(\mathbb{N} \cup \Sigma)^+$ then we have to consider the ambiguity of terminal-preserving morphisms (i.e., more precisely, of substitutions):

Example 4.4. Let $\alpha := 1 \cdot \mathtt{a} \cdot 2 \cdot 2 \cdot \mathtt{b} \cdot 3$.

Let the substitution σ be given by $\sigma(1) := \mathtt{a}$, $\sigma(2) := \varepsilon$, $\sigma(3) := \mathtt{b}$. Then σ and, thus, the word $\sigma(\alpha) = \mathtt{a\,a\,b\,b}$ are unambiguous with respect to α.

If, for any substitution $\sigma : (\mathbb{N} \cup \{\mathtt{a}, \mathtt{b}\})^* \longrightarrow \{\mathtt{a}, \mathtt{b}\}^*$, $\sigma(2) \neq \varepsilon$ then σ and $\sigma(\alpha)$ necessarily are not unambiguous since, in this case, a substitution τ can be given with $\tau(\alpha) = \sigma(\alpha)$ and $\tau(2) = \varepsilon$.

This situation changes as soon as there are at least three distinct letters in the terminal alphabet. Hence, let $\Sigma := \{\mathtt{a}, \mathtt{b}, \mathtt{c}\}$. Then we can easily find unambiguous nonerasing substitutions $\sigma : (\mathbb{N} \cup \Sigma)^* \longrightarrow \Sigma^*$ for α:

Let the substitution σ be given by $\sigma(1) := \mathtt{a}$, $\sigma(2) := \mathtt{a\,c}$, $\sigma(3) := \mathtt{b}$. Then σ and, thus, the word $\sigma(\alpha) = \mathtt{a\,a\,a\,c\,a\,c\,b\,b}$ are unambiguous with respect to α since all morphisms τ with $\tau(\alpha) = \sigma(\alpha)$ and, for some symbol x in α, $\tau(x) \neq \sigma(x)$ are not terminal-preserving. ◇

In addition to this standard type of ambiguity we regard a second definition which disallows any morphism to map a variable onto the empty word. Thus, while Definition 4.1 describes the unambiguity of morphisms in free *monoids*, we now merely regard the unambiguity of morphisms in free *semigroups*:

Definition 4.2 (Weak unambiguity). *Let Σ be an alphabet. Let $\alpha \in (\mathbb{N} \cup \Sigma)^+$, and let $\sigma : (\mathbb{N} \cup \Sigma)^+ \longrightarrow \Sigma^+$ be a (terminal-preserving) morphism. Then both σ and $\sigma(\alpha)$ are called* weakly unambiguous *(with respect to α) provided that there exists no (terminal-preserving) morphism $\tau : (\mathbb{N} \cup \Sigma)^+ \longrightarrow \Sigma^+$ such that $\tau(\alpha) = \sigma(\alpha)$ and, for some $i \in \mathrm{var}(\alpha)$, $\tau(i) \neq \sigma(i)$.*

The following examples briefly illuminate this second (and, within the scope of our thesis, less crucial) type of unambiguity. In order to allow a first comparison of Definition 4.1 and Definition 4.2, we regard the same patterns as in previous examples. Our first example is derived from Example 4.3:

Example 4.5. Let $\alpha_{\mathtt{ab}}^{\mathrm{tf}} := 1 \cdot 1 \cdot 2 \cdot 2 \cdot 3 \cdot 3$.
Let the morphism σ be given by $\sigma(1) := \mathtt{a\,b}$, $\sigma(2) := \mathtt{b}$, $\sigma(3) := \mathtt{b}$. Then, unlike the insights described in Example 4.3, both σ and $\sigma(\alpha_{\mathtt{ab}}^{\mathrm{tf}}) = \mathtt{a\,b\,a\,b}^5$ are weakly unambiguous with respect to $\alpha_{\mathtt{ab}}^{\mathrm{tf}}$. This is caused by the shape of $\sigma(1)$ and the fact that $|\sigma(2)| = |\sigma(3)| = 1$. ◇

The second example examines the pattern and a particular substitution introduced in Example 4.4:

Example 4.6. Let $\alpha := 1 \cdot \mathtt{a} \cdot 2 \cdot 2 \cdot \mathtt{b} \cdot 3$.
Let the substitution σ be given by $\sigma(1) := \mathtt{a\,b}$, $\sigma(2) := \mathtt{a\,b}$, $\sigma(3) := \mathtt{a\,b}$. Then both σ and $\sigma(\alpha) = \mathtt{a\,b\,a\,a\,b\,a\,b\,b\,a\,b}$ are weakly unambiguous with respect to α, although $\sigma(2) \neq \varepsilon$ (cf. our statement on σ not being unambiguous in Example 4.4). ◇

The last example on weak unambiguity again deals with our "standard" pattern and word given in Chapter 1 and discussed in Example 4.2:

Example 4.7. Let $\alpha := 1 \cdot 2 \cdot 3 \cdot 4 \cdot 1 \cdot 4 \cdot 3 \cdot 2$.
Let $w := \mathtt{a\,b\,a\,b}^2\,\mathtt{a\,b}^3\,\mathtt{a\,b}^4\,\mathtt{a\,b\,a\,b}^4\,\mathtt{a\,b}^3\,\mathtt{a\,b}^2$. Then, as already described in Chapter 1, there exist different nonerasing morphisms σ, τ with $\sigma(\alpha) = w = \tau(\alpha)$ such as

$$\overbrace{\mathtt{a\,b}}^{\sigma(1)}\,\overbrace{\mathtt{a\,b\,b}}^{\sigma(2)}\,\overbrace{\mathtt{a\,b\,b\,b}}^{\sigma(3)}\,\overbrace{\mathtt{a\,b\,b\,b\,b}}^{\sigma(4)}\,\overbrace{\mathtt{a\,b}}^{\sigma(1)}\,\overbrace{\mathtt{a\,b\,b\,b\,b}}^{\sigma(4)}\,\overbrace{\mathtt{a\,b\,b\,b}}^{\sigma(3)}\,\overbrace{\mathtt{a\,b\,b}}^{\sigma(2)}$$

$$\underbrace{\mathtt{a\,b\,a}}_{\tau(1)}\,\underbrace{\mathtt{b\,b}}_{\tau(2)}\,\underbrace{\mathtt{a\,b\,b\,b\,a}}_{\tau(3)}\,\underbrace{\mathtt{b\,b\,b\,b}}_{\tau(4)}\,\underbrace{\mathtt{a\,b\,a}}_{\tau(1)}\,\underbrace{\mathtt{b\,b\,b\,b}}_{\tau(4)}\,\underbrace{\mathtt{a\,b\,b\,b\,a}}_{\tau(3)}\,\underbrace{\mathtt{b\,b}}_{\tau(2)}$$

Consequently, w is not weakly unambiguous with respect to α, either. ◇

The general relation between our two ways of understanding unambiguity follows directly by definition – for any pattern $\alpha \in (\mathbb{N} \cup \Sigma)^+$, the set of weakly unambiguous morphisms (and weakly unambiguous words) with respect to α is a superset of the set of unambiguous nonerasing morphisms (and unambiguous words generated by nonerasing morphisms):

Proposition 4.1. *Let Σ be an alphabet. Let $\alpha \in (\mathbb{N} \cup \Sigma)^+$, and let $\sigma : (\mathbb{N} \cup \Sigma)^* \longrightarrow \Sigma^*$ be an (if applicable, terminal-preserving) nonerasing morphism. Then $\sigma(\alpha)$ is weakly unambiguous if it is unambiguous. In general, the converse of this statement does not hold true.*

It is an immediate consequence of the pattern $\alpha_{\mathtt{ab}}^{\mathrm{tf}} = 1 \cdot 1 \cdot 2 \cdot 2 \cdot 3 \cdot 3$ and the word $w = \mathtt{a\,b\,a\,b}^5$ that Proposition 4.1 cannot state the equivalence of unambiguity and weak unambiguity for nonerasing morphisms and their images since w is weakly unambiguous with respect to $\alpha_{\mathtt{ab}}^{\mathrm{tf}}$ (cf. Example 4.6), but not unambiguous (cf. Example 4.4). If we, in turn, consider morphisms that assign the empty word to at least one variable in the pattern – a view which does not correspond to our main focus since we largely seek for *injective* morphisms that are *unambiguous* (see Chapter 1) – then there are unambiguous morphisms which (by definition) cannot be weakly unambiguous, as shown by the pattern $\alpha = 1 \cdot 2 \cdot 2$ and the word $w = \mathtt{a}$ (cf. Example 4.1). We now conclude our formal view on *un*ambiguity and proceed with our notations on ambiguity.

We call $\sigma(\alpha)$ *ambiguous (with respect to α)* if it is not unambiguous (or, if applicable, weakly unambiguous), but we use this term in an informal context only. If we wish to address an ambiguous word formally, then we simply refer to the negation of the terms for unambiguity introduced above (cf., e. g., Example 4.7). Additionally, we regard three distinct types of (potentially) ambiguous words, the first of which – for the sake of a more concise presentation and with regard to our needs in Chapter 5.2.1 – momentarily remains restricted to terminal-free patterns.

Intuitively, this first kind of ambiguity describes those words w which, with regard to a pattern $\alpha = i_1 \cdot i_2 \cdot [\dots] \cdot i_n \in \mathbb{N}^+$, $n \in \mathbb{N}$, and for *every* k, $1 \leq k \leq n$, contain a nonempty subword w_k such that w_k *must* be generated by the variable i_k – no matter which morphism σ with $\sigma(\alpha) = w$ is considered. Thus, for every such σ and for the variable i with $i = i_k$, $\sigma(i)$ must contain w_k, and, moreover, there must be a $j \in \mathbb{N}$ such that i_k under every possible morphism mapping α onto w always generates the jth occurrence of w_k in w. Consequently, in a sense, some parts of such a word are "unambiguous" since they can only be generated by a unique occurrence of a variable, and for each occurrence of each variable there must be such a subword in w. Hence, if we use a hatched area for the w_k and empty area for those subwords not belonging to some w_k, such a word graphically looks as follows (recall that $|\alpha| = n$):

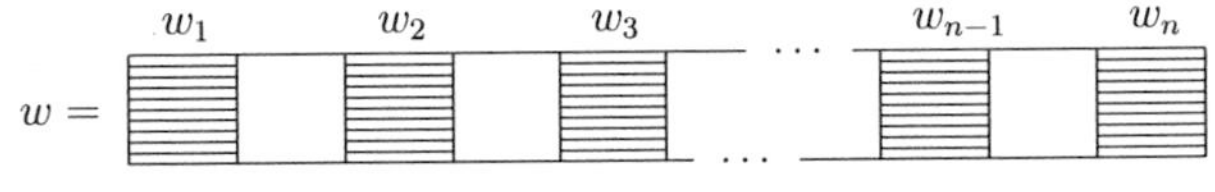

w_1 w_2 w_3 ... w_{n-1} w_n

$w =$ ...

Thus, there must be exactly n subwords w_k in w, and these w_k must be nonempty and non-overlapping. We proceed with the formal definition:

Definition 4.3 (Moderate ambiguity of common morphisms). *Let Σ be an alphabet. Let $\alpha := i_1 \cdot i_2 \cdot [\dots] \cdot i_n$, $n \in \mathbb{N}$, $i_k \in \mathbb{N}$, $1 \leq i_k \leq n$, and $w \in \Sigma^+$. Then w is called* moderately ambiguous (with respect to α)† *provided that there exist $l_2, l_3, \dots, l_n \in \mathbb{N}$ and $r_1, r_2, \dots, r_{n-1} \in \mathbb{N}$ with*

$$1 \leq r_1 < l_2 \leq r_2 < l_3 \leq r_3 < l_4 \leq \ldots \leq r_{n-2} < l_{n-1} \leq r_{n-1} < l_n \leq |w|$$

†Note that, in [78], we call a similar type of words "significant for α".

such that there is at least one morphism $\sigma : \mathbb{N}^* \longrightarrow \Sigma^*$ *with* $\sigma(\alpha) = w$ *and, for every such morphism* σ *and for every* k, $1 \leq k \leq n-1$,

- $|\,\sigma(i_1 \cdot i_2 \cdot [\ldots] \cdot i_k)\,| < l_{k+1}$ *and*
- $|\,\sigma(i_1 \cdot i_2 \cdot [\ldots] \cdot i_k)\,| \geq r_k$.

Additionally, we designate each morphism σ *with* $\sigma(\alpha) = w$ *as* moderately ambiguous (with respect to α) *as well.*

The following criterion allows to decide on the question of whether a given morphism is moderately ambiguous with respect to a given pattern, and it provides an alternative intuitive understanding of moderate ambiguity:

Lemma 4.1. *Let* Σ *be an alphabet. Let* $\alpha := i_1 \cdot i_2 \cdot [\ldots] \cdot i_n$, $n \in \mathbb{N}$, $i_k \in \mathbb{N}$, $1 \leq i_k \leq n$, *and let* $\sigma : \mathbb{N}^* \longrightarrow \Sigma^*$ *be a morphism. Then* σ *is moderately ambiguous with respect to* α *if and only if there do not exist morphisms* $\tau, \tau' : \mathbb{N}^* \longrightarrow \Sigma^*$ *such that*

- $\tau(\alpha) = \tau'(\alpha) = \sigma(\alpha)$ *and*
- *for some* s, $1 \leq s \leq n-1$, $|\,\tau(i_1 \cdot i_2 \cdot [\ldots] \cdot i_s)\,| \geq |\,\tau'(i_1 \cdot i_2 \cdot [\ldots] \cdot i_{s+1})\,|$.

Proof. We begin with the *only if* direction. Hence, let σ be moderately ambiguous with respect to α. Then there are appropriate l_k, $2 \leq k \leq n$ and $r_{k'}$, $1 \leq k' \leq n-1$ satisfying the requirements noted in Definition 4.3. Now assume to the contrary that there exist morphisms τ, τ' satisfying $\tau(\alpha) = \tau'(\alpha) = \sigma(\alpha)$ and

$$|\,\tau(i_1 \cdot i_2 \cdot [\ldots] \cdot i_s)\,| \geq |\,\tau'(i_1 \cdot i_2 \cdot [\ldots] \cdot i_{s+1})\,|.$$

Then by definition

$$\begin{aligned} l_{s+1} &> |\,\tau(i_1 \cdot i_2 \cdot [\ldots] \cdot i_s)\,| \\ &\geq |\,\tau'(i_1 \cdot i_2 \cdot [\ldots] \cdot i_{s+1})\,| \\ &\geq r_{s+1}. \end{aligned}$$

This contradicts the condition $l_{s+1} \leq r_{s+1}$.

We proceed with the *if* direction and we proof it by contraposition. Hence, let σ be not moderately ambiguous with respect to α. Then, by definition there exists an s, $1 \leq s \leq m-1$, which can*not*, at a time, satisfy the following two requirements:

- there are $r_s, l_{s+1}, r_{s+1} \in \mathbb{N}$ with $1 \leq r_s < l_{s+1} \leq r_{s+1} \leq n$ and
- for each morphism τ with $\tau(\alpha) = \sigma(\alpha)$, $|\tau(i_1 \cdot i_2 \cdot [\ldots] \cdot i_s)| < l_{s+1}$, $|\tau(i_1 \cdot i_2 \cdot [\ldots] \cdot i_s)| \geq r_s$ and $|\,\tau(i_1 \cdot i_2 \cdot [\ldots] \cdot i_{s+1})\,| \geq r_{s+1}$.

Hence, there exist morphisms τ, τ' with $\tau(\alpha) = \tau'(\alpha) = \sigma(\alpha)$ that do *not* satisfy

$$r_s \leq |\,\tau(i_1 \cdot i_2 \cdot [\ldots] \cdot i_s)\,| < l_{s+1} \leq r_{s+1} \leq |\,\tau'(i_1 \cdot i_2 \cdot [\ldots] \cdot i_{s+1})$$

and therefore – turning this into a positive statement – it is

$$|\,\tau(i_1 \cdot i_2 \cdot [\ldots] \cdot i_s)\,| \geq |\,\tau'(i_1 \cdot i_2 \cdot [\ldots] \cdot i_{s+1})\,|.$$

This proves the lemma. □

We now briefly illustrate the actual elements of Definition 4.3. Thus, let w be moderately ambiguous with respect to $\alpha = i_1 \cdot i_2 \cdot [\dots] \cdot i_n \in \mathbb{N}^+$, $n \in \mathbb{N}$. Then, by

- $\mathtt{A}_j \in \Sigma$, $1 \leq j \leq |w|$, and
- a notation "$\forall\,\sigma : \sigma(i_k)$", $1 \leq k \leq n$ which says that, for every morphism σ with $\sigma(\alpha) = w$, the hereby marked subword w_k of w results from the image of i_k under σ,

the word w reads as follows:

$$\begin{aligned} w \;=\;& \overbrace{\mathtt{A}_1\,\mathtt{A}_2[\dots]\,\mathtt{A}_{r_1}}^{w_1,\text{ i. e. }\forall\sigma:\sigma(i_1)}\,\mathtt{A}_{r_1+1}\,\mathtt{A}_{r_1+2}[\dots]\,\mathtt{A}_{l_2-1}\,\overbrace{\mathtt{A}_{l_2}\,\mathtt{A}_{l_2+1}[\dots]\,\mathtt{A}_{r_2}}^{w_2,\text{ i. e. }\forall\sigma:\sigma(i_2)}\,[\dots] \\ & \overbrace{\mathtt{A}_{l_{n-1}}\,\mathtt{A}_{l_{n-1}+1}[\dots]\,\mathtt{A}_{r_{n-1}}}^{w_{n-1},\text{ i. e. }\forall\sigma:\sigma(i_{n-1})}\,\mathtt{A}_{r_{n-1}+1}\,\mathtt{A}_{r_{n-1}+2}[\dots]\,\mathtt{A}_{l_n-1}\,\overbrace{\mathtt{A}_{l_n}\,\mathtt{A}_{l_n+1}[\dots]\,\mathtt{A}_{|w|}}^{w_n,\text{ i. e. }\forall\sigma:\sigma(i_n)}\;. \end{aligned}$$

Since, in order to satisfy the requirements on l_k and r_k, each w_k must be nonempty we can directly conclude:

Proposition 4.2. *Let Σ be an alphabet, and let $\alpha \in \mathbb{N}^+$. Let $\sigma : \mathbb{N}^* \longrightarrow \Sigma^*$ be a morphism with, for some $i \in \mathrm{var}(\alpha)$, $\sigma(i) = \varepsilon$. Then neither σ nor $\sigma(\alpha)$ are moderately ambiguous with respect to α.*

With regard to nonerasing morphisms, the relation between unambiguity and moderate ambiguity is an immediate consequence of the respective definitions. If a nonerasing morphism σ is unambiguous for a pattern α then we can define $r_1 := 1$, $l_n := |\sigma(i_1 \cdot i_2 \cdot [\dots] \cdot i_{n-1})| + 1$ and $l_k := r_k := |\sigma(i_1 \cdot i_2 \cdot [\dots] \cdot i_{k-1})| + 1$, $2 \leq k \leq n-1$. This choice satisfies the conditions of Definition 4.3 and, hence, we may state:

Proposition 4.3. *Let Σ be an alphabet. Let $\alpha \in \mathbb{N}^+$, and let $\sigma : \mathbb{N}^* \longrightarrow \Sigma^*$ be a nonerasing morphism. Then $\sigma(\alpha)$ is moderately ambiguous if it is unambiguous. In general, the converse of this statement does not hold true.*

The correctness of the second claim of Proposition 4.3 is substantiated by Example 4.9 (see below).

We proceed with some examples on Definition 4.3, the first of which again addresses our favourite pattern:

Example 4.8. Once more, let $\alpha := 1 \cdot 2 \cdot 3 \cdot 4 \cdot 1 \cdot 4 \cdot 3 \cdot 2$.

Let the morphism σ be given by $\sigma(1) := \mathtt{a}$, $\sigma(2) := \mathtt{a\,b}$, $\sigma(3) := \mathtt{b}$, $\sigma(4) := \mathtt{b\,a}$. Then σ and, thus, the word $\sigma(\alpha) = \mathtt{a\,a\,b\,b\,b\,a\,a\,b\,a\,b\,a\,b}$ are moderately ambiguous with respect to α because they are unambiguous (cf. Proposition 4.3 and Example 4.2).

Let the morphism σ_{c} be given by $\sigma_{\mathrm{c}}(i) := \mathtt{a\,b}^i$, $i \in \mathbb{N}$. Then σ_{c} and, thus, the word $w := \sigma_{\mathrm{c}}(\alpha) = \mathtt{a\,b\,a\,b}^2\,\mathtt{a\,b}^3\,\mathtt{a\,b}^4\,\mathtt{a\,b\,a\,b}^4\,\mathtt{a\,b}^3\,\mathtt{a\,b}^2$ are not moderately ambiguous with respect to α as w can also be generated, e. g., by the morphism τ with $\tau(1) := \mathtt{a\,b\,a\,b}^2$, $\tau(2) := \varepsilon$, $\tau(3) := \mathtt{a\,b}^3\,\mathtt{a\,b}^2$ and $\tau(4) := \mathtt{b}^2$ (cf. Proposition 4.2 and Example 4.2). $\diamond$

While Example 4.8 mainly illustrates Proposition 4.2 and Proposition 4.3, the following two examples are crucial for the understanding of moderate unambiguity:

Example 4.9. Let $\alpha := 1 \cdot 2 \cdot 2 \cdot 3 \cdot 1 \cdot 4 \cdot 3 \cdot 4$.

Let the morphism σ be given by $\sigma(1) := \mathtt{a\,a}$, $\sigma(2) := \mathtt{a\,b}$, $\sigma(3) := \mathtt{b\,a}$, $\sigma(4) := \mathtt{b}$. Then there is exactly one different morphism τ with $w := \sigma(\alpha) = \tau(\alpha)$, namely $\tau(1) := \mathtt{a\,a}$, $\tau(2) := \mathtt{a\,b}$, $\tau(3) := \mathtt{b}$, $\tau(4) := \mathtt{a\,b}$. If we now write α as $i_1 \cdot i_2 \cdot i_3 \cdot i_4 \cdot i_5 \cdot i_6 \cdot i_7 \cdot i_8$ then we can define, e. g., $r_1 := 1$, $l_2 = r_2 := 3$, $l_3 = r_3 := 5$, $l_4 = r_4 := 7$, $l_5 = r_5 := 9$, $l_6 = r_6 := 11$, $l_7 = r_7 := 12$, $l_8 := 14$.

$$\alpha = i_1 = 1 \cdot i_2 = 2 \cdot i_3 = 2 \cdot i_4 = 3 \cdot i_5 = 1 \cdot i_6 = 4 \cdot i_7 = 3 \cdot i_8 = 4$$

	$\sigma(1)$		$\sigma(2)$		$\sigma(2)$		$\sigma(3)$		$\sigma(1)$		$\sigma(4)$	$\sigma(3)$		$\sigma(4)$
$w =$	a	a	a	b	a	b	b	a	a	a	b	b	a	b
	$\tau(1)$		$\tau(2)$		$\tau(2)$		$\tau(3)$	$\tau(1)$		$\tau(4)$		$\tau(3)$	$\tau(4)$	
Pos. in w:	1	2	3	4	5	6	7	8	9	10	11	12	13	14
l_k/r_k	r_1		l_2/r_2		l_3/r_3		l_4/r_4		l_5/r_5		l_6/r_6	l_7/r_7		l_8

This satisfies the conditions in Definition 4.3. Thus, w is moderately ambiguous (and so are σ and τ). ◇

As mentioned above, Example 4.9 additionally underlines the correctness of the second claim in Proposition 4.3.

Example 4.10. Let $\alpha := 1 \cdot 2 \cdot 3 \cdot 1 \cdot 3 \cdot 2$.

Let the morphism σ be given by $\sigma(1) := \mathtt{a\,b}^2$, $\sigma(2) := \mathtt{a\,b}^3$, $\sigma(3) := \mathtt{a\,b}^6$. It can be easily verified that there are exactly four morphisms τ with $\tau(\alpha) = \sigma(\alpha) =: w$ and, for some $i \in \text{var}(\alpha)$, $\tau(i) \neq \sigma(i)$, none of which maps a variable in α onto the empty word. With regard to the question of whether $\sigma(\alpha)$ is moderately ambiguous, we compare σ with one selected τ':

$$w = \overbrace{\mathtt{abb}}^{\sigma(1)}\overbrace{\mathtt{abbb}}^{\boxed{\sigma(2)}}\overbrace{\mathtt{abbbbbb}}^{\sigma(3)}\overbrace{\mathtt{abb}}^{\sigma(1)}\overbrace{\mathtt{abbbbbb}}^{\sigma(3)}\overbrace{\mathtt{abbb}}^{\sigma(2)}$$

$$w = \underbrace{\mathtt{abbabbb}}_{\tau'(1)}\underbrace{\mathtt{abbb}}_{\boxed{\tau'(2)}}\underbrace{\mathtt{bbb}}_{\tau'(3)}\underbrace{\mathtt{abbabbb}}_{\tau'(1)}\underbrace{\mathtt{bbb}}_{\tau'(3)}\underbrace{\mathtt{abbb}}_{\tau'(2)}$$

We now write α as $i_1 \cdot i_2 \cdot i_3 \cdot i_4 \cdot i_5 \cdot i_6$. Then $|\sigma(i_1 \cdot i_2)| = 7 = |\tau'(i_1)|$. Therefore, according to Lemma 4.1, w is not moderately ambiguous with respect to α – although all morphisms τ are nonerasing, and even, with regard to the variable i_2 used for our argumentation, it is $\tau'(i_2) = \tau'(2) = \sigma(2)$. ◇

As a consequence of the particular restrictions on the morphisms generating a moderately ambiguous word w with respect to a pattern α we can regard the subword w_k in w as "linked" with the kth symbol in α. While Definition 4.3 merely is based on the existence of such w_k in a moderately ambiguous word, our learning theoretical considerations in Chapter 5.2.1 require some knowledge on the maximum length the w_k can have. Formally, this leads to the concept of subwords that are associated with corresponding occurrences of variables:

Definition 4.4 (Subwords associated with indices)**.** *Let Σ be an alphabet. Let $\alpha := i_1 \cdot i_2 \cdot [\dots] \cdot i_n$, $n \in \mathbb{N}$, $i_k \in \mathbb{N}$, $1 \leq i_k \leq n$, and let $w \in \Sigma^+$ be a word that is moderately ambiguous with respect to α. Let the indices $l_2, l_3, \dots, l_n \in \{1, 2, \dots, |\sigma(\alpha)|\}$ and $r_1, r_2, \dots, r_{n-1} \in \{1, 2, \dots, |\sigma(\alpha)|\}$, the existence of which is granted by Definition 4.3, be chosen in such a way that for every k, $1 \leq k \leq n-1$,*

- *l_{k+1} and r_k satisfy the conditions of Definition 4.3,*
- *there exists a σ with $\sigma(\alpha) = w$ and $|\,\sigma(i_1 \cdot i_2 \cdot [\,\dots] \cdot i_k)\,| = l_{k+1} - 1$ and*
- *there exists a σ with $\sigma(\alpha) = w$ and $|\,\sigma(i_1 \cdot i_2 \cdot [\,\dots] \cdot i_k)\,| = r_k$.*

Then – for $w = \mathtt{A}_1\,\mathtt{A}_2[\,\dots]\,\mathtt{A}_m \in \Sigma^+$, $m \in \mathbb{N}$, and with $l_1 := 1$, $r_n := m$ – the subword $w_{k,\max} := \mathtt{A}_{l_k}\,\mathtt{A}_{l_k+1}[\,\dots]\,\mathtt{A}_{r_k}$, $1 \le k \le n$, is said to be associated with k (with respect to α and w)*.*

Since in Definition 4.4

- the l_k are chosen minimally among all candidates satisfying the requirements of Definition 4.3 and
- the r_k are chosen maximally among all candidates satisfying the requirements of Definition 4.3

we can now complete our above illustration of a moderately ambiguous word w with respect to $\alpha = i_1 \cdot i_2 \cdot [\dots] \cdot i_n \in \mathbb{N}^+$, $n \in \mathbb{N}$. Thus, with a notation "$\exists\,\sigma : \sigma(i_{k-1}) \wedge \exists\sigma : \sigma(i_k)$", $2 \le k \le n$, which says that, for a morphism σ with $\sigma(\alpha) = w$, the hereby marked subword of w results from $\sigma(i_{k-1})$ and, for a second morphism σ' with $\sigma'(\alpha) = w$, results from $\sigma'(i_k)$, we can exactly identify the subword which is associated with some k:

$$w \;=\; \overbrace{\mathtt{A}_1\,\mathtt{A}_2[\,\dots]\,\mathtt{A}_{r_1}}^{w_{1,\max}}\;\underbrace{\mathtt{A}_{r_1+1}\,\mathtt{A}_{r_1+2}[\,\dots]\,\mathtt{A}_{l_2-1}}_{\exists\sigma:\sigma(i_1)\,\wedge\,\exists\sigma:\sigma(i_2)}\;\overbrace{\mathtt{A}_{l_2}\,\mathtt{A}_{l_2+1}[\,\dots]\,\mathtt{A}_{r_2}}^{w_{2,\max}}\;[\,\dots]$$

$$\overbrace{\mathtt{A}_{l_{n-1}}\,\mathtt{A}_{l_{n-1}+1}[\,\dots]\,\mathtt{A}_{r_{n-1}}}^{w_{n-1,\max}}\;\underbrace{\mathtt{A}_{r_{n-1}+1}\,\mathtt{A}_{r_{n-1}+2}[\,\dots]\,\mathtt{A}_{l_n-1}}_{\exists\sigma:\sigma(i_{n-1})\,\wedge\,\exists\sigma:\sigma(i_n)}\;\overbrace{\mathtt{A}_{l_n}\,\mathtt{A}_{l_n+1}[\,\dots]\,\mathtt{A}_{|w|}}^{w_{n-1,\max}}\;.$$

We now identify all associated subwords for the pattern and the word already examined in Example 4.9:

Example 4.11. Let $\alpha := 1 \cdot 2 \cdot 2 \cdot 3 \cdot 1 \cdot 4 \cdot 3 \cdot 4$.

Let $w :=$ `aaababbaaabbab`. Then Example 4.9 states that there are exactly two morphisms σ, τ with $\sigma(\alpha) = w = \tau(\alpha)$, namely $\sigma(1) :=$ `aa`, $\sigma(2) :=$ `ab`, $\sigma(3) :=$ `ba`, $\sigma(4) :=$ `b` and $\tau(1) :=$ `aa`, $\tau(2) :=$ `ab`, $\tau(3) :=$ `b`, $\tau(4) :=$ `ab`. Consequently, with $\alpha = i_1 \cdot i_2 \cdot i_3 \cdot i_4 \cdot i_5 \cdot i_6 \cdot i_7 \cdot i_8$, we have

- $r_1 = |\sigma(1)| = |\tau(1)| = 2 = \max\{1, 2\}$,
- $l_2 = |\sigma(1)| + 1 = |\tau(1)| + 1 = 3 = \min\{3, 4\}$,
- $r_2 = |\sigma(1 \cdot 2)| = |\tau(1 \cdot 2)| = 4 = \max\{3, 4\}$,
- $l_3 = |\sigma(1 \cdot 2)| + 1 = |\tau(1 \cdot 2)| + 1 = 5 = \min\{5, 6\}$,
- $r_3 = |\sigma(1 \cdot 2 \cdot 2)| = |\tau(1 \cdot 2 \cdot 2)| = 6 = \max\{5, 6\}$,
- $l_4 = |\sigma(1 \cdot 2 \cdot 2)| + 1 = |\tau(1 \cdot 2 \cdot 2)| + 1 = 7 = \min\{7\}$,

- $r_4 = \min\{|\sigma(1 \cdot 2 \cdot 2 \cdot 3)|, |\tau(1 \cdot 2 \cdot 2 \cdot 3)|\} = 7 = \max\{7\}$,
- $l_5 = \max\{|\sigma(1 \cdot 2 \cdot 2 \cdot 3)|, |\tau(1 \cdot 2 \cdot 2 \cdot 3)|\} + 1 = 9 = \min\{9\}$,
- $r_5 = \min\{|\sigma(1 \cdot 2 \cdot 2 \cdot 3 \cdot 1)|, |\tau(1 \cdot 2 \cdot 2 \cdot 3 \cdot 1)|\} = 9 = \max\{9\}$,
- $l_6 = \max\{|\sigma(1 \cdot 2 \cdot 2 \cdot 3 \cdot 1)|, |\tau(1 \cdot 2 \cdot 2 \cdot 3 \cdot 1)|\} + 1 = 11 = \min\{11\}$,
- $r_6 = |\sigma(1 \cdot 2 \cdot 2 \cdot 3 \cdot 1 \cdot 4)| = |\tau(1 \cdot 2 \cdot 2 \cdot 3 \cdot 1 \cdot 4)| = 11 = \max\{11\}$,
- $l_7 = |\sigma(1 \cdot 2 \cdot 2 \cdot 3 \cdot 1 \cdot 4)| + 1 = |\tau(1 \cdot 2 \cdot 2 \cdot 3 \cdot 1 \cdot 4)| + 1 = 12 = \min\{12\}$,
- $r_7 = \min\{|\sigma(1 \cdot 2 \cdot 2 \cdot 3 \cdot 1 \cdot 4 \cdot 3)|, |\tau(1 \cdot 2 \cdot 2 \cdot 3 \cdot 1 \cdot 4 \cdot 3))|\} = 12 = \max\{12\}$,
- $l_8 = \max\{|\sigma(1 \cdot 2 \cdot 2 \cdot 3 \cdot 1 \cdot 4 \cdot 3)|, |\tau(1 \cdot 2 \cdot 2 \cdot 3 \cdot 1 \cdot 4 \cdot 3)|\} + 1 = 14 = \min\{14\}$.

The subsequent illustration summarises the necessary elements of this argumentation:

$$\alpha = i_1 = 1 \cdot i_2 = 2 \cdot i_3 = 2 \cdot i_4 = 3 \cdot i_5 = 1 \cdot i_6 = 4 \cdot i_7 = 3 \cdot i_8 = 4$$

(overbraces)	$\sigma(1)$		$\sigma(2)$		$\sigma(2)$		$\sigma(3)$		$\sigma(1)$		$\sigma(4)$	$\sigma(3)$		$\sigma(4)$
w =	a	a	a	b	a	b	b	a	a	a	b	b	a	b
(underbraces)	$\tau(1)$		$\tau(2)$		$\tau(2)$		$\tau(3)$	$\tau(1)$		$\tau(4)$		$\tau(3)$	$\tau(4)$	
Pos. in w:	1	2	3	4	5	6	7	8	9	10	11	12	13	14
l_k/r_k		r_1	l_2	r_2	l_3	r_3	l_4/r_4		l_5/r_5		l_6/r_6	l_7/r_7		l_8

Thus, the respective associated subwords in w look as follows:

(overbraces)	$w_{1,\max}$		$w_{2,\max}$		$w_{3,\max}$		$w_{4,\max}$		$w_{5,\max}$		$w_{6,\max}$	$w_{7,\max}$		$w_{8,\max}$
w =	a	a	a	b	a	b	b	a	a	a	b	b	a	b
Pos. in w:	1	2	3	4	5	6	7	8	9	10	11	12	13	14

In particular, this example explains why we cannot, w. l. o. g., reduce the idea of associated subwords to something like "subwords associated to variables" (which should be much easier to explain and, moreover, to grasp). As shown above, $w_{1,\max} = \mathtt{a\,a} \neq \mathtt{a} = w_{5,\max}$, but evidently $i_1 = 1 = i_5$. Thus, we simply cannot guarantee that the associated subwords of two different occurrences of a single variable are the same, and therefore we have to link the subword to the index k in our slightly cumbersome definition. For our later considerations on moderate ambiguity (see Chapter 5.2), however, this nontrivial and potentially counter-intuitive aspect is rather marginal. ◇

Definition 4.3 facilitates a refinement of Proposition 4.3, which says that, in any word w that is both unambiguous and moderately ambiguous, the sum of the lengths of all associated subwords equals the length of w:

Proposition 4.4. *Let Σ be an alphabet. Let $\alpha := i_1 \cdot i_2 \cdot [\dots] \cdot i_n$, $n \in \mathbb{N}$, $i_k \in \mathbb{N}$, $1 \leq i_k \leq n$, and let $w \in \Sigma^+$ be a word that is moderately ambiguous with respect to α. Let the indices $l_2, l_3, \dots, l_n \in \{1, 2, \dots, |\sigma(\alpha)|\}$ and $r_1, r_2, \dots, r_{n-1} \in \{1, 2, \dots, |\sigma(\alpha)|\}$ be chosen in such a way that they correspond to the requirements established in Definition 4.4. Then w is unambiguous with respect to α if and only if, for every k, $1 \leq k \leq n-1$, $r_k = l_{k+1} - 1$.*

While, within the present chapter, we restrict the concept of moderate ambiguity to terminal-free patterns, our next definition deals with the ambiguity of substitutions. Thus, this second type of ambiguity to be named explicitly is exclusively applicable to those patterns that contain at least one terminal symbol:

Definition 4.5 (Terminal-comprising ambiguity). *Let Σ be an alphabet, and let $\alpha := \beta_0\, v_1\, \beta_1\, v_2\, \beta_2\, [\dots]\, \beta_{n-1}\, v_n\, \beta_n \in (\mathbb{N} \cup \Sigma)^+ \setminus \mathbb{N}^+$ with $n \in \mathbb{N}$, $\beta_0, \beta_n \in \mathbb{N}^*$, $\beta_1, \beta_2, \dots \beta_{n-1} \in \mathbb{N}^+$ and $v_1, v_2, \dots v_n \in \Sigma^+$. Then a word $w \in \Sigma^*$ is said to be* terminal-comprisingly ambiguous (with respect to α) *if there exist substitutions $\sigma, \tau : (\mathbb{N} \cup \Sigma)^* \longrightarrow \Sigma^*$ with $\sigma(\alpha) = w = \tau(\alpha)$ such that, for some k, $0 \leq k \leq n$, $|\sigma(\beta_k)| \neq |\tau(\beta_k)|$.*

Additionally, we designate each substitution σ with $\sigma(\alpha) = w$ as terminal-comprisingly ambiguous (with respect to α) *as well.*

In order to examine Definition 4.5 in a more concrete manner we again give a number of examples:

Example 4.12. Let $\alpha := 1 \cdot 1 \cdot \mathtt{a} \cdot 2 \cdot 2 \cdot 3 \cdot 3 \cdot \mathtt{b} \cdot 4$. Thus, it is $\alpha = \beta_0\, v_1\, \beta_1\, v_2\, \beta_2$ with $\beta_0 = 1 \cdot 1$, $\beta_1 = 2 \cdot 2 \cdot 3 \cdot 3$, $\beta_2 = 4$ and $v_1 = \mathtt{a}$, $v_2 = \mathtt{b}$.

Let $w := \mathtt{abbababb}$. Then the substitutions σ with $\sigma(1) := \sigma(4) := \varepsilon$, $\sigma(2) := \mathtt{b}$, $\sigma(3) := \mathtt{ab}$ and τ with $\tau(1) := \tau(2) := \tau(3) := \varepsilon$, $\tau(4) := \mathtt{bababb}$ yield $\sigma(\alpha) = w = \tau(\alpha)$, but nevertheless it is $|\sigma(\beta_1)| = 6 \neq 0 = |\tau(\beta_1)|$. Thus, w, σ and τ are terminal-comprisingly ambiguous with respect to α.

Let the substitution σ be given by $\sigma(1) := \sigma(2) := \mathtt{ab}$, $\sigma(3) := \mathtt{abb}$, $\sigma(4) := \mathtt{b}$. Then σ and, thus, the word $\sigma(\alpha) = \mathtt{ababaababababbabbbb}$ are not terminal-comprisingly ambiguous with respect to α as they are unambiguous. Note that this fact nicely contrasts with the insights presented in Example 4.4 on the ambiguity of morphic images of the pattern $\alpha' := 1 \cdot \mathtt{a} \cdot 2 \cdot 2 \cdot \mathtt{b} \cdot 3$ where the substitution assigns a nonempty word to $\beta_1' = 2 \cdot 2$.

Let the substitution σ be given by $\sigma(1) = \sigma(4) := \varepsilon$, $\sigma(2) = \sigma(3) := \mathtt{c}$. Then σ and, thus, the word $\sigma(\alpha) = \mathtt{acccb}$ are not terminal-comprisingly ambiguous with respect to α although they are not unambiguous. The only substitutions τ, τ' with $\tau \neq \tau'$, $\tau \neq \sigma \neq \tau'$ and $\tau(\alpha) = \tau'(\alpha) = \sigma(\alpha)$ are given by $\tau(1) = \tau(3) = \tau(4) := \varepsilon$, $\tau(2) := \mathtt{cc}$ and $\tau(1) = \tau(2) = \tau(4) := \varepsilon$, $\tau(3) := \mathtt{cc}$. Hence, it is $|\sigma(\beta_0)| = |\tau(\beta_0)| = \tau'(\beta_0)| = 0$, $|\sigma(\beta_1)| = |\tau(\beta_1)| = |\tau'(\beta_1)| = 4$ and $|\sigma(\beta_2)| = |\tau(\beta_2)| = |\tau'(\beta_2)| = 0$. ◇

From Definitions 4.1 and 4.5 and the pattern $\alpha = 1 \cdot 1 \cdot \mathtt{a} \cdot 2 \cdot 2 \cdot 3 \cdot 3 \cdot \mathtt{b} \cdot 4$ and word $w = \mathtt{acccb}$ given in Example 4.12 we can directly conclude:

Proposition 4.5. *Let Σ be an alphabet. Let $\alpha \in (\mathbb{N} \cup \Sigma)^+ \setminus \mathbb{N}^+$, and let $\sigma : (\mathbb{N} \cup \Sigma)^* \longrightarrow \Sigma^*$ be a substitution. Then $\sigma(\alpha)$ is not unambiguous if it is terminal-comprisingly ambiguous. In general, the converse of this statement does not hold true.*

In anticipation of our results in Chapter 5 and Chapter 6 we state that terminal-comprising ambiguity shows a number of counter-intuitive and discontinuous properties that are much harder to understand than that of "standard" ambiguity. This is a consequence of the following fact proven by the pattern $\alpha = 1 \cdot 1 \cdot \mathtt{a} \cdot 2 \cdot 2 \cdot 3 \cdot 3 \cdot \mathtt{b} \cdot 4$ and the word $w = \mathtt{abbababb}$ discussed in Example 4.12:

Proposition 4.6. *Let Σ be an alphabet. Then there is a substitution $\sigma : (\mathbb{N} \cup \Sigma)^* \longrightarrow \Sigma^*$ and an $\alpha := \beta_0\, v_1\, \beta_1\, v_2\, \beta_2\, [\dots]\, \beta_{n-1}\, v_n\, \beta_n \in (\mathbb{N} \cup \Sigma)^+ \setminus \mathbb{N}^+$ with $n \in \mathbb{N}$, $\beta_0, \beta_n \in \mathbb{N}^*$, $\beta_1, \beta_2, \dots \beta_{n-1} \in \mathbb{N}^+$, $v_1, v_2, \dots v_n \in \Sigma^+$ such that*

1. *for every k, $0 \leq k \leq n$, $\sigma(\beta_k) = \varepsilon$ or $\sigma(\beta_k)$ is unambiguous with respect to β_k, and*
2. *$\sigma(\alpha)$ is not unambiguous with respect to α.*

Thus, the substitution σ referred to in Proposition 4.6 must be terminal-comprisingly ambiguous. Consequently, Definition 4.5 mainly addresses the *difference* between the ambiguity of common and terminal-preserving morphisms.

We continue our introduction of types of ambiguity with the third (and last) one. It describes a potential property of ambiguous words that in the examples in the present chapter is used frequently (cf., e.g., Examples 4.2 and 4.3) and that in the context of the subsequent chapters – particularly of those dealing with applications – turns out to be sort of the "disastrous" ambiguity since corresponding words do not provide any reliable proof for the existence of certain variables in their preimage. However, in spite of these emerging unpleasant consequences of the said kind of ambiguity, we prefer a more sober name:

Definition 4.6 (Substantial ambiguity). *Let Σ be an alphabet. Let $\alpha \in (\mathbb{N} \cup \Sigma)^+$ and $i \in \text{var}(\alpha)$. Then a word $w \in \Sigma^+$ is called* substantially ambiguous for i (in α) *provided that there exist (terminal-preserving) morphisms $\sigma, \tau : (\mathbb{N} \cup \Sigma)^* \longrightarrow \Sigma^*$ with $\sigma(\alpha) = w = \tau(\alpha)$, $\sigma(i) \neq \varepsilon$ and $\tau(i) = \varepsilon$.*

Additionally, we designate each (terminal-preserving) morphism σ with $\sigma(\alpha) = w$ and $\sigma(i) \neq \varepsilon$ as substantially ambiguous for i (in α) *as well.*

Finally we say that a word w (or a morphism σ) is substantially ambiguous (with respect to α) *if there exists a variable $i \in \text{var}(\alpha)$ such that w (or σ, respectively) is substantially ambiguous for i in α.*

Note that Definition 4.6 – unlike Definitions 4.3 and 4.5 – does not label *every* morphism σ with $\sigma(\alpha) = w$ as substantially ambiguous in the case that w is substantially ambiguous, but only those with $\sigma(i) \neq \varepsilon$. This arbitrary decision is meant to support the intuition according to which a morphism σ that *seems* to reflect the existence of the variable i in α looses this property if there exists a morphism τ with $\tau(\alpha) = \sigma(\alpha)$ and $\tau(i) = \varepsilon$ (see Chapter 1). Of course, we shall not consider such a morphism τ capable to prove any occurrence of i in α, and therefore it is not covered by the terminology of Definition 4.6.

The relation between weak unambiguity – ignoring morphisms that map a variable onto the empty word – and substantial ambiguity – requiring the existence of such morphisms – follows directly by definition:

Proposition 4.7. *Let Σ be an alphabet. Let $\alpha \in (\mathbb{N} \cup \Sigma)^+$, and let $\sigma : (\mathbb{N} \cup \Sigma)^* \longrightarrow \Sigma^*$ be an (if applicable, terminal-preserving) nonerasing morphism. Then $\sigma(\alpha)$ is substantially ambiguous if it is weakly unambiguous, but not unambiguous. In general, the converse of this statement does not hold true.*

On the one hand, it is an evident necessary condition for a word being moderately ambiguous that it is not substantially ambiguous (cf. Proposition 4.2). On the other hand, the pattern $\alpha = 1 \cdot 2 \cdot 3 \cdot 1 \cdot 3 \cdot 2$ and the word $w = \mathtt{a}\,\mathtt{b}^2\,\mathtt{a}\,\mathtt{b}^3\,\mathtt{a}\,\mathtt{b}^6\,\mathtt{a}\,\mathtt{b}^2\,\mathtt{a}\,\mathtt{b}^6\,\mathtt{a}\,\mathtt{b}^3$ (discussed in Example 4.10) prove that this condition is not characteristic:

Proposition 4.8. *Let Σ be an alphabet. Let $\alpha \in \mathbb{N}^+$, and let $\sigma : \mathbb{N}^* \longrightarrow \Sigma^*$ be a nonerasing morphism. Then $\sigma(\alpha)$ is not moderately ambiguous if it is substantially ambiguous. In general, the converse of this statement does not hold true.*

Before we introduce our research questions on the ambiguity of morphisms, we now give a short summary of the concepts introduced above, which moreover contains some additional interpreting remarks.

Our first definition describes the most fundamental category, namely the *unambiguity* of a word $w \in \Sigma^+$ with respect to a pattern α, i. e. there is exactly one morphism $\sigma : (\mathbb{N} \cup \Sigma)^* \longrightarrow \Sigma^*$ with $\sigma(\alpha) = w$ (cf. Definition 4.1). Concerning those morphisms and words that are not unambiguous we have several terms for expressing different "degrees"† of ambiguity. Although we consider the corresponding categories to be of intrinsic interest, this terminology is strongly motivated by special needs involved in the application of our results to problems for pattern languages. Since we are mainly interested in the question of whether or not all variables in a pattern are required for morphically generating the word under consideration we identify two types of ambiguous words that show particular corresponding properties. First, a *substantially ambiguous* word is generated by both a morphism σ that assigns a nonempty word to a variable i in the respective pattern and a morphism τ which maps i onto the *empty* word (cf. Definition 4.6). Thus, as claimed in Chapter 1 and as to be substantiated in some of the subsequent application chapters (see Chapters 6.1.1, 6.2.1 and 6.3.1), w does not allow to draw any reliable conclusions on the existence of i in the pattern. Second, a *moderately ambiguous* word w – which actually may be ambiguous or unambiguous – for every occurrence i_k of any variable i in α contains a *nonempty* "unambiguous subword" w_k, i. e., for every morphism σ with $\sigma(\alpha) = w$, w_k *must* be generated by i_k (cf. Definition 4.3). Consequently, a moderately ambiguous word must be generated by nonerasing morphisms only, and therefore it cannot be substantially ambiguous. Note that there are ambiguous words which are neither substantially nor moderately ambiguous (see Example 4.10 and Proposition 4.8); hence, our concepts do not lead to a partition of the set of all ambiguous words.

In addition to these elementary types of (un-)ambiguity, the present chapter introduces two more definitions, the first of which addresses a particular restricted *range* for the morphisms: the *weak unambiguity* limits any considerations on the subject to morphic images in the *free semigroup* Σ^+ so that it merely considers nonerasing morphisms; thus, a word w is weakly unambiguous with respect to a pattern α if and only if there is exactly one morphism $\sigma : (\mathbb{N} \cup \Sigma)^+ \longrightarrow \Sigma^+$ with $\sigma(\alpha) = w$ (cf. Definition 4.2). Finally, our second additional definition is concerned with a special kind of ambiguity resulting from a *domain* for the morphisms which contains terminal symbols: It describes a situation where, for a pattern $\alpha \in (\mathbb{N} \cup \Sigma)^+ \setminus \mathbb{N}^+$ and a word $w \in \Sigma^+$, there are substitutions σ, τ with $\sigma(\alpha) = w = \tau(\alpha)$ and, moreover, a variable $i \in \mathrm{var}(\alpha)$, a letter $\mathtt{A} \in \Sigma$ and $k \in \mathbb{N}$ such that the kth occurrence of $\mathtt{A}$ in w results, on the one hand, from

†Note that we do not use this term in the sense of Mateescu and Salomaa [54], who understand the degree of ambiguity of a word w with respect to a pattern α as the *number* of morphisms σ with $\sigma(\alpha) = w$. This approach, however, does not match with our vaguely information theoretical point of view, which requires qualitative rather than quantitative criteria.

$\sigma(i)$ and, on the other hand, from no $\tau(i')$, $i' \in \text{var}(\alpha)$. If two morphisms in such a way disagree on the question of whether a certain occurrence of a letter in their common image results from the image of a terminal symbol or a variable then we speak of *terminal-comprising ambiguity* (cf. Definition 4.5). The latter definition mainly aims at the requirements of Chapter 6, which examines the extensibility of the results in Chapter 5 to terminal-preserving morphisms. In this context note again that the definition of moderate ambiguity, within the present chapter, due to mere presentational reasons is restricted to terminal-free patterns. A comprehensive definition (covering substitutions as well) is given as soon as it is needed, i. e. in Chapter 6 (see Definition 6.1).

4.2 Research questions

Using the definitions introduced in the previous Chapter 4.1, the present thesis systematically examines selected basic questions on the ambiguity of morphisms. Since hardly any related results can be found in literature so far, we mainly wish to investigate the *(non-)existence* of moderately ambiguous or unambiguous morphisms with respect to arbitrary patterns rather than to discuss the (un-)ambiguity of arbitrary morphisms in detail; still, we address ourselves to the latter task whenever appropriate or necessary. In other words, our analysis is comprehensive with regard to the patterns in $\mathbb{N}^+$ and (as far as possible) in $(\mathbb{N} \cup \Sigma)^+$ for arbitrary alphabets Σ, but it is not comprehensive concerning the set of all morphisms $\sigma : \mathbb{N}^* \longrightarrow \Sigma^*$ (or, if applicable, $\sigma : (\mathbb{N} \cup \Sigma)^* \longrightarrow \Sigma^*$), i. e. we do not attempt to exactly distinguish all unambiguous morphisms from all ambiguous ones with respect to arbitrary patterns. Nevertheless, our proofs for positive results are always constructive so that they provide numerous examples of how exactly unambiguity or moderate ambiguity can be achieved. Throughout the subsequent chapters, as a consequence of our information theoretical motivation, we normally focus on the ambiguity of morphisms σ that are nonerasing (or even injective) and on alphabets Σ with at least two distinct letters.

Consequently, with regard to any such alphabet Σ and *common morphisms* $\sigma : \mathbb{N}^* \longrightarrow \Sigma^*$, the most elementary question to be discussed below reads as follows:

Question 4.1. *Let Σ be an alphabet, $|\Sigma| \geq 2$, and let $\alpha \in \mathbb{N}^+$. Is there a (preferably nonerasing) morphism $\sigma : \mathbb{N}^* \longrightarrow \Sigma^*$ which is unambiguous with respect to α?*

From Example 4.1 we can immediately derive that there are patterns which cannot unambiguously be mapped onto any word. Hence, Question 4.1 – which is mainly discussed in Chapter 5.4 – does not have an answer in the affirmative for all patterns. Our main result on unambiguity characterises those patterns for which there exists an unambiguous nonerasing morphism (see Corollary 5.14).

Concerning the weaker requirement of moderate ambiguity ("weaker" according to Proposition 4.3), we pose the analogous question:

Question 4.2. *Let Σ be an alphabet, $|\Sigma| \geq 2$, and let $\alpha \in \mathbb{N}^+$. Is there a nonerasing morphism $\sigma : \mathbb{N}^* \longrightarrow \Sigma^*$ which is moderately ambiguous with respect to α?*

Note that the restriction to nonerasing morphisms due to Definitionen 4.3 is mandatory. Question 4.2 is examined in Chapter 5.2. The corresponding main result is given in Corollary 5.6.

The most convenient answer to these two questions of course might be to find a single morphism which is unambiguous or moderately ambiguous for all patterns in a preferably large set:

Question 4.3. *Let Σ be an alphabet, $|\Sigma| \geq 2$, and let $\Pi \subseteq \mathbb{N}^+$ be a set of patterns. Is there a (preferably nonerasing) morphism $\sigma : \mathbb{N}^* \longrightarrow \Sigma^*$ such that, for every pattern $\alpha \in \Pi$, σ is unambiguous or moderately ambiguous with respect to α?*

Results related to Question 4.3 can be found, e. g., in Theorem 5.3, Proposition 5.4 and Theorem 5.8.

Concerning weak unambiguity, i. e. the unambiguity in free semigroups without the identity element ε, the present thesis merely contains a number of brief notes. They deal with the following question:

Question 4.4. *Let Σ be an alphabet, $|\Sigma| \geq 2$. Is there a morphism $\sigma : \mathbb{N}^+ \longrightarrow \Sigma^+$ such that, for every pattern $\alpha \in \mathbb{N}^+$, σ is weakly unambiguous with respect to α? Is there an injective morphism $\sigma : \mathbb{N}^+ \longrightarrow \Sigma^+$ such that, for every pattern $\alpha \in \mathbb{N}^+$, σ is weakly unambiguous with respect to α?*

Question 4.4 is answered in Chapter 5.3. Additionally, due to Proposition 4.1, each positive result on the existence of unambiguous morphisms can be immediately interpreted as an insight on weak unambiguity, as well.

In Chapter 6 we mainly examine to which extent our results on the ambiguity of common morphisms can be extended to *terminal-preserving morphisms* (i. e., normally, substitutions). Consequently, we discuss the following problem:

Question 4.5. *Let Σ be an alphabet, and let $\sigma : \mathbb{N}^* \longrightarrow \Sigma^*$ be a morphism that is moderately ambiguous (or even unambiguous) with respect to an arbitrary pattern or set of patterns. Let $\sigma' : (\mathbb{N} \cup \Sigma)^* \longrightarrow \Sigma^*$ be a substitution satisfying, for every $j \in \mathbb{N}$, $\sigma'(j) = \sigma(j)$. For which patterns $\alpha \in (\mathbb{N} \cup \Sigma)^+$ is σ' moderately ambiguous (or unambiguous)?*

A major result addressing Question 4.5 is given by Theorem 6.2.

Technically, we conduct our analysis on the ambiguity of substitutions rather by a problem on the avoidability of terminal-comprising ambiguity:

Question 4.6. *Let Σ be an alphabet, and let $\alpha \in (\mathbb{N} \cup \Sigma)^+$. Is there a substitution $\sigma : (\mathbb{N} \cup \Sigma)^* \longrightarrow \Sigma^*$ which is not terminal-comprisingly ambiguous with respect to α?*

Question 4.6 largely determines our research in Chapter 6, and, particularly, in Chapter 6.3 (though we tend to consider both terminal-comprising and substantial ambiguity in this part of the thesis). A major result derived from this problem is presented, e. g., in Theorem 6.10.

Finally, the substantial ambiguity of morphisms is not a subject of a particular question. Instead of this, if possible, we treat any statement on substantial ambiguity simply as a more precise formulation of the fact that the morphisms under considerations are not moderately ambiguous (this option follows from Proposition 4.8) or not unambiguous (which holds by definition). Thereby, we frequently can draw conclusions on the crucial properties of E-pattern languages discussed in Chapters 3.2.1 and 3.3. Consequently, no particular chapter is dedicated to substantial ambiguity, but it is dealt with whenever appropriate.

Chapter 5

The ambiguity of common morphisms

In the present chapter, we examine the existence of moderately ambiguous (cf. Chapter 5.2), weakly unambiguous (cf. Chapter 5.3) and unambiguous (cf. Chapter 5.4) morphisms $\sigma : \mathbb{N}^* \longrightarrow \Sigma^*$ with respect to arbitrary patterns in $\mathbb{N}^+$. While the main direction of our research is determined by Questions 4.2 and 4.1, we additionally tackle Questions 4.3 and 4.4, and we present numerous more specific insights into the ambiguity of common morphisms not covered by any labelled questions – such as some brief, but nontrivial remarks on the problem of whether the specific morphisms introduced within this thesis can be improved in a particular way.

With regard to the application of our results to problems for pattern languages, our current focus on common morphisms implies that the class of terminal-free E-pattern languages (or, concerning weak unambiguity, the class of terminal-free NE-pattern languages) is the natural counterpart of our combinatorial considerations. Hence, in Chapter 5.2.1, we shall apply our insights into the existence of moderately ambiguous morphisms to the problem of the learnability of terminal-free E-pattern languages (cf. Question 3.1).

As a decisive prerequisite of our research, however, we first turn our attention to Example 4.1 again. Therein, among others, the pattern $1 \cdot 2$ is presented for which there is no unambiguous or moderately ambiguous morphism. Obviously, this pattern is morphically imprimitive and we can effortlessly prove our first main result stating that *all* morphically imprimitive patterns have this property as long as we restrict ourselves to nonerasing morphisms:

Theorem 5.1. *Let Σ be an alphabet. Let $\alpha \in \mathbb{N}^+$ be morphically imprimitive. Then every nonerasing morphism $\sigma : \mathbb{N}^* \longrightarrow \Sigma^*$ is substantially ambiguous with respect to α.*

Proof. If α is morphically imprimitive then, by definition, there exist a pattern $\beta \in \mathbb{N}^+$ and morphisms ϕ, ϕ' with $|\beta| < |\alpha|$, $\phi(\alpha) = \beta$ and $\phi'(\beta) = \alpha$. If we now consider the morphism $\psi := \phi' \circ \phi$ then $\psi(\alpha) = \alpha$ and, for some $i \in \text{var}(\alpha)$, $\psi(i) = \varepsilon$ (due to $|\beta| < |\alpha|$ and, hence, $\phi(i) = \varepsilon$). For any nonerasing morphism σ, we define the morphism $\tau : \mathbb{N}^* \longrightarrow \Sigma^*$ by $\tau := \sigma \circ \psi$. Consequently, $\tau(\alpha) = \sigma(\alpha)$ and $\tau(i) = \varepsilon \neq \sigma(i)$. Thus, σ is substantially ambiguous with respect to α. □

Theorem 5.1 implies that, whenever nonerasing morphisms are considered, we have to restrict our further search for moderately ambiguous or unambiguous morphisms to morphically primitive patterns. Due to this potential importance of the question of whether a pattern is morphically primitive, we give a characterisation of morphically imprimitive patterns in the subsequent chapter. The corresponding result shall prove to be a crucial tool for several of our subsequent argumentations and, additionally, it is used in Chapter 5.1.1 for deepening our knowledge on the shortest generators of terminal-free E-pattern languages, i. e. on the succinct patterns.

5.1 A crucial partition of word semigroups

We now present a necessary and sufficient structural condition deciding on whether or not a terminal-free pattern is morphically primitive. Due to Theorem 5.1, the corresponding criterion provides a valuable tool for several proofs on the (un-)ambiguity of morphisms. Furthermore, we shall use it for demonstrating that the partition of word semigroups into morphically primitive and morphically imprimitive strings is crucial for numerous prominent formal concepts related to finite strings and morphisms.

Before we can pursue our goal, however, we have to note a number of useful minor observations on (endo-)morphisms mapping two terminal-free patterns onto each other, which alternatively can also be interpreted as statements on fixed points of morphisms. The first of the resulting lemmata says that the composition of these morphisms either is trivial for a pattern α among the two patterns under consideration or there is at least one variable in α showing a special property:

Lemma 5.1. *Let $\alpha, \beta \in \mathbb{N}^+$, and let $\phi, \psi : \mathbb{N}^* \longrightarrow \mathbb{N}^*$ be morphisms with $\phi(\alpha) = \beta$ and $\psi(\beta) = \alpha$. Then either, for every $j \in \mathrm{var}(\alpha)$, $\psi(\phi(j)) = j$ or there exists a $j' \in \mathrm{var}(\alpha)$ such that $|\psi(\phi(j'))| \geq 2$ and $j' \in \mathrm{var}(\psi(\phi(j')))$.*

We call any variable j' satisfying these two conditions an *anchor variable (with respect to ϕ and ψ).*

Proof. Let $\alpha := j_1 \cdot j_2 \cdot [\dots] \cdot j_m$, $m \in \mathbb{N}$, $j_1, j_2, \dots, j_m \in \mathbb{N}$. Then $\beta = \phi(j_1) \cdot \phi(j_2) \cdot [\dots] \cdot \phi(j_m)$. Let j_{k_0} be the leftmost variable such that $\psi(\phi(j_{k_0})) \neq j_{k_0}$. Now assume to the contrary there is no anchor variable in α. Then $\psi(\phi(j_{k_0}))$ necessarily equals ε as otherwise $\psi(\beta) \neq \alpha$. Hence, $|\psi(\phi(j_1)) \cdot \psi(\phi(j_2)) \cdot [\dots] \cdot \psi(\phi(j_{k_0}))| = k_0 - 1$, and obviously, as there is no anchor variable in α, $|\psi(\phi(j_1)) \cdot \psi(\phi(j_2)) \cdot [\dots] \cdot \psi(\phi(j_k))| \leq k - 1$ for every $k > k_0$. Consequently, $|\psi(\beta)| < |\alpha|$ and therefore $\psi(\beta) \neq \alpha$. This contradiction proves the lemma. □

Additionally, if a pattern contains an anchor variable, then it must also contain a variable which is mapped onto ε:

Lemma 5.2. *Let $\alpha, \beta \in \mathbb{N}^+$, and let $\phi, \psi : \mathbb{N}^* \longrightarrow \mathbb{N}^*$ be morphisms with $\phi(\alpha) = \beta$ and $\psi(\beta) = \alpha$. Then either, for every $j \in \mathrm{var}(\alpha)$, $\psi(\phi(j)) = j$ or there exists a $j' \in \mathrm{var}(\alpha)$ such that $\psi(\phi(j')) = \varepsilon$.*

Proof. Directly from Lemma 5.1. □

Using Lemma 5.2, we can draw an important conclusion for morphically primitive patterns. As stated by the following lemma, these patterns do not contain any anchor variable:

Lemma 5.3. *Let $\alpha, \beta \in \mathbb{N}^+$, α morphically primitive. Let $\phi, \psi : \mathbb{N}^* \longrightarrow \mathbb{N}^*$ be morphisms with $\phi(\alpha) = \beta$ and $\psi(\beta) = \alpha$. Then, for every $j \in \mathrm{var}(\alpha)$, $\psi(\phi(j)) = j$.*

Proof. Assume to the contrary that there exists a $j' \in \mathrm{var}(\alpha)$ with $\psi(\phi(j')) \neq j'$. Then, according to Lemma 5.2, there is a $j'' \in \mathrm{var}(\alpha)$ such that $\psi(\phi(j'')) = \varepsilon$. We now regard the morphism $\phi : \mathbb{N}^* \longrightarrow \mathbb{N}^*$ given by

$$\phi'(j) := \begin{cases} \varepsilon \ , & j = j'' \ , \\ j \ , & \text{else} \ , \end{cases}$$

$j \in \mathrm{var}(\alpha)$, and we define $\alpha' := \phi'(\alpha)$. Hence, $|\alpha'| < |\alpha|$. Moreover – since, for every $k \in \mathrm{var}(\phi(j''))$, $\psi(k) = \varepsilon$ – we can state $\psi(\phi(\alpha')) = \alpha$, and therefore α and α' are morphically coincident. This contradicts the condition of α being morphically primitive. □

Finally, we now can present our characteristic criterion for morphically imprimitive patterns mentioned above:

Theorem 5.2. *A pattern $\alpha \in \mathbb{N}^+$ is morphically imprimitive if and only if there exists a decomposition*

$$\alpha = \beta_0\,\gamma_1\,\beta_1\,\gamma_2\,\beta_2\,[\ldots]\,\beta_{n-1}\,\gamma_n\,\beta_n$$

for an $n \in \mathbb{N}$, arbitrary $\beta_k \in \mathbb{N}^*$ and $\gamma_k \in \mathbb{N}^+$, $k \leq n$, such that

(i) *for every k, $1 \leq k \leq n$, $|\gamma_k| \geq 2$,*

(ii) *for every k, $1 \leq k \leq n$, and for every k', $0 \leq k' \leq n$, $\mathrm{var}(\gamma_k) \cap \mathrm{var}(\beta_{k'}) = \emptyset$, and*

(iii) *for every k, $1 \leq k \leq n$, there exists a $j_k \in \mathrm{var}(\gamma_k)$ such that $|\gamma_k|_{j_k} = 1$ and, for every k', $1 \leq k' \leq n$, if $j_k \in \mathrm{var}(\gamma_{k'})$ then $\gamma_k = \gamma_{k'}$.*

Proof. We first prove the *if* part of the theorem. Hence, let $\alpha \in \mathbb{N}^+$ be a pattern such that there exists a decomposition satisfying conditions (i), (ii), and (iii). We show that then there exists a pattern $\delta \in \mathbb{N}^+$ such that δ and α are morphically coincident and $|\delta| < |\alpha|$.

We define

$$\delta := \beta_0 \cdot j_1 \cdot \beta_1 \cdot j_2 \cdot \beta_2 \cdot [\ldots] \cdot \beta_{n-1} \cdot j_n \cdot \beta_n$$

with j_k derived from condition (iii) for every $k \leq n$. Then $|\delta| < |\alpha|$ because of condition (i).

We now show that α and δ are coincident. As a first morphism we define

$$\phi(j) := \begin{cases} \gamma_k \ , & \text{there is a } k,\ 1 \leq k \leq n, \text{ with } j = j_k \ , \\ j \ , & \text{else} \ , \end{cases}$$

$j \in \mathrm{var}(\delta)$. Because of conditions (ii) and (iii), ϕ is a well-defined morphism; obviously, $\phi(\delta) = \alpha$. The second morphism reads

$$\psi(j) := \begin{cases} \varepsilon \ , & \text{there is a } k,\ 1 \leq k \leq n, \text{ with } j \in \mathrm{var}(\gamma_k) \text{ and } j \neq j_k \ , \\ j \ , & \text{else} \ , \end{cases}$$

$j \in \mathrm{var}(\alpha)$. Consequently, $\psi(\alpha) = \delta$ and therefore α and δ are coincident. Thus, α is morphically imprimitive.

For the *only if* part assume that $\alpha \in \mathbb{N}^+$ is morphically imprimitive. We show that this assumption implies the existence of a decomposition of α satisfying conditions (i), (ii), and (iii): If α is morphically imprimitve then there exist morphisms $\phi, \psi : \mathbb{N}^* \longrightarrow \mathbb{N}^*$ and a morphically primitive pattern $\delta \in \mathbb{N}^+$ such that $|\delta| < |\alpha|$, $\phi(\delta) = \alpha$, and $\psi(\alpha) = \delta$. This leads to

Claim 1. For every $j \in \mathrm{var}(\delta)$, $\phi(j) \neq \varepsilon$.

Proof (Claim 1). Since δ, α, ϕ and ψ satisfy the conditions of Lemma 5.3 we may conclude that, for every $j \in \mathrm{var}(\delta)$, $\psi(\phi(j)) = j \neq \varepsilon$. Consequently, $\phi(j) \neq \varepsilon$. □ (Claim 1)

In addition, Lemma 5.3 also is the decisive tool for the proof of

Claim 2. For every $j \in \mathrm{var}(\delta)$, there is a $j' \in \mathrm{var}(\alpha)$ such that $j' \in \mathrm{var}(\phi(j))$ and $|\delta|_j = |\alpha|_{j'}$.

Proof (Claim 2). Assume to the contrary that there is a $j \in \mathrm{var}(\delta)$ such that, for every $i \in \mathrm{var}(\alpha)$, $i \notin \mathrm{var}(\phi(j))$ or $|\alpha|_i \neq |\delta|_j$. Consequently, for every $i \in \mathrm{var}(\phi(j))$, $|\alpha|_i > |\delta|_j$ since necessarily $|\alpha|_i \geq |\delta|_j$. Therefore, for every $i \in \mathrm{var}(\phi(j))$, $j \notin \mathrm{var}(\psi(i))$, and, thus, $\psi(\phi(j)) \neq j$. This contradicts Lemma 5.3. □ (Claim 2)

We now regard the following subsets of $\mathrm{var}(\delta)$: $X_1 := \{j \in \mathrm{var}(\delta) \mid |\phi(j)| = 1\}$ and $X_2 := \mathrm{var}(\delta) \setminus X_1$. This partition of $\mathrm{var}(\delta)$ leads to a particular decomposition of δ:

$$\begin{aligned}\delta \;=\; & \bar{\beta}_0 \cdot \underbrace{\bar{\jmath}_1 \cdot \bar{\jmath}_2 \cdot [\ldots] \cdot \bar{\jmath}_{s_1}}_{\bar{\gamma}_1} \cdot \\ & \bar{\beta}_1 \cdot \underbrace{\bar{\jmath}_{s_1+1} \cdot \bar{\jmath}_{s_1+2} \cdot [\ldots] \cdot \bar{\jmath}_{s_2}}_{\bar{\gamma}_2} \cdot \bar{\beta}_2 \cdot \\ & [\ldots] \cdot \\ & \bar{\beta}_{m-1} \cdot \underbrace{\bar{\jmath}_{s_{m-1}+1} \cdot \bar{\jmath}_{s_{m-1}+2} \cdot [\ldots] \cdot \bar{\jmath}_{s_m}}_{\bar{\gamma}_m} \cdot \bar{\beta}_m\end{aligned}$$

with

- $m \in \mathbb{N}$,
- for every k, $1 \leq k \leq m$, $|\bar{\gamma}_k| =: p_k \geq 1$,
- for every i, $1 \leq i \leq m$, $s_i := \sum_{k=1}^{i} p_k$,
- for every l, $1 \leq l \leq s_m$, $\bar{\jmath}_l \in X_2$, and
- $\bar{\beta}_0, \bar{\beta}_m \in X_1^*$, $\bar{\beta}_1, \bar{\beta}_2, \ldots, \bar{\beta}_{m-1} \in X_1^+$.

This decomposition of δ is unique, and, by ϕ, it induces an appropriate decomposition of $\alpha = \phi(\delta)$:

$$\begin{aligned}
\alpha = {} & \underbrace{\phi(\bar\beta_0)}_{\beta_0}\cdot\underbrace{\phi(\bar\jmath_1)}_{\gamma_1}\underbrace{\varepsilon}_{\beta_1}\underbrace{\phi(\bar\jmath_2)}_{\gamma_2}\underbrace{\varepsilon}_{\beta_2}[\ldots]\underbrace{\varepsilon}_{\beta_{s_1-1}}\underbrace{\phi(\bar\jmath_{s_1})}_{\gamma_{s_1}}\cdot\\
& \underbrace{\phi(\bar\beta_1)}_{\beta_{s_1}}\cdot\underbrace{\phi(\bar\jmath_{s_1+1})}_{\gamma_{s_1+1}}\underbrace{\varepsilon}_{\beta_{s_1+1}}\underbrace{\phi(\bar\jmath_{s_1+2})}_{\gamma_{s_1+2}}\underbrace{\varepsilon}_{\beta_{s_1+2}}[\ldots]\underbrace{\varepsilon}_{\beta_{s_2-1}}\underbrace{\phi(\bar\jmath_{s_2})}_{\gamma_{s_2}}\cdot\underbrace{\phi(\bar\beta_2)}_{\beta_{s_2}}\cdot\\
& [\ldots]\cdot\\
& \underbrace{\phi(\bar\beta_{m-1})}_{\beta_{s_{m-1}}}\cdot\underbrace{\phi(\bar\jmath_{s_{m-1}+1})}_{\gamma_{s_{m-1}+1}}\underbrace{\varepsilon}_{\beta_{s_{m-1}+1}}\underbrace{\phi(\bar\jmath_{s_{m-1}+2})}_{\gamma_{s_{m-1}+2}}\underbrace{\varepsilon}_{\beta_{s_{m-1}+2}}[\ldots]\underbrace{\varepsilon}_{\beta_{s_m-1}}\underbrace{\phi(\bar\jmath_{s_m})}_{\gamma_{s_m}}\cdot\underbrace{\phi(\bar\beta_m)}_{\beta_{s_m}}.
\end{aligned}$$

Then, for this decomposition of α, Claim 1 and the definition of $X_2 \subseteq \text{var}(\delta)$ prove condition (i). Condition (ii) follows from Claim 2 and the statement that, for every $k \leq m$, $\bar\beta_k \in X_1^*$. Finally, condition (iii) is satisfied because of Claim 2 and the fact that the above decomposition is induced by a morphism, leading to $\gamma_k = \phi(\bar\jmath_k) = \phi(\bar\jmath_{k'}) = \gamma_{k'}$ for every k, k' with $\bar\jmath_k = \bar\jmath_{k'}$. □

Note that Theorem 5.2 does not imply a new decidability result. It is a simple and well-known fact that morphic (semi-)coincidence can be tested algorithmically (cf. our notes on the membership problem for pattern languages in Chapter 3.2.1).

Apart from the intrinsic interest involved in the decomposition introduced by Theorem 5.2 and its use for several of our subsequent proofs, it provides a simple method for checking whether a pattern is morphically imprimitive. This is demonstrated by the following example:

Example 5.1. Obviously, any pattern α, $|\alpha| \geq 2$, necessarily is morphically imprimitive if there is a variable $i \in \mathbb{N}$ such that $|\alpha|_i = 1$.

Our favourite example $\alpha := 1 \cdot 2 \cdot 3 \cdot 4 \cdot 1 \cdot 4 \cdot 3 \cdot 2$ and the pattern $\alpha_1 := 1 \cdot 2 \cdot 2 \cdot 1 \cdot 2$ are morphically primitive because no variable for every of its occurrences has the same "environment" (i.e. a suitable γ) of length greater or equal 2 so that this environment does not share any of its variables with any potential β; thus, there is no decomposition of these patterns satisfying the conditions of Theorem 5.2.

Contrary to this, the pattern $\alpha_2 := 1 \cdot 2 \cdot 3 \cdot 3 \cdot 1 \cdot 2 \cdot 3$ is morphically imprimitive with $\beta_0 = \varepsilon$, $\gamma_1 = 1 \cdot 2$, $\beta_1 = 3 \cdot 3$, $\gamma_2 = 1 \cdot 2$ and $\beta_2 = 3$. Note that this obligatory decomposition of a morphically imprimitive pattern does not have to be unique.

Furthermore, it now can be verified easily that most examples in Chapter 4 are morphically primitive – which, by the way, because of Theorem 5.1 is a necessity.

Finally, there are numerous examples for morphically primitive patterns given in the subsequent chapters, e. g. in the proof for Proposition 5.1. ◇

Referring to Theorem 5.2, we can also effortlessly prove the following closure property of the sets of morphically primitive and imprimitive patterns:

Corollary 5.1. *Neither the set* PRIM $\subset \mathbb{N}^+$ *of all morphically primitive patterns nor the set* IMPRIM $\subset \mathbb{N}^+$ *of all morphically imprimitive patterns is closed under concatenation.*

Proof. Let $\alpha := 1 \cdot 2 \cdot 2 \cdot 1 \cdot 2$ and $\beta := 2$. Then, by Theorem 5.2, $\alpha, \beta \in$ PRIM, but $\alpha \cdot \beta \in$ IMPRIM.

Let $\alpha := 1 \cdot 2$ and $\beta := 2 \cdot 1$. Then, by Theorem 5.2, $\alpha, \beta \in$ IMPRIM, but $\alpha \cdot \beta \in$ PRIM. □

Corollary 5.1 is needed at a later stage of our thesis.

Furthermore, the criterion in Theorem 5.2 is also known to be characteristic for another prominent set of strings. According to Head [28], a string is decomposable in the way described by Theorem 5.2 if and only if it is a fixed point of a nontrivial morphism. Hence, the concept of fixed points and that of morphically imprimitive patterns are equivalent:

Corollary 5.2. *A pattern $\alpha \in \mathbb{N}^+$ is morphically imprimitive if and only if there exists a nontrivial morphism $\phi : \mathrm{var}(\alpha)^* \longrightarrow \mathrm{var}(\alpha)^*$ such that α is a fixed point of ϕ.*

Proof. Directly from Theorem 5.2 and Head [28]. □

For additional information on fixed points of morphisms, Hamm and Shallit [26] and Levé and Richomme [46] can be consulted.

Note that, alternatively, we can easily give a direct proof for Corollary 5.2 without making use of the fact that Theorem 5.2 and Head [28] deal with the same structural characterisation for morphically primitive patterns and fixed points of morphisms. In this case, of course, there is no need to discuss the correctness of Theorem 5.2 as it then follows from Corollary 5.2 and [28]. Nevertheless, we prefer to deal with our independently developed direct proof for Theorem 5.2 as our argumentation contains worthwile preparatory thoughts for our subsequent reasoning; furthermore, the adaptation of the results in [28] to our notations is a cumbersome task which we therefore do not wish to leave to the reader.

The subsequent chapter complements our preliminary notes on morphic (im-)primitivity by an immediate conclusion on terminal-free E-pattern languages.

5.1.1 Application: A characterisation of succinctness

As explained above, the decomposition introduced by Theorem 5.2 characterises two combinatorial concepts related to strings and morphisms: the morphically imprimitive patterns and the fixed points of morphisms. In addition to this, it can be effortlessly concluded from Corollary 3.3 that it also has major value for E-pattern languages, because the morphically primitive patterns exactly correspond to the succinct patterns:

Corollary 5.3. *Let $\alpha \in \mathbb{N}^+$, and let Σ be an alphabet, $|\Sigma| \geq 2$. Then α is succinct on Σ if and only if α is morphically primitive.*

Proof. Directly from Corollary 3.3. □

If we wish to know whether each terminal-free E-pattern language L contains an unambiguous word with respect to at least one pattern α satisfying $L = L_{\mathrm{E}}(\alpha)$ (a question posed by Reidenbach [76, 81] and to be answered by Corollary 5.12) then we can refer to Corollary 5.3. It implies that our comprehensive negative result given in Theorem 5.1 on the substantial ambiguity of all nonerasing morphisms with respect to any morphically imprimitive pattern fortunately does not affect any E-pattern *language* since, by definition, for every such language a succinct generating pattern can be given.

Summarising our previous results, we now can state that several elementary partitions of free semigroups considered so far are identical – a string is morphically imprimitive if and only if it is a fixed point of a nontrivial morphism if and only if it is prolix. Thus, dealing with the complement of these sets, the equivalences identified so far read as follows:

Corollary 5.4. *Let $\alpha \in \mathbb{N}^+$. Then the following statements are equivalent:*

- *α is morphically primitive.*
- *α is not a fixed point of any nontrivial morphism $\phi : \mathrm{var}(\alpha)^* \longrightarrow \mathrm{var}(\alpha)^*$.*
- *α is succinct on any alphabet Σ with $|\Sigma| \geq 2$.*

Proof. Directly from Corollary 5.2 and Corollary 5.3. □

Consequently, each of these sets is characterised by the fact that none of its strings can be decomposed in the way described by Theorem 5.2.

Corollary 5.4 demonstrates that morphic primitivity indeed is a central concept when dealing with morphisms in free monoids, and numerous of our subsequent result shall substantiate this claim. Therefore, at several stages of this thesis, we can extend Corollary 5.4 by adding alternative ways of characterising morphic primitivity.

5.2 Moderately ambiguous morphisms

We proceed with the first main part of our considerations on the (un-)ambiguity of common morphisms $\sigma : \mathbb{N}^* \longrightarrow \Sigma^*$. To this end, we mostly deal with Question 4.2 and additionally with Question 4.3, which ask for the existence of moderately ambiguous morphisms (cf. Definition 4.3) for arbitrary terminal-free patterns. Hence, for any pattern $\alpha \in \mathbb{N}^+$, we wish to find a morphism σ such that, for every occurrence i_k of every variable in $\mathrm{var}(\alpha)$, the word $\sigma(\alpha)$ contains a subword w_{i_k} which, with regard to any morphism $\tau : \mathbb{N}^* \longrightarrow \Sigma^*$ satisfying $\tau(\alpha) = \sigma(\alpha)$, must be generated by i_k. In Chapter 5.2.1, we apply our respective results to the problem of the learnability of terminal-free E-pattern languages as described by Question 3.1.

Concerning the morphically imprimitive patterns we already know all the necessary facts for presenting a final answer to Question 4.2. Theorem 5.1 states that each nonerasing morphism is substantially ambiguous with respect to each morphically primitive pattern, and therefore, as each moderately ambiguous morphism necessarily is not substantially ambiguous (cf. Proposition 4.8), we cannot find any morphically imprimitive pattern for which Question 4.2 has an answer in the affirmative:

Corollary 5.5. *Let Σ be an alphabet. Let $\alpha \in \mathbb{N}^+$ be morphically imprimitive. Then there is no nonerasing morphism $\sigma : \mathbb{N}^* \longrightarrow \Sigma^*$ such that σ is moderately ambiguous with respect to α.*

Proof. Directly from Theorem 5.1 and Proposition 4.8. □

Hence, we have to turn our attention to morphically primitive patterns. To this end, we introduce the following type of morphisms, the shape of which shall be decisive for a major part of the remainder of our thesis:

Definition 5.1 (Set 3-SEG of morphisms). *Let* Σ *be an alphabet,* $\{\mathtt{a}, \mathtt{b}\} \subseteq \Sigma$*. Let* 3-SEG *denote the set of all morphisms* $\sigma_{\text{3-seg}} : \mathbb{N}^* \longrightarrow \Sigma^*$ *where, for every* $j \in \mathbb{N}$*, there exist* $\mathtt{A}, \mathtt{B}, \mathtt{C}, \mathtt{D} \in \{\mathtt{a}, \mathtt{b}\}$, $\mathtt{A} \neq \mathtt{B}$, $\mathtt{C} \neq \mathtt{D}$, *such that*

$$\sigma_{\text{3-seg}}(j) = \mathtt{A}\,\mathtt{B}^{3j}\,\mathtt{A}\ \mathtt{a}\,\mathtt{b}^{3j+1}\,\mathtt{a}\ \mathtt{C}\,\mathtt{D}^{3j+2}\,\mathtt{C}\,.$$

Clearly, each morphism $\sigma_{\text{3-seg}} \in$ 3-SEG is injective on $\mathbb{N}^+$; more precisely, it is a *biprefix code* (cf. Berstel and Perrin [10]). Additionally note that our decision to introduce the full set 3-SEG instead of just one single morphism representing it (such as the simple morphism σ introduced in Example 5.2, see below) – which leads to a more involved appearance of our subsequent formal reasoning – is not so much motivated by the needs of the present chapter, but rather by those of our considerations on unambiguous morphisms given in Chapter 5.4.

Before we proceed with interpreting notes on the structure of the morphisms in 3-SEG, we briefly illustrate Definition 5.1 by presenting the most elementary morphism contained in this set:

Example 5.2. For every $j \in \mathbb{N}$, let the morphism $\sigma : \mathbb{N}^* \longrightarrow \{\mathtt{a}, \mathtt{b}\}^*$ be given by

$$\sigma(j) := \mathtt{a}\,\mathtt{b}^{3j}\,\mathtt{a}\ \mathtt{a}\,\mathtt{b}^{3j+1}\,\mathtt{a}\ \mathtt{a}\,\mathtt{b}^{3j+2}\,\mathtt{a}\,.$$

Then $\sigma \in$ 3-SEG. ◇

For our subsequent argumentation, it is essential that each morphism $\sigma_{\text{3-seg}} \in$ 3-SEG maps each variable onto a word that consists of three *segments* $\mathtt{a}\,\mathtt{b}^m\,\mathtt{a}$, $m \in \mathbb{N}$. Since, for any pattern $\alpha \in \mathbb{N}^+$, each of these segments in $\sigma_{\text{3-seg}}(\alpha)$ solely is generated by the occurrences of some particular variable j in α, we can unequivocally call each word $\mathtt{A}\,\mathtt{B}^{3j+p}\,\mathtt{A}$, $\mathtt{A}, \mathtt{B} \in \{\mathtt{a}, \mathtt{b}\}$, $\mathtt{A} \neq \mathtt{B}$, $p \in \{0, 1, 2\}$, a *segment of* $\sigma_{\text{3-seg}}(j)$.

We now demonstrate that each morphism $\sigma_{\text{3-seg}} \in$ 3-SEG is moderately ambiguous with respect to every morphically primitive pattern α. The proof for this main achievement of the present Chapter 5.2, however, shall require a complex reasoning. We begin with a list of simple combinatorial observations which impose a number of strong restrictions on any morphism τ with $\tau(\alpha) = \sigma_{\text{3-seg}}(\alpha)$. These first formal remarks exclusively deal with those variables j in α for which $\tau(j)$ contains any segment of $\sigma_{\text{3-seg}}(j)$ (note that a straightforward reasoning proves the existence of such variables for every morphism τ with $\tau(\alpha) = \sigma_{\text{3-seg}}(\alpha)$). Since the subsequent lemmata are needed for an unobstructed understanding of our main argumentation in Lemma 5.8, it is important to keep them in mind. Additionally, they shall also be used for our reasoning in Chapter 5.4.

Our first lemma says that, for any morphism τ with $\tau(\alpha) = \sigma_{\text{3-seg}}(\alpha)$ and for any variable j in α, $\tau(j)$ does not contain any segment of $\sigma_{\text{3-seg}}(j)$ twice (or even more than twice):

Lemma 5.4. *Let* $\alpha \in \mathbb{N}^+$*, and let* $\sigma_{\text{3-seg}} \in$ 3-SEG*. Let* $\tau : \mathbb{N}^* \longrightarrow \{\mathtt{a}, \mathtt{b}\}^*$ *be any morphism with* $\tau(\alpha) = \sigma_{\text{3-seg}}(\alpha)$*. Then, for every* $j \in \text{var}(\alpha)$*, for every* $p \in \{0, 1, 2\}$ *and for every* $\mathtt{A}, \mathtt{B} \in \{\mathtt{a}, \mathtt{b}\}$ *with* $\mathtt{A} \neq \mathtt{B}$*,* $\tau(j) \neq \ldots\ \mathtt{A}\,\mathtt{B}^{3j+p}\,\mathtt{A}\ \ldots\ \mathtt{A}\,\mathtt{B}^{3j+p}\,\mathtt{A}\ \ldots$.

Proof. Assume to the contrary that there is a $j \in \text{var}(\alpha)$ with $|\tau(j)|_{\mathtt{A}\mathtt{B}^{3j+p}\mathtt{A}} \geq 2$. Then $|\tau(\alpha)|_{\mathtt{A}\mathtt{B}^{3j+p}\mathtt{A}} \geq 2|\alpha|_j > |\alpha|_j = |\,\sigma_{\text{3-seg}}(\alpha)|_{\mathtt{A}\mathtt{B}^{3j+p}\mathtt{A}}$. This contradicts the condition $\tau(\alpha) = \sigma_{\text{3-seg}}(\alpha)$. □

In order to address the segments more precisely we henceforth use the following terms: For any variable $j \in \mathbb{N}$ and $\mathtt{A}, \mathtt{B}, \mathtt{C}, \mathtt{D} \in \{\mathtt{a}, \mathtt{b}\}$, $\mathtt{A} \neq \mathtt{B}$, $\mathtt{C} \neq \mathtt{D}$, we call $\mathtt{A}\,\mathtt{B}^{3j}\,\mathtt{A}$ the *left segment*, $\mathtt{a}\,\mathtt{b}^{3j+1}\,\mathtt{a}$ the *inner segment* and $\mathtt{C}\,\mathtt{D}^{3j+2}\,\mathtt{C}$ the *right segment* of $\sigma_{\text{3-seg}}(j)$. Additionally, we extend this terminology to segments of the image of any (sub-)pattern $\gamma \in \mathbb{N}^+$: A word $w \in \{\mathtt{a}, \mathtt{b}\}^+$ is a segment of $\sigma_{\text{3-seg}}(\gamma)$ if and only if there exists a variable $j \in \text{var}(\gamma)$ such that w is a segment of $\sigma_{\text{3-seg}}(j)$.

The next lemma says that if, for any morphism τ with $\tau(\alpha) = \sigma_{\text{3-seg}}(\alpha)$ and for any variable j in α, $\tau(j)$ contains the left and the inner segment (or the inner and the right segment) of $\sigma_{\text{3-seg}}(j)$ then these segments must occur in the "natural" order:

Lemma 5.5. *Let $\alpha \in \mathbb{N}^+$, and let $\sigma_{\text{3-seg}} \in$ 3-SEG. Let $\tau : \mathbb{N}^* \longrightarrow \{\mathtt{a}, \mathtt{b}\}^*$ be any morphism with $\tau(\alpha) = \sigma_{\text{3-seg}}(\alpha)$. For every $j \in \text{var}(\alpha)$, for every $p \in \{0, 1\}$ and for every $\mathtt{A}, \mathtt{B}, \mathtt{C}, \mathtt{D} \in \{\mathtt{a}, \mathtt{b}\}$, $\mathtt{A} \neq \mathtt{B}$, $\mathtt{C} \neq \mathtt{D}$, if $\tau(j) = \ldots\ \mathtt{A}\,\mathtt{B}^{3j+p}\,\mathtt{A}\ \ldots$ and $\tau(j) = \ldots\ \mathtt{C}\,\mathtt{D}^{3j+p+1}\,\mathtt{C}\ \ldots$ then $\tau(j) = \ldots\ \mathtt{A}\,\mathtt{B}^{3j+p}\,\mathtt{A}\ \mathtt{C}\,\mathtt{D}^{3j+p+1}\,\mathtt{C}\ \ldots$.*

Proof. Because of the equality of $\tau(\alpha)$ and $\sigma_{\text{3-seg}}(\alpha)$, it is obvious that $|\tau(\alpha)|_{\mathtt{A}\mathtt{B}^{3j+p}\mathtt{A}} = |\tau(\alpha)|_{\mathtt{C}\mathtt{D}^{3j+p+1}\mathtt{C}} = |\tau(\alpha)|_{\mathtt{A}\mathtt{B}^{3j+p}\mathtt{A}\mathtt{C}\mathtt{D}^{3j+p+1}\mathtt{C}} = |\alpha|_j$. Thus, for *every occurrence* of $\mathtt{A}\,\mathtt{B}^{3j+p}\,\mathtt{A}$ and $\mathtt{C}\,\mathtt{D}^{3j+p+1}\,\mathtt{C}$ in $\tau(\alpha)$, $\tau(\alpha) = \ldots\ \mathtt{A}\,\mathtt{B}^{3j+p}\,\mathtt{A}\ \mathtt{C}\,\mathtt{D}^{3j+p+1}\,\mathtt{C}\ \ldots$. Since $\tau(j) = \ldots\ \mathtt{A}\,\mathtt{B}^{3j+p}\,\mathtt{A}\ \ldots$ and $\tau(j) = \ldots\ \mathtt{C}\,\mathtt{D}^{3j+p+1}\,\mathtt{C}\ \ldots$, Lemma 5.5 follows immediately. □

Furthermore, it follows similarly that if, for any morphism τ with $\tau(\alpha) = \sigma_{\text{3-seg}}(\alpha)$ and for any variable j in α, $\tau(j)$ contains the left and the right segment of $\sigma_{\text{3-seg}}(j)$ then it must also contain the inner segment of $\sigma_{\text{3-seg}}(j)$ and, again, these segments must occur in the canonical order:

Lemma 5.6. *Let $\alpha \in \mathbb{N}^+$, and let $\sigma_{\text{3-seg}} \in$ 3-SEG. Let $\tau : \mathbb{N}^* \longrightarrow \{\mathtt{a}, \mathtt{b}\}^*$ be any morphism with $\tau(\alpha) = \sigma_{\text{3-seg}}(\alpha)$. For every $j \in \text{var}(\alpha)$ and for every $\mathtt{A}, \mathtt{B}, \mathtt{C}, \mathtt{D} \in \{\mathtt{a}, \mathtt{b}\}$, $\mathtt{A} \neq \mathtt{B}$, $\mathtt{C} \neq \mathtt{D}$, if $\tau(j) = \ldots\ \mathtt{A}\,\mathtt{B}^{3j}\,\mathtt{A}\ \ldots$ and $\tau(j) = \ldots\ \mathtt{C}\,\mathtt{D}^{3j+2}\,\mathtt{C}\ \ldots$ then $\tau(j) = \ldots\ \mathtt{A}\,\mathtt{B}^{3j}\,\mathtt{A}\ \mathtt{a}\,\mathtt{b}^{3j+1}\,\mathtt{a}\ \mathtt{C}\,\mathtt{D}^{3j+2}\,\mathtt{C}\ \ldots$.*

Proof. Directly from a straightforward adaptation of the proof of Lemma 5.5. □

We conclude our list of preliminary observations on the impact of some $j \in \text{var}(\alpha)$ for which $\tau(j)$ contains any segment $s = \mathtt{A}\,\mathtt{B}^{3j+p}\,\mathtt{A}$, $p \in \{0, 1, 2\}$, with an immediate consequence of Lemma 5.4. It is based on the fact that, for any n, $1 \leq n \leq |\alpha|_j$, the nth *occurrence* of j in α under both σ and τ necessarily generates the nth *occurrence* of s in $\sigma_{\text{3-seg}}(\alpha) = \tau(\alpha)$. Thus, in addition to what is stated in Lemmata 5.4, 5.5, 5.6, we can also use a variable j with the said property for drawing conclusions about the images under τ of neighbours of j in α. In anticipation of the requirements of the subsequent main Lemma 5.8 and for the sake of a more concise presentation, we focus on a variable $j \in \text{var}(\alpha)$ for which $\tau(j)$ contains the *inner* segment of $\sigma_{\text{3-seg}}(j)$:

Lemma 5.7. *Let $\alpha \in \mathbb{N}^+$, and let $\sigma_{\text{3-seg}} \in$ 3-SEG. Let, for some variable $j \in \text{var}(\alpha)$ and $\alpha_1, \alpha_2 \in \mathbb{N}^*$, $\alpha = \alpha_1 \cdot j \cdot \alpha_2$. Let $\tau : \mathbb{N}^* \longrightarrow \{\mathtt{a}, \mathtt{b}\}^*$ be any morphism with $\tau(\alpha) = \sigma_{\text{3-seg}}(\alpha)$. If, for some $w_1, w_2 \in \{\mathtt{a}, \mathtt{b}\}^*$, $\tau(j) = w_1\ \mathtt{a}\,\mathtt{b}^{3j+1}\,\mathtt{a}\ w_2$ then, for some $\mathtt{A}, \mathtt{B}, \mathtt{C}, \mathtt{D} \in \{\mathtt{a}, \mathtt{b}\}$, $\mathtt{A} \neq \mathtt{B}$, $\mathtt{C} \neq \mathtt{D}$,*

- *$\tau(\alpha_1)\, w_1 = \sigma(\alpha_1)\ \mathtt{A}\,\mathtt{B}^{3j}\,\mathtt{A}$ and*
- *$w_2\, \tau(\alpha_2) = \mathtt{C}\,\mathtt{D}^{3j+2}\,\mathtt{C}\ \sigma(\alpha_2)$.*

Proof. Straightforward from Lemma 5.4. □

We now proceed with the crucial lemma on the fact that each morphism $\sigma_{3\text{-seg}} \in$ 3-SEG is moderately ambiguous with respect to every morphically primitive pattern α. It shows that, for *every* morphism τ with $\tau(\alpha) = \sigma_{3\text{-seg}}(\alpha)$ and for *every* variable $j \in \text{var}(\alpha)$, $\tau(j)$ must contain the inner segment of $\sigma_{3\text{-seg}}(j)$ and the letter to the left and the letter to the right of this segment:

Lemma 5.8. *Let $\alpha \in \mathbb{N}^+$ be a morphically primitive pattern, and let $\sigma_{3\text{-seg}} \in$ 3-SEG. Then, for every morphism $\tau : \mathbb{N}^* \longrightarrow \{\mathtt{a}, \mathtt{b}\}^*$ with $\tau(\alpha) = \sigma_{3\text{-seg}}(\alpha)$ and for every $i \in$ $\text{var}(\alpha)$, $\tau(i) = \ldots \; \mathtt{E} \; \mathtt{a}\,\mathtt{b}^{3i+1}\,\mathtt{a} \; \mathtt{F} \; \ldots$, $\mathtt{E}, \mathtt{F} \in \{\mathtt{a}, \mathtt{b}\}$.*

Proof. We argue by contraposition; in other words, we show that if there exists a morphism τ with $\tau(\alpha) = \sigma_{3\text{-seg}}(\alpha)$ and, for some $i \in \text{var}(\alpha)$, $\tau(i) \neq \ldots \; \mathtt{E} \; \mathtt{a}\,\mathtt{b}^{3i+1}\,\mathtt{a} \; \mathtt{F} \ldots$ then α is morphically imprimitive. To this end, we use Theorem 5.2.

We start with a partition of $\text{var}(\alpha)$ into subsets X_1, X_2, X_3 depending on any morphism τ satisfying the said conditions. From an informal point of view, this partition is given as follows: First, let X_1 be the set of all variables i in α such that $\tau(i)$ contains the inner segment of $\sigma_{3\text{-seg}}(i)$, of the left and right segment of $\sigma_{3\text{-seg}}(i)$ at least one letter and at least one segment of some $\sigma_{3\text{-seg}}(j)$, $j \neq i$. Second, let X_2 be the set of all variables i in α such that $\tau(i)$ does not contain any letter of at least one segment of $\sigma_{3\text{-seg}}(i)$. Third (and last), let X_3 be the set of all variables i in α such that $\tau(i)$ contains the inner segment of $\sigma_{3\text{-seg}}(i)$ and at least one letter of the left and right segment of $\sigma_{3\text{-seg}}(i)$, but no segment of some $\sigma_{3\text{-seg}}(j)$, $j \neq i$.

Since $\tau(\alpha) = \sigma_{3\text{-seg}}(\alpha)$ and thus, for every $i \in \text{var}(\alpha)$, $\tau(i)$ is a subword of $\sigma_{3\text{-seg}}(\alpha)$ this vague definition of X_1, X_2 and X_3 results in several evident restrictions on the images under τ of the variables in α (cf. Lemmata 5.4, 5.5, 5.6, 5.7) such that the introduced subsets of $\text{var}(\alpha)$ read formally:

$$
\begin{aligned}
X_1 := \{ i \in \text{var}(\alpha) \;\mid\; & \tau(i) = \ldots \; \mathtt{A}\,\mathtt{B}^{3j+2}\,\mathtt{A} \; \mathtt{C}\,\mathtt{D}^{3i}\,\mathtt{C} \; \mathtt{a}\,\mathtt{b}^{3i+1}\,\mathtt{a} \; \mathtt{E} \; \ldots \; \underline{\text{or}} \\
& \tau(i) = \ldots \; \mathtt{F} \; \mathtt{a}\,\mathtt{b}^{3i+1}\,\mathtt{a} \; \mathtt{A}\,\mathtt{B}^{3i+2}\,\mathtt{A} \; \mathtt{C}\,\mathtt{D}^{3j}\,\mathtt{C} \; \ldots \; , \\
& \mathtt{A}, \mathtt{B}, \mathtt{C}, \mathtt{D}, \mathtt{E}, \mathtt{F} \in \{\mathtt{a}, \mathtt{b}\}, \; \mathtt{A} \neq \mathtt{B}, \; \mathtt{C} \neq \mathtt{D}, j \in \mathbb{N} \, \}, \\
X_2 := \{ i \in \text{var}(\alpha) \;\mid\; & \tau(i) \neq \ldots \; \mathtt{E} \; \mathtt{a}\,\mathtt{b}^{3i+1}\,\mathtt{a} \; \mathtt{F} \; \ldots \; , \; \mathtt{E}, \mathtt{F} \in \{\mathtt{a}, \mathtt{b}\} \, \}, \\
X_3 := \{ i \in \text{var}(\alpha) \;\mid\; & \tau(i) = \ldots \; \mathtt{E} \; \mathtt{a}\,\mathtt{b}^{3i+1}\,\mathtt{a} \; \mathtt{F} \; \ldots \; \underline{\text{and}} \\
& \tau(i) \neq \ldots \; \mathtt{A}\,\mathtt{B}^{3j+2}\,\mathtt{A} \; \mathtt{C}\,\mathtt{D}^{3i}\,\mathtt{C} \; \ldots \; \underline{\text{and}} \\
& \tau(i) \neq \ldots \; \mathtt{A}\,\mathtt{B}^{3i+2}\,\mathtt{A} \; \mathtt{C}\,\mathtt{D}^{3j}\,\mathtt{C} \; \ldots \; , \\
& \mathtt{A}, \mathtt{B}, \mathtt{C}, \mathtt{D}, \mathtt{E}, \mathtt{F} \in \{\mathtt{a}, \mathtt{b}\}, \; \mathtt{A} \neq \mathtt{B}, \; \mathtt{C} \neq \mathtt{D}, j \in \mathbb{N} \, \}.
\end{aligned}
$$

Directly from the definition, it can be verified that $X_1 \cap X_2 = X_1 \cap X_3 = X_2 \cap X_3 = \emptyset$ and $X_1 \cup X_2 \cup X_3 = \text{var}(\alpha)$. According to the condition on our proof, there is a variable $i \in \text{var}(\alpha)$ with $\tau(i) \neq \ldots \mathtt{E}\,\mathtt{a}\,\mathtt{b}^{3i+1}\,\mathtt{a}\,\mathtt{F} \ldots$ and therefore $X_2 \neq \emptyset$. Note that our subsequent argumentation in Claim 3 shows that this leads to $X_1 \neq \emptyset$.

As we now wish to show that $X_2 \neq \emptyset$ implies α being morphically imprimitive, we need to find an appropriate decomposition of α satisfying the three conditions of Theorem 5.2. We start our argumentation with the following one:

$$\alpha = \bar{\beta}_0 \, \bar{\gamma}_1 \, \bar{\beta}_1 \, \bar{\gamma}_2 \, \bar{\beta}_2 \, [\ldots] \, \bar{\beta}_{\bar{m}-1} \, \bar{\gamma}_{\bar{m}} \, \bar{\beta}_{\bar{m}}$$

with $\bar{m} \geq 1$ and

- $\bar{\beta}_0, \bar{\beta}_{\bar{m}} \in X_3^*$ and $\bar{\beta}_k \in X_3^+$, $1 \leq k \leq \bar{m}-1$, and
- $\bar{\gamma}_k \in (X_1 \cup X_2)^+$, $1 \leq k \leq \bar{m}$.

Note that $\bar{m} \geq 1$ is guaranteed because of $X_2 \neq \emptyset$.

Obviously, this decomposition is unique, and it satisfies condition (ii) of Theorem 5.2 since X_1, X_2 and X_3 are disjoint:

Claim 1. For every k, k', $1 \leq k, k' \leq \bar{m}$, $\text{var}(\bar{\gamma}_k) \cap \text{var}(\bar{\beta}_{k'}) = \emptyset$.

Concerning condition (i) of Theorem 5.2, we need to examine the given decomposition of α in a bit more detail. The subsequent claim says that, for every $\bar{\gamma}_k$, $\tau(\bar{\gamma}_k)$ "almost" corresponds to $\sigma_{\text{3-seg}}(\bar{\gamma}_k)$, i.e. $\tau(\bar{\gamma}_k)$ contains *at least* $3|\bar{\gamma}_k| - 2$ complete segments of $\sigma_{\text{3-seg}}(\bar{\gamma}_k)$ (and potentially some letters of two other segments) and *at most* $3|\bar{\gamma}_k|$ complete segments of $\sigma_{\text{3-seg}}(\bar{\gamma}_k)$ and, moreover, for every variable $i \in \text{var}(\bar{\gamma}_k)$ the inner segment of $\sigma_{\text{3-seg}}(i)$ as often as i is contained in $\bar{\gamma}_k$, and it does not contain any segment of $\sigma_{\text{3-seg}}(i')$ if $i' \notin \text{var}(\bar{\gamma}_k)$:

Claim 2. For every $\bar{\gamma}_k = i_1 \cdot i_2 \cdot [\ldots] \cdot i_s$, $s \in \mathbb{N}$, $i_1, i_2, \ldots, i_s \in \mathbb{N}$, and for every $i' \in \text{var}(\alpha)$,

$$\begin{aligned}
\tau(\bar{\gamma}_k) &= \ldots \mathtt{E}\ \mathtt{a}\,\mathtt{b}^{3i_1+1}\,\mathtt{a}\ \mathtt{A}\,\mathtt{B}^{3i_1+2}\,\mathtt{A}\ \mathtt{C}\,\mathtt{D}^{3i_2}\,\mathtt{C}\,[\ldots]\ \mathtt{a}\,\mathtt{b}^{3i_s+1}\,\mathtt{a}\ \mathtt{G}\ldots \ \underline{\text{and}}\\
\tau(\bar{\gamma}_k) &\neq \ldots \mathtt{G}\ \mathtt{a}\,\mathtt{b}^{3i_1+1}\,\mathtt{a}\ \mathtt{A}\,\mathtt{B}^{3i_1+2}\,\mathtt{A}\,[\ldots]\ \mathtt{C}\,\mathtt{D}^{3i_s+2}\,\mathtt{C}\ \mathtt{E}\,\mathtt{F}^{3i'}\,\mathtt{E}\ldots \ \underline{\text{and}}\\
\tau(\bar{\gamma}_k) &\neq \ldots \mathtt{E}\,\mathtt{F}^{3i'+2}\,\mathtt{E}\ \mathtt{A}\,\mathtt{B}^{3i_1}\,\mathtt{A}\ \mathtt{a}\,\mathtt{b}^{3i_1+1}\,\mathtt{a}\,[\ldots]\ \mathtt{a}\,\mathtt{b}^{3i_s+1}\,\mathtt{a}\ \mathtt{G}\ldots,
\end{aligned}$$

$\mathtt{A}, \mathtt{B}, \mathtt{C}, \mathtt{D}, \mathtt{E}, \mathtt{F}, \mathtt{G} \in \{\mathtt{a}, \mathtt{b}\}$ *with* $\mathtt{A} \neq \mathtt{B}$, $\mathtt{C} \neq \mathtt{D}$, $\mathtt{E} \neq \mathtt{F}$.

For any subpattern δ of α satisfying the statement in Claim 2 we say that $\tau(\delta)$ *corresponds to $\sigma_{\text{3-seg}}(\delta)$ (apart from a negligible prefix and suffix).*

Claim 2 follows from the fact that every $\bar{\gamma}_k$ is surrounded by $\bar{\beta}_{k-1} \in X_3^*$ and $\bar{\beta}_k \in X_3^*$. Thus, with Lemma 5.7 applied to the variables in X_3, $\tau(\bar{\gamma}_k)$ is fixed by $\tau(\bar{\beta}_{k-1})$ and $\tau(\bar{\beta}_k)$: as these two subwords of $\tau(\alpha)$ by definition correspond to $\sigma_{\text{3-seg}}(\bar{\beta}_{k-1})$ and $\sigma_{\text{3-seg}}(\bar{\beta}_k)$, respectively, $\tau(\bar{\gamma}_k)$ must also correspond to $\sigma_{\text{3-seg}}(\bar{\gamma}_k)$. Consequently – and since by definition, for no $\delta \in X_1^+$, $\tau(\delta)$ corresponds to $\sigma_{\text{3-seg}}(\delta)$ – $\bar{\gamma}_k \notin X_1^+$.

We proceed our argumentation on condition (i) of Theorem 5.2 being satisfied for the regarded decomposition by a closer look at the images under τ of those subpatterns δ of α which exclusively consist of variables in X_2. In this regard we can see that $\tau(\delta)$ necessarily does not correspond to $\sigma_{\text{3-seg}}(\delta)$:

Claim 3. For every $\delta = i_1 \cdot i_2 \cdot [\ldots] \cdot i_t$, $t \in \mathbb{N}$, $i_1, i_2, \ldots, i_t \in X_2$,

$$\tau(\delta) \neq \ldots \mathtt{E}\ \mathtt{a}\,\mathtt{b}^{3i_1+1}\,\mathtt{a}\ \mathtt{A}\,\mathtt{B}^{3i_1+2}\,\mathtt{A}\ \mathtt{C}\,\mathtt{D}^{3i_2}\,\mathtt{C}\,[\ldots]\ \mathtt{a}\,\mathtt{b}^{3i_t+1}\,\mathtt{a}\ \mathtt{F}\ldots,$$

$\mathtt{A}, \mathtt{B}, \mathtt{C}, \mathtt{D}, \mathtt{E}, \mathtt{F} \in \{\mathtt{a}, \mathtt{b}\}$ *with* $\mathtt{A} \neq \mathtt{B}$, $\mathtt{C} \neq \mathtt{D}$.

The correctness of Claim 3 follows from a straightforward combinatorial examination of the definition of X_2. Thus, and from Claim 2, it is $\bar{\gamma}_k \notin X_2^+$ and therefore $\bar{\gamma}_k$ must consist of variables in X_1 and of variables in X_2:

Claim 4. For every k, $1 \leq k \leq \bar{m}$, $|\bar{\gamma}_k| \geq 2$.

Hence, condition (i) of Theorem 5.2 is satisfied for the above decomposition.

With regard to condition (iii), however, the decomposition possibly requires some modifications. We wish to have a decomposition where there is *exactly* one occurrence of an $i \in X_1$ in each $\bar{\gamma}_k$ since this variable is meant to serve as the variable j_k referred to in condition (iii) of Theorem 5.2. For the given decomposition, however, we can only conclude that there is *at least* one occurrence of an $i \in X_1$ in each $\bar{\gamma}_k$. Therefore we transform it into a specific decomposition where every $\bar{\gamma}_k$ contains exactly one $i \in X_1$. To this end, we apply two different types of operations, namely a *splitting* of certain $\bar{\gamma}_k$ and a *redefinition* of X_1 and X_3.

We first split every $\bar{\gamma}_k$ that contains more than one occurrence of a variable from X_1, and we do so by identifying all so-called *splitting points* in $\bar{\gamma}_k$. Intuitively, these splitting points lead to a maximum $s \in \mathbb{N}$ for which there exists a decomposition $\bar{\gamma}_k = \bar{\gamma}_{k,1}\bar{\gamma}_{k,2}[\ldots]\bar{\gamma}_{k,s}$ such that, for every k', $1 \leq k' \leq s$, $\tau(\bar{\gamma}_{k,k'})$ corresponds to $\sigma_{3\text{-seg}}(\bar{\gamma}_{k,k'})$ "as far as possible". Formally, a splitting point is an inner substring δ of $\bar{\gamma}_k$, i. e. $\bar{\gamma}_k = \bar{\gamma}_{k,l}\delta\bar{\gamma}_{k,r}$ with $\bar{\gamma}_{k,l} = i_1 \cdot i_2 \cdot [\ldots] \cdot i_p$ and $\bar{\gamma}_{k,r} = i_{p+1} \cdot i_{p+2} \cdot [\ldots] \cdot i_{p+q}$, $p, q \in \mathbb{N}$, $i_1, i_2, \ldots, i_{p+q} \in X_1 \cup X_2$, that satisfies one of the following conditions:

1. $\delta = \varepsilon$ and

$$\begin{aligned} \tau(\bar{\gamma}_{k,l}) &= \ldots \mathtt{E\ a b}^{3i_1+1}\mathtt{a\ A B}^{3i_1+2}\mathtt{A}\,[\ldots]\ \mathtt{C D}^{3i_p}\mathtt{C\ a b}^{3i_p+1}\mathtt{a\ F} \ldots, \\ &\quad \mathtt{A,B,C,D,E,F} \in \{\mathtt{a},\mathtt{b}\},\ \mathtt{A} \neq \mathtt{B},\ \mathtt{C} \neq \mathtt{D},\ \text{and} \\ \tau(\bar{\gamma}_{k,r}) &= \ldots \mathtt{E\ a b}^{3i_{p+1}+1}\mathtt{a\ A B}^{3i_{p+1}+2}\mathtt{A}\,[\ldots]\ \mathtt{C D}^{3i_{p+q}}\mathtt{C\ a b}^{3i_{p+q}+1}\mathtt{a\ F} \ldots, \\ &\quad \mathtt{A,B,C,D,E,F} \in \{\mathtt{a},\mathtt{b}\},\ \mathtt{A} \neq \mathtt{B},\ \mathtt{C} \neq \mathtt{D},\ \underline{\text{or}} \end{aligned}$$

2. $\delta = i'$, $i' \in \mathbb{N}$, and

$$\begin{aligned} \tau(\bar{\gamma}_{k,l}) &= \ldots \mathtt{G\ a b}^{3i_1+1}\mathtt{a\ A B}^{3i_1+2}\mathtt{A}\,[\ldots]\ \mathtt{a b}^{3i_p+1}\mathtt{a\ C D}^{3i_p+2}\mathtt{C\ E F}^{3i'}\mathtt{E} \ldots, \\ &\quad \mathtt{A,B,C,D,E,F,G} \in \{\mathtt{a},\mathtt{b}\},\ \mathtt{A} \neq \mathtt{B},\ \mathtt{C} \neq \mathtt{D},\ \mathtt{E} \neq \mathtt{F},\ \text{and} \\ \tau(\bar{\gamma}_{k,r}) &= \ldots \mathtt{E F}^{3i'+2}\mathtt{E\ A B}^{3i_{p+1}}\mathtt{A\ a b}^{3i_{p+1}+1}\mathtt{a}\,[\ldots]\ \mathtt{C D}^{3i_{p+q}}\mathtt{C\ a b}^{3i_{p+q}+1}\mathtt{a\ G} \ldots, \\ &\quad \mathtt{A,B,C,D,E,F,G} \in \{\mathtt{a},\mathtt{b}\},\ \mathtt{A} \neq \mathtt{B},\ \mathtt{C} \neq \mathtt{D},\ \mathtt{E} \neq \mathtt{F}\,. \end{aligned}$$

For a better understanding of the definition of a splitting point, recall Claim 2 and Claim 3. Furthermore, these claims are sufficient for verifying the following facts:

- A $\bar{\gamma}_k$ with only one occurrence of a variable from X_1 does not contain any splitting point.
- For every splitting point δ of type 2, i. e. $\delta = i' \in \mathbb{N}$, necessarily $i' \in X_2$.
- For two splitting points δ, δ', necessarily $\bar{\gamma}_k \neq \ldots\ \delta\,\delta'\ \ldots$.

After *all* of the splitting points have been identified in $\bar{\gamma}_k$, for each of them the following *splitting operation* is performed:

1. If $|\delta| = 0$ then δ is renamed to $\dot{\beta}$.
2. If $|\delta| = 1$ then a $\dot{\beta} = \varepsilon$ is inserted to the right of δ, i. e. $\bar{\gamma}_k := \bar{\gamma}_{k,l}\,\delta\,\dot{\beta}\,\bar{\gamma}_{k,r}$.

Note that, in case 2, we can arbitrarily choose to insert $\dot{\beta}$ to the left or to the right of δ, but it is essential to do this for all splitting points in the same way. This will be relevant for our argumentation on the crucial Claim 6.

When this has been accomplished for all splitting points then we regard the following decomposition of α:

$$\alpha = \hat{\beta}_0\, \hat{\gamma}_1\, \hat{\beta}_1\, \hat{\gamma}_2\, \hat{\beta}_2\, [\dots]\, \hat{\beta}_{\hat{m}-1}\, \hat{\gamma}_{\hat{m}}\, \hat{\beta}_{\hat{m}}$$

with $\hat{m} \geq 1$ and

- $\hat{\beta}_k \in X_3^*$, $0 \leq k \leq \hat{m}$, where, for every $1 \leq k' \leq \hat{m} - 1$, $\hat{\beta}_{k'} = \varepsilon$ if and only if at exactly the position of $\hat{\beta}_{k'}$ a $\dot{\beta}$ has been inserted by a splitting operation, i. e. in this case $\hat{\beta}_{k'}$ simply is a renaming of a $\dot{\beta}$, and

- $\hat{\gamma}_k \in (X_1 \cup X_2)^+$, $1 \leq k \leq \hat{m}$.

Consequently, if in some $\bar{\gamma}_k$ there is, e. g., exactly one splitting point, i. e. $\bar{\gamma}_k = \bar{\gamma}_{k,l}\, \delta\, \bar{\gamma}_{k,r}$, then, for some $h < \hat{m}$, the splitting operation leads to $\hat{\gamma}_h = \bar{\gamma}_{k,l}$ and $\hat{\gamma}_{h+1} = \bar{\gamma}_{k,r}$ (in case of $|\delta| = 0$) or $\hat{\gamma}_h = \bar{\gamma}_{k,l}\, \delta$ and $\hat{\gamma}_{h+1} = \bar{\gamma}_{k,r}$ (in case of $|\delta| = 1$). Additionally note that $\hat{m} \geq 1$ again follows from $X_2 \neq \emptyset$.

After the splitting operations we can record:

Claim 5. For every k, $1 \leq k \leq \hat{m}$, $\hat{\gamma}_k$ contains exactly one occurrence of an $i \in X_1$.

Claim 5 follows from Claim 2, Claim 3 and the definition of the splitting points.

Moreover, the resulting decomposition has a second crucial property:

Claim 6. For every k, k', $1 \leq k, k' \leq \hat{m}$, if $\mathrm{var}(\hat{\gamma}_k) \cap \mathrm{var}(\hat{\gamma}_{k'}) \cap X_1 \neq \emptyset$ then $\hat{\gamma}_k = \hat{\gamma}_{k'}$.

Proof (Claim 6). If $|\hat{\gamma}_k| = |\hat{\gamma}_{k'}| = 1$ then Claim 6 trivially holds true. Therefore we restrict ourselves to the case $|\hat{\gamma}_k| \geq 2$ or $|\hat{\gamma}_{k'}| \geq 2$. Now assume to the contrary that there are k, k', $1 \leq k, k' \leq \hat{m}$, with $\mathrm{var}(\hat{\gamma}_k) \cap \mathrm{var}(\hat{\gamma}_{k'}) \cap X_1 \neq \emptyset$ and $\hat{\gamma}_k \neq \hat{\gamma}_{k'}$. Because of Claim 5 we can write $\hat{\gamma}_k$ as $\hat{\gamma}_k = i_1 \cdot i_2 \cdot [\dots] \cdot i_p \cdot i_x \cdot i_{p+1} \cdot i_{p+2} \cdot [\dots] \cdot i_{p+q}$ with $p, q \in \mathbb{N}_0$, $i_x \in X_1$, $i_1, i_2, \dots, i_{p+q} \in X_2$ and $\hat{\gamma}_{k'}$ as $\hat{\gamma}_{k'} = i'_1 \cdot i'_2 \cdot [\dots] \cdot i'_r \cdot i_x \cdot i'_{r+1} \cdot i'_{r+2} \cdot [\dots] \cdot i'_{r+s}$ with $r, s \in \mathbb{N}_0$, $i'_1, i'_2, \dots, i'_{r+s} \in X_2$. Note that our condition $|\hat{\gamma}_k| \geq 2$ or $|\hat{\gamma}_{k'}| \geq 2$ implies $p + q + r + s \geq 1$.

We now assume, first, that $p = r$ and $q = s$ (we shall examine the case where there is $p \neq r$ or $q \neq s$ later) and, second, w. l. o. g. that $t \in \mathbb{N}$ is the largest number with $i_t \neq i'_t$ and $t \leq p$. The latter assumption does not restrict our reasoning since, for the case that the only different variables in $\hat{\gamma}_k, \hat{\gamma}_{k'}$ are to the right of i_x, an analogous argumentation can be applied. Under these two assumptions, we now examine Claim 3, which says that, for *every* y, $0 \leq y \leq p - t$, $\tau(i_t \cdot i_{t+1} \cdot [\dots] \cdot i_{t+y})$ does not correspond to $\sigma_{\text{3-seg}}(i_t \cdot i_{t+1} \cdot [\dots] \cdot i_{t+y})$ (and of course $\tau(i'_t \cdot i'_{t+1} \cdot [\dots] \cdot i'_{t+y})$ does not correspond to $\sigma_{\text{3-seg}}(i'_t \cdot i'_{t+1} \cdot [\dots] \cdot i'_{t+y})$) as all of the variables under consideration are in X_2. More precisely, we may conclude that, again for every y, $0 \leq y \leq p - t$, $\tau(i_{t+y+1} \cdot i_{t+y+2} \cdot [\dots] \cdot i_x)$ contains the right segment of $\sigma_{\text{3-seg}}(i_{t+y})$ (and, additionally, $\tau(i'_{t+y+1} \cdot i'_{t+y+2} \cdot [\dots] \cdot i_x)$ contains the right segment of $\sigma_{\text{3-seg}}(i'_{t+y})$) since, otherwise, there would have been a splitting point somewhere between i_t and i_x (and between i'_t and i_x) – this statement can be verified by a closer look at the definition of both types of splitting points, where the condition for the left subpattern $\bar{\gamma}_{k,l}$ in the case of $\tau(\bar{\gamma}_{k,l})$ containing the right segment of

$\sigma_{3\text{-seg}}(i_{t+y})$ would have led to the insertion of the said splitting point. Thus, with $y := 0$, this implies for some $\mathtt{A}, \mathtt{B}, \mathtt{C}, \mathtt{D}, \mathtt{G} \in \{\mathtt{a}, \mathtt{b}\}$, $\mathtt{A} \neq \mathtt{B}$, $\mathtt{C} \neq \mathtt{D}$,

$$\begin{aligned} \tau(i_{t+1} \cdot i_{t+2} \cdot [\ldots] \cdot i_x) &= \ldots \mathtt{A}\,\mathtt{B}^{3i_t+2}\,\mathtt{A}\;\mathtt{C}\,\mathtt{D}^{3i_{t+1}}\,\mathtt{C}\,[\ldots]\;\mathtt{a}\,\mathtt{b}^{3i_x+1}\,\mathtt{a}\;\mathtt{G}\ldots \\ &= \tau(i'_{t+1} \cdot i'_{t+2} \cdot [\ldots] \cdot i_x)\,. \end{aligned}$$

On the other hand, we know that, for the same $\mathtt{C}, \mathtt{D}, \mathtt{G}$ as above and some $\mathtt{E}, \mathtt{F} \in \{\mathtt{a}, \mathtt{b}\}$, $\mathtt{E} \neq \mathtt{F}$,

$$\sigma_{3\text{-seg}}(i'_t \cdot i'_{t+1} \cdot i'_{t+2} \cdot [\ldots] \cdot i_x) = \ldots\;\mathtt{E}\,\mathtt{F}^{3i'_t+2}\,\mathtt{E}\;\mathtt{C}\,\mathtt{D}^{3i_{t+1}}\,\mathtt{C}\,[\ldots]\;\mathtt{a}\,\mathtt{b}^{3i_x+1}\,\mathtt{a}\;\mathtt{G}\ldots.$$

However – since $i_x \in X_1$ and, hence, $\tau(i_x)$ contains the inner segment of $\sigma_{3\text{-seg}}(i_x)$ – we know that $\tau(i_x)$ and $\sigma_{3\text{-seg}}(i_x)$ generate the *same occurrence* of the subword $\mathtt{a}\,\mathtt{b}^{3i_x+1}\,\mathtt{a}$ in $\tau(\alpha) = \sigma_{3\text{-seg}}(\alpha)$. Thus, roughly speaking, if we compare $\tau(i'_t \cdot i'_{t+1} \cdot [\ldots] \cdot i_x)$ with $\sigma_{3\text{-seg}}(i'_t \cdot i'_{t+1} \cdot [\ldots] \cdot i_x)$ then we deal with the same part of $\sigma_{3\text{-seg}}(\alpha)$ (cf. our remarks introducing Lemma 5.7). Consequently, we can conclude from Lemma 5.7 that the conditions $\tau(\alpha) = \sigma_{3\text{-seg}}(\alpha)$ and $i_x \in X_1$ imply $\mathtt{A}\,\mathtt{B}^{3i_t+2}\,\mathtt{A} = \mathtt{E}\,\mathtt{F}^{3i'_t+2}\,\mathtt{E}$ and, more precisely, $\mathtt{A} = \mathtt{E}$, $\mathtt{B} = \mathtt{F}$ and $i_t = i'_t$. This contradicts our assumption $i_t \neq i'_t$.

We proceed with the remaining case $p \neq r$ or $q \neq s$. Due to the same reason as given above and, thus, w. l. o. g., we focus on $p \neq r$. Additionally and again w. l. o. g., we assume $p < r$. If there is a $t \in \mathbb{N}_0$, $t < p$, such that $i_{p-t} \neq i'_{r-t}$ then we can apply exactly the same argument as above. Thus, $i_1 \cdot i_2 \cdot [\ldots] \cdot i_p$ must be a suffix of $i'_1 \cdot i'_2 \cdot [\ldots] \cdot i'_r$. If $\alpha = \hat{\gamma}_k \ldots$ or $\alpha = \ldots\, i_\sharp\, \hat{\gamma}_k \ldots$ with $i_\sharp \neq i'_{r-p}$ then our argumentation again is equivalent to that on the case $p = r$. Hence, $i_\sharp = i'_{r-p}$. Since $i'_{r-p} \in X_2$, $\hat{\beta}_{k-1}$ must have been a splitting point separating i'_{r-p} and i_1, whereas there has not been any splitting point between i'_{r-p} and i'_{r-p+1}. Since $i_1 \cdot i_2 \cdot [\ldots] \cdot i_p$ is a suffix of $i'_1 \cdot i'_2 \cdot [\ldots] \cdot i'_r$ this contradicts the definition of splitting points. □ (Claim 6)

In a final step, we now remove all $\hat{\gamma}_k$ with $|\hat{\gamma}_k| = 1$; this type of $\hat{\gamma}$ can occur, e. g., for $\bar{\gamma}_{k'} = i_1 \cdot i_2 \cdot i_3$ with $i_1, i_3 \in X_1$ and $i_2 \in X_2$. Consequently, for every $\hat{\gamma}_k = i$, $i \in X_1$, we shift i to X_3, or, more precisely, we introduce $X'_3 := X_3 \cup \{i \mid \exists\, k : \hat{\gamma}_k = i\}$ and $X'_1 := X_1 \setminus \{i \mid \exists\, k : \hat{\gamma}_k = i\}$. Note that because of Claim 5 and Claim 6 this *redefinition* operation does not affect any $\hat{\gamma}_k$ with $|\hat{\gamma}_k| \geq 2$.

This leads to the final decomposition of α:

$$\alpha = \beta_0\, \gamma_1\, \beta_1\, \gamma_2\, \beta_2\, [\ldots]\, \beta_{m-1}\, \gamma_m\, \beta_m$$

with $m \geq 1$ and

- $\beta_k \in X'^*_3$, $0 \leq k \leq m$, where, for every $1 \leq k' \leq m-1$, $\beta_{k'} = \varepsilon$ if and only if the variables to the right and to the left of $\beta_{k'}$ have been split by a splitting operation and none of the resulting neighbouring $\hat{\gamma}_{k''}$ has been removed by a shifting operation, and

- $\gamma_k \in (X'_1 \cup X_2)^+$, $1 \leq k \leq m$.

Again, this decomposition is unique, and $m \geq 1$ is granted since X_2 is not redefined and (according to our assumption) $X_2 \neq \emptyset$.

We conclude the proof of Lemma 5.8 with the verification of the conditions in Theorem 5.2:

Claim 7. For every k, $1 \leq k \leq m$, $|\gamma_k| \geq 2$.

Claim 7 is evident since the redefinition operation does not shorten or split any $\hat{\gamma}_k$ with $|\hat{\gamma}_k| \geq 2$. Consequently, the above decomposition conforms with condition (i) of Theorem 5.2. The next claim follows directly from the fact that X_1', X_2 and X_3' are disjoint:

Claim 8. For every k, $1 \leq k \leq m$, and for every k', $0 \leq k' \leq m$, $\mathrm{var}(\gamma_k) \cap \mathrm{var}(\beta_{k'}) = \emptyset$.

Thus, condition (ii) of Theorem 5.2 is satisfied as well. Since, according to the notes on Claim 7, the splitting operation does not modify any $\hat{\gamma}_k$ with $|\hat{\gamma}_k| \geq 2$ we can easily conclude from Claim 6:

Claim 9. For every k, $1 \leq k \leq m$, γ_k contains exactly one $i \in X_1'$ and, for every k', $1 \leq k' \leq m$, if $\mathrm{var}(\gamma_k) \cap \mathrm{var}(\gamma_{k'}) \cap X_1' \neq \emptyset$ then $\gamma_k = \gamma_{k'}$.

This proves that condition (iii) of Theorem 5.2 is satisfied.

Consequently, if there is an $i \in \mathrm{var}(\alpha)$ such that $\tau(i) \neq \ldots \mathtt{E}\ \mathtt{a}\,\mathtt{b}^{3i+1}\,\mathtt{a}\ \mathtt{F} \ldots$ then, according to Theorem 5.2, α is morphically imprimitive. This proves the lemma. □

Referring to, e.g., Lemma 4.1, it is intuitively clear that Lemma 5.8 implies the moderate ambiguity of every morphism $\sigma_{\text{3-seg}} \in$ 3-SEG with respect to any morphically primitive pattern α; additionally, for any variable i in α, it provides deep insights into the core of the subword which is associated with any occurrences of i in α, namely the string $\mathtt{E}\ \mathtt{a}\,\mathtt{b}^{3i+1}\,\mathtt{a}\ \mathtt{F}$, $\mathtt{E}, \mathtt{F} \in \{\mathtt{a}, \mathtt{b}\}$. The latter aspect, however, is rather relevant for our studies in the subsequent Chapter 5.2.1.

Consequently, we can effortlessly complete the proof for the main result of the present chapter:

Theorem 5.3. *Let $\alpha \in \mathbb{N}^+$ be morphically primitive. Then every morphism in 3-SEG is moderately ambiguous with respect to α.*

Proof. Let $\alpha := i_1 \cdot i_2 \cdot [\ldots] \cdot i_n$, $n \in \mathbb{N}$, $i_1, i_2, \ldots, i_n \in \mathbb{N}$, and let $\sigma_{\text{3-seg}}$ be any morphism in 3-SEG. For every k, $1 \leq k \leq n$, for any $\mathtt{A}, \mathtt{B}, \mathtt{C} \in \{\mathtt{a}, \mathtt{b}\}$ and referring to the conditions given in Definition 4.3, we now define

- for $1 \leq k \leq n-1$, $r_k := |\ \sigma_{\text{3-seg}}(i_1 \cdot i_2 \cdot [\ldots] \cdot i_{k-1})\ \mathtt{A}\,\mathtt{B}^{3i_k}\,\mathtt{A}\ \mathtt{a}\,\mathtt{b}^{3i_k+1}\,\mathtt{a}\ \mathtt{C}\ |$ and,
- for $2 \leq k \leq n$, $l_k := |\ \sigma_{\text{3-seg}}(i_1 \cdot i_2 \cdot [\ldots] \cdot i_{k-1})\ \mathtt{A}\,\mathtt{B}^{3i_k}\,\mathtt{A}\ |$.

Then it can be easily verified that

$$1 \leq r_1 < l_2 \leq r_2 < l_3 \leq r_3 < l_4 \leq \ldots \leq r_{n-2} < l_{n-1} \leq r_{n-1} < l_n \leq |\sigma_{\text{3-seg}}(\alpha)|.$$

According to Lemma 5.8, for every k, $1 \leq k \leq n$, for some $\mathtt{A}, \mathtt{C} \in \{\mathtt{a}, \mathtt{b}\}$ and for every morphism $\tau : \mathbb{N}^* \longrightarrow \{\mathtt{a}, \mathtt{b}\}^*$ with $\tau(\alpha) = \sigma_{\text{3-seg}}(\alpha)$, there exist $w_{k,1}, w_{k,2} \in \{\mathtt{a}, \mathtt{b}\}^*$ such that $\tau(\alpha)$ is given as

$$\tau(\alpha) = \tau(i_1 \cdot i_2 \cdot [\ldots] \cdot i_{k-1})\, w_{k,1}\ \mathtt{A}\ \mathtt{a}\,\mathtt{b}^{3i_k+1}\,\mathtt{a}\ \mathtt{C}\ w_{k,2}\, \tau(i_{k+1} \cdot i_{k+2} \cdot [\ldots] \cdot i_n).$$

Additionally, due to Lemma 5.7, we know that, for every $i \in \mathrm{var}(\alpha)$,

- $|\ \tau(i_1 \cdot i_2 \cdot [\ldots] \cdot i_{k-1})\ |_{\mathtt{a}\mathtt{b}^{3i+1}\mathtt{a}} = |\ i_1 \cdot i_2 \cdot [\ldots] \cdot i_{k-1}\ |_i$,

- $|\,\tau(i_{k+1} \cdot i_{k+2} \cdot [\ldots] \cdot i_n)\,|_{\mathtt{a}\,\mathtt{b}^{3i+1}\,\mathtt{a}} = |\,i_{k+1} \cdot i_{k+2} \cdot [\ldots] \cdot i_n\,|_i$ and,
- for every $i' \in \mathrm{var}(\alpha)$ with $i' \neq i$, $|\tau(i')|_{\mathtt{a}\,\mathtt{b}^{3i+1}\,\mathtt{a}} = 0$.

Consequently, for every τ and for every k, $1 \leq k \leq n-1$, we have

- $|\,\tau(i_1 \cdot i_2 \cdot [\ldots] \cdot i_k)\,| < l_{k+1}$ and
- $|\,\tau(i_1 \cdot i_2 \cdot [\ldots] \cdot i_k)\,| \geq r_k$.

Hence, $\sigma_{3\text{-seg}}$ is moderately ambiguous with respect to α. □

Thus, with respect to each morphically primitive pattern α, moderate ambiguity can be achieved by a morphism $\sigma_{3\text{-seg}}$ mapping each variable j onto a word that contains three unique segments, and the inner of these segments is the part which, for all morphisms τ with $\tau(\alpha) = \sigma(\alpha)$, always must be generated by j. This illustrative and at first glance simplifying conclusion – which speaks of the variables in the pattern instead of their particular occurrences (as required by Definition 4.3) – is justified by the easily provable fact that, with regard to any $\sigma_{3\text{-seg}} \in$ 3-SEG, all occurrences of any variable have the same associated subword. Hence the corresponding behaviour of the morphisms in 3-SEG is less complex than, e. g., that of the morphism presented in Example 4.11.

It is not surprising that, in addition to their moderate ambiguity, the morphisms in 3-SEG are even *un*ambiguous with respect to a major set of nontrivial patterns. Concerning the morphism $\sigma \in$ 3-SEG given by $\sigma(j) := \mathtt{a}\,\mathtt{b}^{3j}\,\mathtt{a}\ \mathtt{a}\,\mathtt{b}^{3j+1}\,\mathtt{a}\ \mathtt{a}\,\mathtt{b}^{3j+2}\,\mathtt{a}$, $j \in \mathbb{N}$ this is demonstrated by Freydenberger [20] providing a characterisation of all patterns with respect to which σ is unambiguous. Nevertheless, from this characterisation (or, alternatively, from our favourite pattern $\alpha = 1 \cdot 2 \cdot 3 \cdot 4 \cdot 1 \cdot 4 \cdot 3 \cdot 2$, cf. Chapter 1) it follows immediately, that the morphism σ is not unambiguous for all morphically primitive patterns. Hence, when dealing in Chapter 5.4 with the existence of unambiguous morphisms, we cannot simply expand our present argumentation.

Summarising our main insights into moderate ambiguity, we now can comprehensively answer Question 4.2. More precisely, we can characterise morphic primitivity of patterns by means of the existence of moderately ambiguous morphisms:

Corollary 5.6. *Let $\alpha \in \mathbb{N}^+$, and let Σ be an alphabet, $|\Sigma| \geq 2$. Then α is morphically primitive if and only if there exists an injective morphism $\sigma : \mathbb{N}^* \longrightarrow \Sigma^*$ such that $\sigma(\alpha)$ is moderately ambiguous.*

Proof. Corollary 5.6 immediately follows from Corollary 5.5 and – since each $\sigma_{3\text{-seg}} \in$ 3-SEG is injective – from Theorem 5.3. □

Concerning our preliminary list of characterisations of morphic primitivity given by Corollary 5.4, however, we prefer to add the original statement of Theorem 5.3 which, by the way, provides a first major answer to Question 4.3:

Corollary 5.7. *Let Σ be an alphabet, $|\Sigma| \geq 2$. Let $\alpha \in \mathbb{N}^+$. Then the following statements are equivalent:*

- *α is morphically primitive.*
- *α is not a fixed point of any nontrivial morphism $\phi : \mathrm{var}(\alpha)^* \longrightarrow \mathrm{var}(\alpha)^*$.*

- *α is succinct on Σ.*
- *Every morphism in 3-SEG is moderately ambiguous with respect to α.*

Proof. Directly from Corollary 5.4, Corollary 5.5 and Theorem 5.3. □

By Corollary 5.7, the most fundamental question on the moderate ambiguity of morphisms is answered, and, as our argumentation is based on the morphisms in 3-SEG, we can even rely on a number of concrete suitable morphisms that are moderately ambiguous whenever the pattern under consideration has a moderately ambiguous morphic image at all. Still, it is open whether, in general, moderate ambiguity *requires* to map the variables in a morphically primitive pattern onto a word which consists of three segments. Since, for every $\sigma_{3\text{-seg}} \in$ 3-SEG and for every $\alpha \in \mathbb{N}^+$,

$$|\,\sigma_{3\text{-seg}}(\alpha)| = \sum_{j \in \mathrm{var}(\alpha)} |\alpha|_j \cdot (9(j+1)),$$

which from a practical point of view is rather unsatisfactory, we now wish to briefly discuss whether the complexity of the morphisms in 3-SEG is really necessary.

In fact, from the examples in Chapter 4, we know numerous patterns with respect to which there is a nonerasing or even injective morphism σ such that $\sigma(\alpha)$ is moderately ambiguous and much shorter than, for any $\sigma_{3\text{-seg}} \in$ 3-SEG, the word $\sigma_{3\text{-seg}}(\alpha)$. Furthermore, we do not know any example pattern for which $\sigma_{3\text{-seg}}$ leads to the shortest moderately ambiguous morphic image possible for this pattern. Nevertheless we can give an example which suggests that, in general, our morphisms in 3-SEG are optimally chosen. More precisely, we show that there exists a morphically primitive pattern α and an injective morphism σ mapping each variable in α onto a word which consists of *two* distinct segments such that σ is *not* moderately ambiguous with respect to α. Consequently, the properties of morphisms based on two segments contrast with those of the morphisms based on three segments (which are noted by Theorem 5.3):

Proposition 5.1. *Let the morphism $\sigma_{2\text{-seg}} : \mathbb{N}^* \longrightarrow \{\mathtt{a}, \mathtt{b}\}^*$ be given by $\sigma_{2\text{-seg}}(j) = \mathtt{a}\,\mathtt{b}^{2j}\,\mathtt{a}\ \mathtt{a}\,\mathtt{b}^{2j+1}\,\mathtt{a}$, $j \in \mathbb{N}$. Then there exists a morphically primitive pattern $\alpha \in \mathbb{N}^+$ such that $\sigma_{2\text{-seg}}$ is not moderately ambiguous with respect to α.*

Proof. Let the pattern $\alpha \in \mathbb{N}^+$ be given by

$$\alpha := 1 \cdot 2 \cdot 3 \cdot 1 \cdot 2 \cdot 3 \cdot 4 \cdot 2 \cdot 5 \cdot 4 \cdot 6 \cdot 7 \cdot 5 \cdot 7 \cdot 5 \cdot 8 \cdot 6 \cdot 8 \cdot 6.$$

By Theorem 5.2, it can be verified that α is morphically primitive.

We now give a morphism $\tau : \mathbb{N}^* \longrightarrow \{\mathtt{a}, \mathtt{b}\}^*$ which shows that $\sigma_{2\text{-seg}}(\alpha)$ is not moderately ambiguous; it is defined as follows:

$$\begin{aligned}
\tau(1) &:= \mathtt{a}\,\mathtt{b}^2\,\mathtt{a}\ \mathtt{a}\,\mathtt{b}^3\,\mathtt{a}\ \mathtt{a}\,\mathtt{b}^4\,\mathtt{a}\ \mathtt{a}\,\mathtt{b}^5\,\mathtt{a}\ \mathtt{a}\,\mathtt{b}^3,\\
\tau(2) &:= \mathtt{b}^3\,\mathtt{a}\ \mathtt{a}\,\mathtt{b}^3,\\
\tau(3) &:= \mathtt{b}^4\,\mathtt{a},\\
\tau(4) &:= \mathtt{a}\,\mathtt{b}^8\,\mathtt{a}\ \mathtt{a}\,\mathtt{b}^9\,\mathtt{a}\ \mathtt{a}\,\mathtt{b},\\
\tau(5) &:= \mathtt{b}^2\,\mathtt{a}\ \mathtt{a}\,\mathtt{b}^{10}\,\mathtt{a}\ \mathtt{a}\,\mathtt{b}^{11}\,\mathtt{a},\\
\tau(6) &:= \mathtt{b}^{11}\,\mathtt{a}\ \mathtt{a}\,\mathtt{b}^{13}\,\mathtt{a},\\
\tau(7) &:= \mathtt{a}\,\mathtt{b}^{14}\,\mathtt{a}\ \mathtt{a}\,\mathtt{b}^{13},\\
\tau(8) &:= \mathtt{a}\,\mathtt{b}^{16}\,\mathtt{a}\ \mathtt{a}\,\mathtt{b}^{17}\,\mathtt{a}\ \mathtt{a}\,\mathtt{b}.
\end{aligned}$$

Then τ and $\sigma_{2\text{-seg}}$ generate the same word when applied to α:

$$
\begin{aligned}
\sigma_{2\text{-seg}}(\alpha) \;=\;& \overbrace{\mathtt{a\,b^2\,a\;a\,b^3\,a}}^{\sigma_{2\text{-seg}}(1)}\;\overbrace{\mathtt{a\,b^4\,a\;a\,b^5\,a}}^{\boxed{\sigma_{2\text{-seg}}(2)}}\;\overbrace{\mathtt{a\,b^3\;b^3\,a\;a\,b^3\;b^4\,a}}^{\sigma_{2\text{-seg}}(3)} \\
& \underbrace{\mathtt{a\,b^2\,a\;a\,b^3\,a\;a\,b^4\,a\;a\,b^5\,a\;a\,b^3}}_{\tau(1)}\;\underbrace{\mathtt{b^3\,a\;a\,b^3}}_{\boxed{\tau(2)}}\;\underbrace{\mathtt{b^4\,a}}_{\tau(3)} \\
& \overbrace{\mathtt{a\,b^2\,a\;a\,b^3\,a}}^{\sigma_{2\text{-seg}}(1)}\;\overbrace{\mathtt{a\,b^4\,a\;a\,b^5\,a}}^{\sigma_{2\text{-seg}}(2)}\;\overbrace{\mathtt{a\,b^3\;b^3\,a\;a\,b^3\;b^4\,a}}^{\sigma_{2\text{-seg}}(3)} \\
& \underbrace{\mathtt{a\,b^2\,a\;a\,b^3\,a\;a\,b^4\,a\;a\,b^5\,a\;a\,b^3}}_{\tau(1)}\;\underbrace{\mathtt{b^3\,a\;a\,b^3}}_{\tau(2)}\;\underbrace{\mathtt{b^4\,a}}_{\tau(3)} \\
& \overbrace{\mathtt{a\,b^8\,a\;a\,b^9\,a}}^{\sigma_{2\text{-seg}}(4)}\;\overbrace{\mathtt{a\,b\;b^3\,a\;a\,b^3\;b^2\,a}}^{\sigma_{2\text{-seg}}(2)}\;\overbrace{\mathtt{a\,b^{10}\,a\;a\,b^{11}\,a}}^{\sigma_{2\text{-seg}}(5)} \\
& \underbrace{\mathtt{a\,b^8\,a\;a\,b^9\,a\;a\,b}}_{\tau(4)}\;\underbrace{\mathtt{b^3\,a\;a\,b^3}}_{\tau(2)}\;\underbrace{\mathtt{b^2\,a\;a\,b^{10}\,a\;a\,b^{11}\,a}}_{\tau(5)} \\
& \overbrace{\mathtt{a\,b^8\,a\;a\,b^9\,a}}^{\sigma_{2\text{-seg}}(4)}\;\overbrace{\mathtt{a\,b\;b^{11}\,a\;a\,b^{13}\,a}}^{\sigma_{2\text{-seg}}(6)} \\
& \underbrace{\mathtt{a\,b^8\,a\;a\,b^9\,a\;a\,b}}_{\tau(4)}\;\underbrace{\mathtt{b^{11}\,a\;a\,b^{13}\,a}}_{\tau(6)} \\
& \overbrace{\mathtt{a\,b^{14}\,a\;a\,b^{13}\;b^2\,a}}^{\sigma_{2\text{-seg}}(7)}\;\overbrace{\mathtt{a\,b^{10}\,a\;a\,b^{11}\,a}}^{\sigma_{2\text{-seg}}(5)}\;\overbrace{\mathtt{a\,b^{14}\,a\;a\,b^{13}\;b^2\,a}}^{\sigma_{2\text{-seg}}(7)}\;\overbrace{\mathtt{a\,b^{10}\,a\;a\,b^{11}\,a}}^{\sigma_{2\text{-seg}}(5)} \\
& \underbrace{\mathtt{a\,b^{14}\,a\;a\,b^{13}}}_{\tau(7)}\;\underbrace{\mathtt{b^2\,a\;a\,b^{10}\,a\;a\,b^{11}\,a}}_{\tau(5)}\;\underbrace{\mathtt{a\,b^{14}\,a\;a\,b^{13}}}_{\tau(7)}\;\underbrace{\mathtt{b^2\,a\;a\,b^{10}\,a\;a\,b^{11}\,a}}_{\tau(5)} \\
& \overbrace{\mathtt{a\,b^{16}\,a\;a\,b^{17}\,a}}^{\sigma_{2\text{-seg}}(8)}\;\overbrace{\mathtt{a\,b\;b^{11}\,a\;a\,b^{13}\,a}}^{\sigma_{2\text{-seg}}(6)}\;\overbrace{\mathtt{a\,b^{16}\,a\;a\,b^{17}\,a}}^{\sigma_{2\text{-seg}}(8)}\;\overbrace{\mathtt{a\,b\;b^{11}\,a\;a\,b^{13}\,a}}^{\sigma_{2\text{-seg}}(6)} \\
& \underbrace{\mathtt{a\,b^{16}\,a\;a\,b^{17}\,a\;a\,b}}_{\tau(8)}\;\underbrace{\mathtt{b^{11}\,a\;a\,b^{13}\,a}}_{\tau(6)}\;\underbrace{\mathtt{a\,b^{16}\,a\;a\,b^{17}\,a\;a\,b}}_{\tau(8)}\;\underbrace{\mathtt{b^{11}\,a\;a\,b^{13}\,a}}_{\tau(6)} \\
=\;& \tau(\alpha)\,.
\end{aligned}
$$

We now write α as $i_1 \cdot i_2 \cdot [\,\dots\,] \cdot i_{19}$. Referring to Definition 4.3, $|\,\sigma_{2\text{-seg}}(i_1 \cdot i_2)| = 22$ implies $r_2 \leq 22$, whereas $|\tau(i_1)| = 26$ leads to $l_2 \geq 27$. Thus, $r_2 < l_2$, which contradicts the definition of moderate ambiguity. □

Due to Proposition 5.1 it is of course not at all suprising that the morphism σ given by $\sigma(j) := \mathtt{a}\,\mathtt{b}^j\,\mathtt{a}$, $j \in \mathbb{N}$ – mapping each variable onto a word which consists of a *single* segment – also fails in guaranteeing moderate ambiguity. This is demonstrated by our standard pattern $\alpha := 1 \cdot 2 \cdot 3 \cdot 4 \cdot 1 \cdot 4 \cdot 3 \cdot 2$ (or, alternatively, the pattern given in the above proof of Proposition 5.1). Additionally, concerning the prominent suffix code σ_{c} given by $\sigma_{\mathrm{c}}(j) := \mathtt{a}\,\mathtt{b}^j$, $j \in \mathbb{N}$, which also assigns sort of a segment to the variables, we can simply refer to our explanations in Example 4.2.

Hence, we conclude from these notes on the moderate ambiguity of selected less complex morphisms that any morphism which, for all morphically primitive patterns, is moderately ambiguous has to map every variable onto a word that consists of three distinct segments, or it must implement a different principle which is not based on segments (provided that such a principle exists). Alternatively, there might of course, for every morphically primitive pattern, exist a way to find a simple *tailor-made* moderately ambiguous morphism. If this is possible, i. e. if there is no pattern α for which some $\sigma_{3\text{-seg}}(\alpha)$ is the shortest moderately ambiguous word, then we expect the corresponding proof to be extraordinarily cumbersome.

Concluding our considerations on moderate ambiguity, we now proceed with an application of our above results to inductive inference of terminal-free E-pattern languages over alphabets with at least *three* distinct letters.

5.2.1 Application: Inductive inference of the class of terminal-free E-pattern languages

In the present chapter, we examine the learnability of the full class of terminal-free E-pattern languages, i. e., roughly speaking, the computability of a terminal-free pattern from the words in its E-pattern language (cf. Chapters 2.5 and 3.3). Our main result provides a comprehensive answer to Question 3.1. The corresponding proof technique is largely based on the existence of *telltales* (cf. Theorems 3.12 and 3.13), which may be interpreted as a finite set of words that contain sufficient *information* about their common generating pattern (and, thus, the language to be inferred) for separating it from all patterns generating proper sublanguages. As these telltales consist of particular *moderately* ambiguous words, our considerations confirm the suggestion by Theorem 3.16 and the proof for Theorem 3.15 (as presented in [81]) that there is a close connection between the ambiguity of morphisms and the inferrability of pattern languages. In other words, it strengthens our claim according to which a word with a *restricted* ambiguity contains reliable information about its generating pattern (cf. our notes in Chapter 1), so that, in a combinatorial context, the structure-preserving capacity of a morphism essentially depends on its ambiguity (and not so much on its injectivity, as it is the case in coding theory).

We begin our considerations by a definition which identifies a special type of morphisms. These morphisms show a property which, by Theorem 3.16, in general is necessary for generating a *meaningful word*[†] in a telltale of a terminal-free E-pattern language:

Definition 5.2 (Telltale candidate morphism). *Let* Σ *be an alphabet,* $|\Sigma| \geq 2$. *Then a morphism* $\sigma : \mathbb{N}^* \longrightarrow \Sigma^*$ *is said to be a* telltale candidate for a variable $j \in \mathbb{N}$ (in a pattern $\alpha \in \mathrm{Pat}_{\mathrm{tf}}$) *provided that there exists an* $\mathtt{A} \in \Sigma$ *such that* $|\sigma(j)|_{\mathtt{A}} = 1$ *and* $|\sigma(\alpha)|_{\mathtt{A}} = |\alpha|_j$. *Additionally, we call* σ *a* telltale candidate (with respect to α) *if there exists a* $j \in \mathrm{var}(\alpha)$ *such that* σ *is a telltale candidate for* j.

Hence, for every telltale candidate σ with respect to a pattern α, there is a variable j in α such that $\sigma(j)$ contains a unique letter $\mathtt{A}$ – i. e. $|\sigma(j)|_{\mathtt{A}} = 1$ and, for every $k \in \mathrm{var}(\alpha) \setminus \{j\}$, $\mathtt{A} \notin \mathrm{term}(\sigma(k))$.

Using this definition, we can semi-formally paraphrase the characterisation of telltales for terminal-free E-pattern languages (with respect to $\mathrm{ePAT}_{\mathrm{tf}}$) given by Theorem 3.16 as follows: A set $T_\alpha := \{w_1, w_2, \ldots w_n\}$ is a telltale for the E-pattern language $L_{\mathrm{E}}(\alpha)$ of a succinct (i. e. morphically primitive, cf. Corollary 5.3) pattern $\alpha \in \mathrm{Pat}_{\mathrm{tf}}$ if and only if, for every $j \in \mathrm{var}(\alpha)$, there exists an index i such that every morphism σ satisfying $\sigma(\alpha) = w_i$ is a telltale candidate for j.

[†]We do not declare this term formally since we use it for illustration purposes only. From a formal point of view, within the present chapter, one might consider a word w a meaningful part of a telltale for a language L if and only if, e. g., there exists a set W of words such that W is not a telltale for L, but $W \cup \{w\}$ is a telltale for L with respect to some indexed family $\mathcal{L} = (L_i)_{i \in \mathbb{N}}$ comprising L. It can be easily seen that our subsequent considerations exclusively deal with words which, in the given sense, are meaningful.

In other words, for every variable in the pattern there must be a telltale candidate for this particular variable, the ambiguity of which is restricted in a certain way. This, in turn, implies that, from our above version of the characterisation of telltales, a sufficient condition for the learnability of the class of terminal-free E-pattern languages incorporating the concept of telltale candidates and the property of moderate ambiguity can be derived with little effort:

Theorem 5.4. *Let Σ be an alphabet, $|\Sigma| \geq 2$. Then $\mathrm{ePAT}_{\mathrm{tf},\Sigma}$ is inferrable from positive data if, for every succinct pattern $\alpha := i_1 \cdot i_2 \cdot [\dots] \cdot i_n \in \mathrm{Pat}_{\mathrm{tf}}$, $n \in \mathbb{N}$, $i_1, i_2, \dots, i_n \in \mathbb{N}$, and for every $i \in \mathrm{var}(\alpha)$, there exists a morphism $\sigma_i : \mathbb{N}^* \longrightarrow \Sigma^*$ such that*

(1) *σ_i is moderately ambiguous with respect to α and,*

(2) *whenever, for some k, $1 \leq k \leq n$, $i_k = i$ then the subword $w_{i,k,\max}$ associated with k contains a letter $\mathtt{A} \in \Sigma$ with $|\sigma_i(\alpha)|_{\mathtt{A}} = |\alpha|_i$.*

Proof. Theorem 5.4 can be easily derived from Theorem 3.16. Nevertheless, we give an independent complete proof, so as to illustrate the use of the decidability of the inclusion problem for $\mathrm{ePAT}_{\mathrm{tf},\Sigma}$ (cf. Theorems 3.6, 3.7) when dealing with learnability questions:

Let α be an arbitrary succinct terminal-free pattern. Then we define the set $T_\alpha \subseteq L_{\mathrm{E},\Sigma}(\alpha)$ by $T_\alpha := \{\sigma_i(\alpha) \mid i \in \mathrm{var}(\alpha)\}$. We now show that T_α is a telltale for $L_{\mathrm{E},\Sigma}(\alpha)$ with respect to $\mathrm{ePAT}_{\mathrm{tf},\Sigma}$. For that purpose assume $T_\alpha \subseteq L_{\mathrm{E},\Sigma}(\beta) \subseteq L_{\mathrm{E},\Sigma}(\alpha)$ for some $\beta \in \mathrm{Pat}_{\mathrm{tf}}$. Then (due to Theorem 3.6) there exists a morphism $\phi : \mathbb{N}^* \longrightarrow \mathbb{N}^*$ such that $\phi(\alpha) = \beta$.

Directly from conditions (1) and (2) it follows that, for every $i \in \mathrm{var}(\alpha)$, each morphism τ_i with $\tau_i(\alpha) = \sigma_i(\alpha)$ maps i onto a word comprising $w_{i,k,\max}$. Therefore, in particular, we may conclude that, for each morphism τ_i' with $\tau_i' \circ \phi(\alpha) = \tau_i'(\phi(\alpha)) = \tau_i'(\beta) = \sigma_i(\alpha)$, the image of i under $\tau_i' \circ \phi$ has this property. Hence, in order to generate the unique letter $\mathtt{A}$ by τ_i', $\phi(i)$ must contain a unique variable $\hat{\jmath}_i$, i. e., for some $\gamma_1, \gamma_2 \in \mathbb{N}^*$, $\phi(i) = \gamma_1 \cdot \hat{\jmath}_i \cdot \gamma_2$ with $|\phi(\alpha)|_{\hat{\jmath}_i} = |\alpha|_i$.

We now define a morphism $\psi : \mathbb{N}^* \longrightarrow \mathbb{N}^*$ as follows:

$$\psi(j) := \begin{cases} i &, \quad j = \hat{\jmath}_i \text{ for some } i \in \mathrm{var}(\alpha)\,, \\ \varepsilon &, \quad \text{else}\,, \end{cases}$$

$j \in \mathrm{var}(\beta)$. Then $\psi(\beta) = \alpha$ and thus, according to Theorem 3.6, $L_{\mathrm{E},\Sigma}(\beta) \supseteq L_{\mathrm{E},\Sigma}(\alpha)$. Consequently, there is no $\beta \in \mathrm{Pat}_{\mathrm{tf}}$ such that $T_\alpha \subseteq L_{\mathrm{E},\Sigma}(\beta) \subset L_{\mathrm{E},\Sigma}(\alpha)$, and therefore T_α is a telltale for $L_{\mathrm{E},\Sigma}(\alpha)$ with respect to $\mathrm{ePAT}_{\mathrm{tf},\Sigma}$.

Hence, if the conditions (1) and (2) are satisfied for every succinct terminal-free pattern, then – as the inclusion is decidable for $\mathrm{ePAT}_{\mathrm{tf},\Sigma}$, see Theorem 3.7 – the conditions of Theorem 3.13 are satisfied for each language $L \in \mathrm{ePAT}_{\mathrm{tf},\Sigma}$, and therefore $\mathrm{ePAT}_{\mathrm{tf},\Sigma}$ is inferrable from positive data in that case. □

Theorem 5.4 formally confirms our above remarks: In order to construct a telltale, we can rely on telltale candidates that are moderately ambiguous in such a way, that a respective unique letter $\mathtt{A}$ is contained in an appropriate associated subword. Fortunately, the previous chapter – and, in particular, Theorem 5.3 – contains deep knowledge on the moderate ambiguity of the morphisms in 3-SEG (cf. Definition 5.1) which, more precisely, says that for every succinct (i. e. morphically primitive, cf. Corollary 5.3) pattern $\alpha \in \mathrm{Pat_{tf}}$ there exist various moderately ambiguous morphisms such as $\sigma_{3\text{-seg}} : \mathbb{N}^* \longrightarrow \{\mathtt{a}, \mathtt{b}\}^*$ given by $\sigma_{3\text{-seg}}(j) := \mathtt{a\,b}^{3j}\,\mathtt{a}\ \mathtt{a\,b}^{3j+1}\,\mathtt{a}\ \mathtt{a\,b}^{3j+2}\,\mathtt{a}$, $j \in \mathbb{N}$ (cf. Example 5.2). These morphisms in 3-SEG, however, are no telltale candidates. Therefore we now transform them into telltale candidates without disturbing their moderate ambiguity. Since, on the one hand, we only need morphic images over a two letter alphabet for guaranteeing moderate ambiguity and, on the other hand, we solely have to deal with pattern languages over an alphabet with at least *three* distinct letters (due to Theorem 3.15), we can simply pick the morphism $\sigma_{3\text{-seg}} \in$ 3-SEG given above and insert a unique letter $\mathtt{A} := \mathtt{c}$ into the subwords associated with all occurrences of any fixed variable. Hence, we regard the following morphisms:

Definition 5.3 ("Telltale morphism" $\sigma_{\mathrm{tt,3\text{-}seg},i}$). *Let Σ be an alphabet, $\{\mathtt{a}, \mathtt{b}, \mathtt{c}\} \subseteq \Sigma$. Then, for every $i, j \in \mathbb{N}$, the morphism $\sigma_{\mathrm{tt,3\text{-}seg},i} : \mathbb{N}^* \longrightarrow \Sigma^*$ is given by*

$$\sigma_{\mathrm{tt,3\text{-}seg},i}(j) := \begin{cases} \mathtt{a\,b}^{3j-2}\,\mathtt{a}\ \mathtt{a\,b}^{3j-1}\,\mathtt{a}\ \mathtt{a\,b}^{3j}\,\mathtt{a} & , \quad i \neq j, \\ \mathtt{a\,b}^{3j-2}\,\mathtt{a}\ \mathtt{c}\ \mathtt{a\,b}^{3j-1}\,\mathtt{a}\ \mathtt{a\,b}^{3j}\,\mathtt{a} & , \quad i = j. \end{cases}$$

By definition, for every $i \in \mathbb{N}$ and for every pattern $\alpha \in \mathrm{Pat_{tf}}$ with $i \in \mathrm{var}(\alpha)$, $\sigma_{\mathrm{tt,3\text{-}seg},i}$ is a telltale candidate for i in α.

Furthermore, our argumentation on the moderate ambiguity of every morphism $\sigma_{3\text{-seg}} \in$ 3-SEG with respect to every morphically primitive pattern (cf. Theorem 5.3) after minor and canonical modifications can be applied to the modified shape of the morphisms $\sigma_{\mathrm{tt,3\text{-}seg},i}$:

Lemma 5.9. *Let $\alpha := i_1 \cdot i_2 \cdot [\ldots] \cdot i_n \in \mathrm{Pat_{tf}}$, $n \in \mathbb{N}$, $i_1, i_2, \ldots, i_n \in \mathbb{N}$, be a succinct pattern. Then, for every $i \in \mathrm{var}(\alpha)$, the morphism $\sigma_{\mathrm{tt,3\text{-}seg},i}$ is moderately ambiguous with respect to α and, for every variable i_k with $i_k = i$, the subword $w_{i,k,\max}$ in $\sigma_{\mathrm{tt,3\text{-}seg},i}(\alpha)$ associated with k satisfies $w_{i,k,\max} = \ldots\ \mathtt{a}\ \mathtt{c}\ \mathtt{a\,b}^{3i-1}\,\mathtt{a}\ \mathtt{a}\ \ldots$.*

Proof. According to Corollary 5.3, the succinctness of α implies that α is morphically primitive; furthermore, both $\sigma_{\mathrm{tt,3\text{-}seg},i}$ and any $\sigma_{3\text{-seg}} \in$ 3-SEG have almost the same structure. Therefore, in order to show that $\sigma_{\mathrm{tt,3\text{-}seg},i}$ is moderately ambiguous, we can straightforward adapt the argumentation given in the proof of Lemma 5.8. The proof on the statement describing the shape of the $w_{i,k,\max}$ is analogous to the corresponding statements in the proof of Theorem 5.3. □

Hence, from the specific moderate ambiguity of the $\sigma_{\mathrm{tt,3\text{-}seg},i}$ and the fact that they are telltale candidates, we can immediately conclude that, for every alphabet Σ with $|\Sigma| \geq 3$, for every terminal-free E-pattern language $L \in \mathrm{ePAT_{tf,\Sigma}}$ and for every succinct pattern $\alpha \in \mathrm{Pat_{tf}}$ satisfying $L = L_{\mathrm{E},\Sigma}(\alpha)$, the set $\{\sigma_i(\alpha) \mid i \in \mathrm{var}(\alpha)\}$ is a telltale for L with respect to $\mathrm{ePAT_{tf,\Sigma}}$. Consequently, by Theorem 3.12, this class is inferrable from positive data:

Theorem 5.5. *Let Σ be a finite alphabet, $|\Sigma| \geq 3$. Then $\mathrm{ePAT_{tf,\Sigma}}$ is inferrable from positive data.*

Proof. Directly from Theorem 5.4 and Lemma 5.9. □

This completely answers Question 3.1.

From Theorem 5.5, the characterisation of the learnability of the full class of terminal-free E-pattern languages with respect to the size of the corresponding terminal alphabet follows immediately:

Corollary 5.8. *Let* Σ *be an alphabet. Then* $\mathrm{ePAT}_{\mathrm{tf},\Sigma}$ *is inferrable from positive data if and only if* $|\Sigma| \neq 2$.

Proof. Directly from Proposition 3.14, Theorem 3.15 and Theorem 5.5. □

Consequently, the learnability of terminal-free E-pattern languages changes depending on the alphabet size, and our reasoning demonstrates that this phenomenon is caused by the fact that the ambiguity of telltale candidates mapping patterns onto words over two distinct letters differs from the ambiguity of those telltale candidates which can use three letters. Hence, roughly speaking, we may summarise that two letters are necessary for restricting the ambiguity of the telltale candidates (according to Lemma 5.9), one particular letter is needed for guaranteeing that the morphism indeed is a telltale candidate and, in general, none of these letters can have both roles. Despite of the complexity of the argumentation required, we do not consider this outcome of our combinatorial approach overly surprising. Nevertheless it is remarkable that the stated properties do not at all match with the most elementary insights in coding theory, where the expressive power of morphic images over a ternary alphabet does not exceed that of words over a binary alphabet.

From a language theoretical point of view and referring to Theorem 3.12, Corollary 5.8 necessarily implies that some fundamental intrinsic (topological) properties of $\mathrm{ePAT}_{\mathrm{tf},\Sigma}$ change under the alphabet extension from $|\Sigma| = 2$ to $|\Sigma| = 3$. We do not feel this aspect to be perfectly understood so far. In particular, it is noteworthy that the varying learnability results for $\mathrm{ePAT}_{\mathrm{tf},\Sigma}$ contrast with the continuous behaviour of the equivalence of terminal-free E-pattern languages over the alphabet sizes under consideration (recall that, as shown by Corollary 3.3, two terminal-free patterns generate the same language over a binary alphabet if and only if they generate the same language over any alphabet with more than two letters). Hence, in this regard, Corollary 5.8 is counter-intuitive.

Evidently, our reasoning on Theorem 5.5 shows the learnability of $\mathrm{ePAT}_{\mathrm{tf},\Sigma}$, $|\Sigma| \geq 3$, by a purely combinatorial argument on morphisms in free monoids which, in turn, is sufficient for proving a structural property of $\mathrm{ePAT}_{\mathrm{tf},\Sigma}$, namely the existence of telltales; finally, by Theorem 3.12, this structural property implies the learnability of the class, which implies the existence of an appropriate learning strategy satisfying the conditions of the LIM model. Consequently, our considerations do not yield any particular learning strategy for $\mathrm{ePAT}_{\mathrm{tf},\Sigma}$, and therefore we can merely refer to the general procedure for learnable indexed families that is provided by Angluin [4].

We regard it as a worthwile (though challenging) problem for the future research on the learnability of terminal-free E-pattern languages to find a tailor-made learning strategy which matches with the characteristic of the subject more accurately than the generic strategy. In this regard, an inconsistent learner such as the procedure provided by Lange and Wiehagen [43] on the full class of NE-pattern languages, which contrary to Angluin's approach does not use any test for the membership problem, might be the overall goal of the corresponding considerations. Since the telltales identified by our reasoning in the previous chapter contain rather long words (whereas those for general NE-pattern languages used by Lange and Wiehagen [43] simply consist of the shortest words in the respective language), it even seems inevitable to avoid membership tests as far as possible.

In addition to this, one might wish to seek for shorter telltale words in terminal-free E-pattern languages. We conjecture that Theorem 3.16 and Theorem 5.4 can serve as powerful tools for such a task. However, as our argumentation is based on moderately ambiguous words – which, for selected patterns, cannot be generated by morphisms mapping the variables onto words just consisting of two distinct segments (see our corresponding explanations in Proposition 5.1) – we expect that, in general, our complex morphisms are necessary for generating meaningful words in telltales. We substantiate this claim by a proposition which shows that if we replace $\sigma_{\text{tt,3-seg},i}$ by a morphism $\sigma_{\neg\text{tt,2-seg},i}$ mapping the variables on just two distinct segments then there exists a succinct pattern such that $\sigma_{\neg\text{tt,2-seg},i}$ does not only conflict with the requirements of Theorem 5.4, but demonstrably does not generate a telltale. The latter claim is verified with the aid of Theorem 3.16.

Consequently, in a sense, our insights gained into the telltale-related properties of morphisms largely correspond to the results on moderate ambiguity described in the previous Chapter 5.2. We wish to emphasise that this is not only a consequence of our proof technique, but also of the fact that the necessary restriction of ambiguity of the meaningful words in telltales postulated by Theorem 3.16 shows close connections to our purely combinatorial concept of moderate ambiguity. Nevertheless, due to the third letter potentially contained in any $\sigma_{\text{tt,3-seg},i}(\alpha)$, $\alpha \in \text{Pat}_{\text{tf}}$, we cannot simply refer to the pattern introduced in the proof for Proposition 5.1, but our reasoning requires a more sophisticated example.

Formalising these explanations, our statement reads as follows:

Proposition 5.2. *Let Σ be an alphabet, $\{\mathtt{a},\mathtt{b},\mathtt{c}\} \subseteq \Sigma$, and let, for every $i, j \in \mathbb{N}$, the morphism $\sigma_{\neg\text{tt,2-seg},i} : \mathbb{N}^* \longrightarrow \Sigma^*$ be given by*

$$\sigma_{\neg\text{tt,2-seg},i}(j) := \begin{cases} \mathtt{a}\,\mathtt{b}^{2j}\,\mathtt{a}\;\mathtt{c}\;\mathtt{a}\,\mathtt{b}^{2j+1}\,\mathtt{a} & , \quad i = j, \\ \mathtt{a}\,\mathtt{b}^{2j}\,\mathtt{a}\;\mathtt{a}\,\mathtt{b}^{2j+1}\,\mathtt{a} & , \quad \text{else.} \end{cases}$$

Then there exists a succinct pattern $\alpha \in \text{Pat}_{\text{tf}}$ such that the set $W_\alpha := \{\sigma_{\neg\text{tt,2-seg},j}(\alpha) \mid j \in \text{var}(\alpha)\}$ is not a telltale for $L_\Sigma(\alpha)$ with respect to $\text{ePAT}_{\text{tf},\Sigma}$.

Proof. Let the pattern $\alpha \in \text{Pat}_{\text{tf}}$ be given by

$$\alpha \;:=\; 1\cdot 2\cdot 3\cdot 4\cdot 1\cdot 2\cdot 3\cdot 4\cdot 5\cdot 2\cdot 6\cdot 5\cdot 7\cdot 8\cdot 6\cdot 8\cdot 6\cdot 9\cdot 7\cdot 9\cdot 7\cdot 10\cdot 4\cdot 10\cdot 11\cdot 12\cdot 4\cdot 12\cdot 11.$$

By Theorem 5.2, it can be verified that α is morphically primitive; by Corollary 5.3, this implies that α is succinct.

Furthermore, let the substitution τ_3 be given by

$$
\begin{aligned}
\tau_3(1) &:= \mathtt{ab^{2}a\,ab^{3}a\,ab^{4}a\,ab^{5}a\,ab^{6}a\,c\,ab^{4}},\\
\tau_3(2) &:= \mathtt{b^{3}a\,ab^{4}},\\
\tau_3(3) &:= \varepsilon,\\
\tau_3(4) &:= \mathtt{b^{4}a\,ab^{9}a},\\
\tau_3(5) &:= \mathtt{ab^{10}a\,ab^{11}a\,ab},\\
\tau_3(6) &:= \mathtt{ba\,ab^{12}a\,ab^{13}a},\\
\tau_3(7) &:= \mathtt{b^{13}a\,ab^{15}a},\\
\tau_3(8) &:= \mathtt{ab^{16}a\,ab^{16}},\\
\tau_3(9) &:= \mathtt{ab^{18}a\,ab^{19}a\,ab},\\
\tau_3(10) &:= \mathtt{ab^{20}a\,ab^{21}a\,ab^{4}},\\
\tau_3(11) &:= \mathtt{b^{18}a\,ab^{23}a},\\
\tau_3(12) &:= \mathtt{ab^{24}a\,ab^{25}a\,ab^{4}}.
\end{aligned}
$$

Then τ_3 and $\sigma_{\neg\mathrm{tt},2\text{-seg},3}$ generate the same word when applied to α:

$$
\begin{aligned}
\sigma_{\neg\mathrm{tt},2\text{-seg},3}(\alpha) &= \underbrace{\overbrace{\mathtt{ab^{2}a\,ab^{3}a}}^{\sigma_{\neg\mathrm{tt},2\text{-seg},3}(1)}\,\overbrace{\mathtt{ab^{4}a\,ab^{5}a}}^{\sigma_{\neg\mathrm{tt},2\text{-seg},3}(2)}\,\mathrlap{\overbrace{\phantom{\mathtt{ab^{6}a\,c\,ab^{4}\,b^{3}a}}}^{\sigma_{\neg\mathrm{tt},2\text{-seg},3}(3)}}\mathtt{ab^{6}a\,c\,ab^{4}}}_{\tau_3(1)}\,\underbrace{\mathtt{b^{3}a}\,\mathrlap{\overbrace{\phantom{\mathtt{ab^{4}\,b^{4}a\,ab^{9}a}}}^{\sigma_{\neg\mathrm{tt},2\text{-seg},3}(4)}}\mathtt{ab^{4}}}_{\tau_3(2)}\,\underbrace{\mathtt{b^{4}a\,ab^{9}a}}_{\tau_3(4)}\\
&\quad \underbrace{\overbrace{\mathtt{ab^{2}a\,ab^{3}a}}^{\sigma_{\neg\mathrm{tt},2\text{-seg},3}(1)}\,\overbrace{\mathtt{ab^{4}a\,ab^{5}a}}^{\sigma_{\neg\mathrm{tt},2\text{-seg},3}(2)}\,\mathrlap{\overbrace{\phantom{\mathtt{ab^{6}a\,c\,ab^{4}\,b^{3}a}}}^{\sigma_{\neg\mathrm{tt},2\text{-seg},3}(3)}}\mathtt{ab^{6}a\,c\,ab^{4}}}_{\tau_3(1)}\,\underbrace{\mathtt{b^{3}a}\,\mathrlap{\overbrace{\phantom{\mathtt{ab^{4}\,b^{4}a\,ab^{9}a}}}^{\sigma_{\neg\mathrm{tt},2\text{-seg},3}(4)}}\mathtt{ab^{4}}}_{\tau_3(2)}\,\underbrace{\mathtt{b^{4}a\,ab^{9}a}}_{\tau_3(4)}\\
&\quad \underbrace{\overbrace{\mathtt{ab^{10}a\,ab^{11}a}}^{\sigma_{\neg\mathrm{tt},2\text{-seg},3}(5)}\,\mathrlap{\overbrace{\phantom{\mathtt{ab\,b^{3}a\,ab^{4}}}}^{\sigma_{\neg\mathrm{tt},2\text{-seg},3}(2)}}\mathtt{ab}}_{\tau_3(5)}\,\underbrace{\mathtt{b^{3}a\,ab^{4}}}_{\tau_3(2)}\,\underbrace{\overbrace{\mathtt{ba\,ab^{12}a\,ab^{13}a}}^{\sigma_{\neg\mathrm{tt},2\text{-seg},3}(6)}}_{\tau_3(6)}\\
&\quad \underbrace{\overbrace{\mathtt{ab^{10}a\,ab^{11}a}}^{\sigma_{\neg\mathrm{tt},2\text{-seg},3}(5)}\,\mathrlap{\overbrace{\phantom{\mathtt{ab\,b^{13}a\,ab^{15}a}}}^{\sigma_{\neg\mathrm{tt},2\text{-seg},3}(7)}}\mathtt{ab}}_{\tau_3(5)}\,\underbrace{\mathtt{b^{13}a\,ab^{15}a}}_{\tau_3(7)}\\
&\quad \underbrace{\overbrace{\mathtt{ab^{16}a\,ab^{16}}}^{\sigma_{\neg\mathrm{tt},2\text{-seg},3}(8)}}_{\tau_3(8)}\,\underbrace{\overbrace{\mathtt{ba\,ab^{12}a\,ab^{13}a}}^{\sigma_{\neg\mathrm{tt},2\text{-seg},3}(6)}}_{\tau_3(6)}\,\underbrace{\overbrace{\mathtt{ab^{16}a\,ab^{16}}}^{\sigma_{\neg\mathrm{tt},2\text{-seg},3}(8)}}_{\tau_3(8)}\,\underbrace{\overbrace{\mathtt{ba\,ab^{12}a\,ab^{13}a}}^{\sigma_{\neg\mathrm{tt},2\text{-seg},3}(6)}}_{\tau_3(6)}\\
&\quad \underbrace{\overbrace{\mathtt{ab^{18}a\,ab^{19}a}}^{\sigma_{\neg\mathrm{tt},2\text{-seg},3}(9)}\,\mathrlap{\overbrace{\phantom{\mathtt{ab\,b^{13}a\,ab^{15}a}}}^{\sigma_{\neg\mathrm{tt},2\text{-seg},3}(7)}}\mathtt{ab}}_{\tau_3(9)}\,\underbrace{\mathtt{b^{13}a\,ab^{15}a}}_{\tau_3(7)}\,\underbrace{\overbrace{\mathtt{ab^{18}a\,ab^{19}a}}^{\sigma_{\neg\mathrm{tt},2\text{-seg},3}(9)}\,\mathrlap{\overbrace{\phantom{\mathtt{ab\,b^{13}a\,ab^{15}a}}}^{\sigma_{\neg\mathrm{tt},2\text{-seg},3}(7)}}\mathtt{ab}}_{\tau_3(9)}\,\underbrace{\mathtt{b^{13}a\,ab^{15}a}}_{\tau_3(7)}\\
&\quad \underbrace{\overbrace{\mathtt{ab^{20}a\,ab^{21}a}}^{\sigma_{\neg\mathrm{tt},2\text{-seg},3}(10)}\,\mathrlap{\overbrace{\phantom{\mathtt{ab^{4}\,b^{4}a\,ab^{9}a}}}^{\sigma_{\neg\mathrm{tt},2\text{-seg},3}(4)}}\mathtt{ab^{4}}}_{\tau_3(10)}\,\underbrace{\mathtt{b^{4}a\,ab^{9}a}}_{\tau_3(4)}\,\underbrace{\overbrace{\mathtt{ab^{20}a\,ab^{21}a}}^{\sigma_{\neg\mathrm{tt},2\text{-seg},3}(10)}\,\mathrlap{\overbrace{\phantom{\mathtt{ab^{4}\,b^{18}a\,ab^{23}a}}}^{\sigma_{\neg\mathrm{tt},2\text{-seg},3}(11)}}\mathtt{ab^{4}}}_{\tau_3(10)}\,\underbrace{\mathtt{b^{18}a\,ab^{23}a}}_{\tau_3(11)}\\
&\quad \underbrace{\overbrace{\mathtt{ab^{24}a\,ab^{25}a}}^{\sigma_{\neg\mathrm{tt},2\text{-seg},3}(12)}\,\mathrlap{\overbrace{\phantom{\mathtt{ab^{4}\,b^{4}a\,ab^{9}a}}}^{\sigma_{\neg\mathrm{tt},2\text{-seg},3}(4)}}\mathtt{ab^{4}}}_{\tau_3(12)}\,\underbrace{\mathtt{b^{4}a\,ab^{9}a}}_{\tau_3(4)}\,\underbrace{\overbrace{\mathtt{ab^{24}a\,ab^{25}a}}^{\sigma_{\neg\mathrm{tt},2\text{-seg},3}(12)}\,\mathrlap{\overbrace{\phantom{\mathtt{ab^{4}\,b^{18}a\,ab^{23}a}}}^{\sigma_{\neg\mathrm{tt},2\text{-seg},3}(11)}}\mathtt{ab^{4}}}_{\tau_3(12)}\,\underbrace{\mathtt{b^{18}a\,ab^{23}a}}_{\tau_3(11)}\\
&= \tau_3(\alpha)\,.
\end{aligned}
$$

Moreover, $\tau_3(3) = \varepsilon$ and, for every $j \in \mathbb{N}$ with $j \in \mathrm{var}(\alpha) \setminus \{3\}$ and for every letter $\mathtt{A}$ occurring in $\sigma_{\neg\mathrm{tt},2\text{-seg},j}(3)$, $|\sigma_{\neg\mathrm{tt},2\text{-seg},j}(3)|_{\mathtt{A}} \geq 2$. Thus, with regard to the variable 3, W_α does not satisfy the characteristic criterion given in Theorem 3.16. This proves the lemma. □

Thus, Lemma 5.9 and Proposition 5.2 and their counterparts on moderate ambiguity, namely Theorem 5.3 and Proposition 5.1, suggest that the rather long and very special words generated by the $\sigma_{\mathrm{tt},3\text{-seg},i}$ in general are necessary for drawing unequivocal conclusions on the respective generating pattern under consideration. If there is no option to switch to shorter meaningful words then we expect this to cause major problems for stochastic finite learning of $\mathrm{ePAT_{tf}}$ (as introduced by Rossmanith and Zeugmann [84] with respect to the full class of NE-pattern languages, cf. our brief notes in Chapter 3.3) – even if it is possible to give a learning strategy for the terminal-free E-pattern languages over suitable alphabets that is not based on exhaustive membership tests.

5.3 Weakly unambiguous morphisms

After we have extensively answered the most basic questions on the existence of moderately ambiguous morphisms, we now start a systematic analysis of unambiguity. We first turn our attention to free semigroups which do not include the identity element ε; hence, we deal with weak unambiguity (cf. Definition 4.2). As mentioned in Chapter 4, we merely give a number of notes on this topic which discuss Question 4.4, and the corresponding results can be obtained easily. Nevertheless, the insights provided have an elementary character.

The first part of Question 4.4 can be answered effortlessly:

Proposition 5.3. *Let Σ be an alphabet. Then there is a nonerasing morphism $\sigma : \mathbb{N}^+ \longrightarrow \Sigma^+$ which is weakly unambiguous with respect to every pattern $\alpha \in \mathbb{N}^+$.*

Proof. For every $i \in \mathbb{N}$, let $|\sigma(i)| = 1$. Then, for every $\alpha \in \mathbb{N}^+$, $|\sigma(\alpha)| = |\alpha|$ and, consequently, $\sigma(\alpha)$ is weakly unambiguous. □

Remarkably, this result demonstrates that, unlike moderate ambiguity (cf. Corollary 5.6), the existence of weakly unambiguous morphisms does not depend on the question of whether or not the pattern under consideration is morphically primitive.

With respect to the different types of pattern languages, our current focus on nonerasing morphisms implies that the statements in the present chapter can be rephrased in terms of NE-pattern languages. Consequently, from Proposition 5.3 we can conclude that, in every NE-pattern language generated by a terminal-free pattern α, there exist several weakly unambiguous words, namely all those words w satisfying $|w| = |\alpha|$. It can be easily seen that this insight also holds for general patterns over an arbitrary alphabet Σ. This combinatorial observation, in turn, is known to be of major importance for inductive inference of NE-pattern languages: Due to the fact given in Proposition 5.3 (and its canonical extension to the patterns in Pat_Σ for any alphabet Σ), for every NE-pattern language L a pattern α with $L = L_{\mathrm{NE}}(\alpha)$ can be inferred from the set of all the shortest words in this language (shown by Lange and Wiehagen [43], though the ambiguity of words is not explicitly discussed by the authors) as these words constitute a telltale for

L with respect to nePAT$_\Sigma$. With regard to E-pattern languages, however, the results by Reidenbach [76, 81] as presented in Theorem 3.15 and Theorem 3.16 and discussed in Chapter 5.2.1 show that such an approach necessarily has to fail since, in general, the words w with $|w| = |\alpha|$ are not unambiguous with respect to a pattern α. Consequently, in addition to our results in Chapter 5.2.1, the analysis by Lange and Wiehagen [43] provides another example where a restricted ambiguity of certain words – which moreover are not generated by injective morphisms – is a powerful tool for inductive inference.

Evidently, the morphisms discussed in the proof for Proposition 5.3 are not injective, whereas the morphisms $\sigma_{3\text{-seg}} \in$ 3-SEG introduced in Chapter 5.2, which are moderately ambiguous for all morphically primitive patterns, have this property. Our next result shows, that Proposition 5.3 cannot be strengthened in an analogous manner, i. e. no injective morphism can be weakly unambiguous for all patterns in $\mathbb{N}^+$. Hence, our answer to the second part of Question 4.4 reads as follows:

Theorem 5.6. *Let Σ be a finite alphabet. Then there is no injective morphism $\sigma : \mathbb{N}^+ \longrightarrow \Sigma^+$ which is weakly unambiguous with respect to every pattern $\alpha \in \mathbb{N}^+$.*

Proof. Assume to the contrary that there is such a morphism σ. Since σ is injective, $\sigma(\alpha) \neq \sigma(\beta)$ for every $\alpha \neq \beta$. In particular, this implies that, for every $i, i' \in \mathbb{N}$ with $i \neq i'$, $\sigma(i) \neq \sigma(i')$. Hence – since $\mathbb{N}$ is infinite whereas Σ is finite – there must be a variable $j \in \mathbb{N}$ with $\sigma(j) = w_1 w_2$ for some $w_1, w_2 \in \Sigma^+$. Now, for an arbitrary variable j' satisfying $j' \neq j$, let $\alpha := j \cdot j'$. Then, for the morphism $\tau : \mathbb{N}^+ \longrightarrow \Sigma^+$ given by $\tau(j) := w_1$ and $\tau(j') := w_2\, \sigma(j')$, $\tau(\alpha) = \sigma(\alpha)$, and, thus, $\sigma(\alpha)$ is not weakly unambiguous. This contradicts the assumption. □

The immediate consequences of Theorem 5.6, which do not only have a major impact on the search for injective weakly unambiguous morphisms, but also on that for unambiguous morphisms, are further discussed in Chapter 5.4.

By Theorem 5.6, we conclude this brief chapter on unambiguity in free semigroups. Nevertheless, the subsequent chapters contain numerous results implictly covering weak unambiguity. This is due to the fact that, by definition, each positive statement on unambiguity also holds for weak unambiguity (cf. Proposition 4.1).

5.4 Unambiguous morphisms

We conclude our studies of the ambiguity of common morphisms $\sigma : \mathbb{N}^* \longrightarrow \Sigma^*$ by a detailed examination of unambiguity. Our considerations mainly strive for an answer to Question 4.1 and the relevant parts of Question 4.3.

We begin with the immediate conclusion from Proposition 4.1 that not only, as stated in the previous chapter, every positive insight into unambiguity at the same time is a result on weak ambiguity, but also that, conversely, every negative statement on weak unambiguity also holds for unambiguity. Hence, we can directly note the following consequence of Theorem 5.6:

Corollary 5.9. *Let Σ be a finite alphabet. Then there is no injective morphism $\sigma : \mathbb{N}^* \longrightarrow \Sigma^*$ which is unambiguous with respect to every pattern $\alpha \in \mathbb{N}^+$.*

Proof. Directly from Theorem 5.6 and Proposition 4.1. □

Due to Theorem 5.1, however, the insight presented in Corollary 5.9 is by no means surprising since, in fact, we can immediately conclude from Theorem 5.1 a much stronger statement, according to which there is no possibility to give a nonerasing morphism that is unambiguous with respect to a morphically imprimitive pattern:

Corollary 5.10. *Let Σ be an alphabet, and let $\alpha \in \mathbb{N}^+$ be a morphically imprimitive pattern. Then there is no nonerasing morphism $\sigma : \mathbb{N}^* \longrightarrow \Sigma^*$ such that σ is unambiguous with respect to α.*

Proof. Directly from Theorem 5.1. □

Thus, for every morphically imprimitive pattern there is no unambiguous morphism at all – at least as long as we restrict ourselves to nonerasing morphisms. If this requirement is omitted then we face a fairly intricate situation:

Example 5.3. Let $\alpha_1 := 1 \cdot 1 \cdot 2 \cdot 2 \cdot 3 \cdot 4 \cdot 3 \cdot 4$, $\beta_1 := 1 \cdot 2 \cdot 2 \cdot 1 \cdot 3 \cdot 4 \cdot 3 \cdot 4$, and $\alpha_2 := 1 \cdot 2 \cdot 3 \cdot 3 \cdot 1 \cdot 2$, $\beta_2 := 1 \cdot 2 \cdot 2 \cdot 3 \cdot 3 \cdot 1 \cdot 2 \cdot 2$. The patterns are morphically imprimitive (cf. Theorem 5.2). For α_1 and α_2 there is no unambiguous morphism. Contrary to this, for β_1 and β_2 there exist unambiguous words such as $\mathtt{abba}$ and $\mathtt{baab}$. ◇

As shown in Example 5.3, there are morphically imprimitive patterns for which we can unambiguously map certain scattered subpatterns onto strings in $\{\mathtt{a}, \mathtt{b}\}^*$, whereas for other, quite similar appearing patterns this is impossible. Furthermore, these scattered subpatterns can consist of parts of some β_k as well as parts of some γ_k in the decomposition which according to Theorem 5.2 is obligatory for these patterns. Some first criteria on this extraordinarily challenging problem of finding unambiguous morphisms for morphically imprimitive patterns are provided by Schneider [93].

Concerning our overall goal of finding morphisms which preserve the structure of their preimage (cf. Chapter 1) it is obviously necessary to focus on nonerasing (or even rather: injective) morphisms. Therefore, after the comprehensive negative result on morphically imprimitive patterns given in Corollary 5.10, we now examine morphically primitive patterns (which additionally, from the pattern language point of view, are by far more interesting; cf. Corollary 5.3). But for these patterns, the negative result in Corollary 5.9 holds as well; since we now deal with unambiguity in free monoids (and not with weak unambiguity) we can even prove the opposite of Proposition 5.3 to be true:

Proposition 5.4. *Let Σ be a finite alphabet. Then there is no nonerasing morphism $\sigma : \mathbb{N}^* \longrightarrow \Sigma^*$ which is unambiguous with respect to every morphically primitive $\alpha \in \mathbb{N}^+$.*

Proof. Assume to the contrary that there is such a morphism. Then – since $\mathbb{N}$ is infinite whereas Σ is finite – there exist $j, j' \in \mathbb{N}$, $j \neq j'$, and an $\mathtt{A} \in \Sigma$ with $\sigma(j) = v\,\mathtt{A}$ and $\sigma(j') = v'\,\mathtt{A}$, $v, v' \in \Sigma^*$. For some $k, k' \in \mathbb{N}$, $k \neq k'$, $j \neq k \neq j'$ and $j \neq k' \neq j'$, we then regard the pattern $\alpha := j \cdot k \cdot j \cdot k' \cdot j' \cdot k \cdot j' \cdot k'$. According to Theorem 5.2, α is morphically primitive. Now consider the morphism τ, given by $\tau(j) := v$, $\tau(j') := v'$, $\tau(k) := \mathtt{A}\,\sigma(k)$, $\tau(k') := \mathtt{A}\,\sigma(k')$. Then evidently $\sigma(\alpha) = \tau(\alpha)$, but, e. g., $\sigma(j) \neq \tau(j)$. This is a contradiction. □

Note that Proposition 5.4 answers Question 4.3 in the negative for the full set of morphically primitive patterns. Therefore, and due to Corollary 5.9, we may conclude that if we have a set of patterns $\Pi \subseteq \mathbb{N}^+$ for which there exists a single nonerasing morphism σ such that σ is unambiguous with respect to each $\alpha \in \Pi$ then Π necessarily is a *proper* subset of the set of all morphically primitive patterns. This insights significantly contrasts with the behaviour of moderately ambiguous words as stated by Theorem 5.3.

Consequently, for every morphically primitive pattern, it is necessary to give an individual injective morphism that leads to an unambiguous word – provided that (unlike the outcome for morphically imprimitive patterns; see Corollary 5.10) such a morphism exists. We shall examine this problem now.

Our considerations start with a closer look at the proof of Proposition 5.4, that utilises a particular phenomenon which obviously leads to the undesirable ambiguity of injective morphisms. In this proof, we can observe that, for the abstract morphically primitive example pattern $\alpha = j \cdot k \cdot j \cdot k' \cdot j' \cdot k \cdot j' \cdot k'$, the ambiguity of the given morphism is caused by the fact that all occurrences of certain variables (namely k, k') have *left* neighbours (the variables j, j'), the morphic images of which have the same *suffix* (namely a word $w = \mathtt{A} \in \Sigma^+$ of length 1). Before we further discuss the question of how a tailor-made unambiguous morphism in such a case might look like, we first formalise our understanding of "neighbourship" of variables. Additionally, we anticipate that an analogous proof can be given focussing on sets of *right* neighbours of some variables, the images of which have the same *prefix*, and therefore we regard both left and right neighbours of any variable:

Definition 5.4 (Sets Λ_j and P_j of variables)**.** *Let $\alpha \in \mathbb{N}^+$. For every $j \in \mathrm{var}(\alpha)$, we define the following sets: $\Lambda_j := \{k \mid \alpha = \ldots \cdot k \cdot j \cdot \ldots\}$ and $\mathrm{P}_j := \{k \mid \alpha = \ldots \cdot j \cdot k \cdot \ldots\}$.*

Thus, Λ_j consists of all "left neighbours" of a variable j in a pattern α and P_j of all "right neighbours".

As mentioned above, the identical suffixes of the images of the variables in some Λ_j entail the ambiguity of the corresponding morphism. We call a set $V \subseteq \mathbb{N}$ of variables *(morphically) s-homogeneous* (*with respect to* a morphism σ) if and only if there exists a word $w \in \Sigma^+$ such that, for every $i \in V$, $\sigma(i) = \ldots w$; correspondingly, we say that V is *(morphically) p-homogeneous* (*with respect to* a morphism σ) if and only if there exists a word $w \in \Sigma^+$ such that, for every $i \in V$, $\sigma(i) = w \ldots$. Additionally, V is *s-heterogeneous* if and only if it is not s-homogeneous, and it is *p-heterogeneous* if and only if it is not p-homogeneous. Provided that the context is understood we simply speak of the *heterogeneity* of V.

Our initial conclusions drawn from the proof of Proposition 5.4 now suggest that it might be *sufficient* for the unambiguity of a morphism σ with respect to a morphically primitive pattern α if, for every $j \in \mathrm{var}(\alpha)$, Λ_j is s-heterogeneous w. r. t. σ and P_j is p-heterogeneous w. r. t. σ. Unfortunately, however, there are patterns for which we cannot find any morphism σ showing such a property:

Example 5.4. Let $\alpha := 1 \cdot 2 \cdot 3 \cdot 2 \cdot 1 \cdot 3 \cdot 1$. Note that α is morphically primitive (cf. Theorem 5.2). Thus, $\Lambda_1 = \{2, 3\}$, $\Lambda_2 = \{1, 3\}$ and $\Lambda_3 = \{1, 2\}$. Then, for a binary alphabet Σ, there is no morphism σ such that, at a time, Λ_1, Λ_2 and Λ_3 are morphically s-heterogeneous with respect to σ.

It can be straightforward verified that analogous examples exist for every finite alphabet Σ. ◇

Thus, in general, we cannot guarantee heterogeneity of the Λ_j and P_j introduced in Definition 5.4. On the other hand, as demonstrated by the subsequent example, it is in general not *necessary* that *each* Λ_j and P_j is heterogeneous:

Example 5.5. Let $\alpha := 1 \cdot 2 \cdot 3 \cdot 4 \cdot 1 \cdot 4 \cdot 3 \cdot 2 \cdot 5 \cdot 6 \cdot 5 \cdot 6 \cdot 3 \cdot 6$. Once more, due to Theorem 5.2, α is morphically primitive. Thus, $\Lambda_1 = \{4\}$, $\Lambda_2 = \{1,3\}$, $\Lambda_3 = \{2,4,6\}$, $\Lambda_4 = \{1,3\}$, $\Lambda_5 = \{2,6\}$, $\Lambda_6 = \{3,5\}$. With respect to the morphism $\sigma_1 : \mathbb{N}^* \longrightarrow \{\mathtt{a},\mathtt{b}\}^*$, given by $\sigma_1(i) := \mathtt{a}\,\mathtt{b}^i\,\mathtt{a}$, $i \in \mathbb{N}$, each Λ_j is s-homogeneous, and σ_1 is not unambiguous: for the morphism τ_1, defined by $\tau_1(1) := \mathtt{a}\,\mathtt{b}$, $\tau_1(2) := \mathtt{a}\,\mathtt{a}\,\mathtt{b}^2\,\mathtt{a}$, $\tau_1(3) := \mathtt{a}\,\mathtt{b}^3$, $\tau_1(4) := \mathtt{a}\,\mathtt{a}\,\mathtt{b}^4\,\mathtt{a}$, $\tau_1(5) := \mathtt{a}\,\mathtt{b}^5$ and $\tau_1(6) := \mathtt{a}\,\mathtt{a}\,\mathtt{b}^6\,\mathtt{a}$, it is $\tau_1(\alpha) = \sigma_1(\alpha)$.

Contrary to this, the morphism $\sigma_2 : \mathbb{N}^* \longrightarrow \{\mathtt{a},\mathtt{b}\}^*$ given by

$$\sigma_2(i) := \begin{cases} \mathtt{b}\,\mathtt{a}^i\,\mathtt{b}\,, & i \in \{1,2\}, \\ \mathtt{a}\,\mathtt{b}^i\,\mathtt{a}\,, & \text{else}, \end{cases}$$

is unambiguous with respect to α, although Λ_6 is not s-heterogeneous with respect to σ_2. ◇

If we take a closer look at Example 5.5 then we can see that the unambiguity of σ_2 is due to the fact that the s-heterogeneous set Λ_2 and the s-homogeneous set Λ_6 are non-disjoint: Roughly speaking, if we assume that there is a morphism τ_2 with $\tau_2 \neq \sigma_2$, $\tau_2(\alpha) = \sigma_2(\alpha)$ and $\tau_2(6)$ contains the rightmost letter of $\sigma_2(5)$ (just as τ_1 does when compared to σ_1) then, since the variables 3 and 5 both are neighbours of the variable 6, $\tau_2(3)$ does not contain the rightmost letter of $\sigma_2(3)$, either. Furthermore, with regard to the variables in Λ_2, as $\tau_2(3)$ does not contain the rightmost letter of $\sigma_2(3)$, $\tau_2(2)$ must contain this letter. Consequently, concerning the variable $1 \in \Lambda_2$, $\tau_2(2)$ also contains the rightmost letter of $\sigma_2(1)$. However, since Λ_2 is morphically heterogeneous with respect to σ_2, $\tau_2(2)$ has to start with both the letter $\mathtt{a}$ and the letter $\mathtt{b}$. This of course is a contradiction and therefore there is no morphism τ_2 satisfying the above conditions. Concerning other morphisms τ_2 potentially leading to $\tau_2 = \sigma_2$, we can apply an analogous argumentation – which, in particular, requires the p-heterogeneity of P_1 in order to prevent that for any two variables $i, i' \in \mathrm{var}(\alpha)$, $i \neq i'$, $\tau_2(i)$ contains the *left*most letter of $\sigma_2(i')$. Therefore, due to the heterogeneity of some *selected* Λ_j and P_j, the morphism σ_2 is unambiguous with respect to α.

Introducing classes of variables belonging to non-disjoint Λ_j (or, alternatively, P_j), we now formalise this concept, and we anticipate that the useful effect described above does not only hold for immediately non-disjoint sets, but also for the "transitive closure of non-disjointness":

Definition 5.5 (Equivalence classes $\Lambda_i^{\sim}$ and $\mathrm{P}_i^{\sim}$ of variables). *Let $\alpha \in \mathbb{N}^+$. Let, for every $j \in \mathrm{var}(\alpha)$, the sets Λ_j and P_j be given according to Definition 5.4. With these sets, construct two equivalence relations $\sim_\lambda$ and $\sim_\rho$ on $\mathrm{var}(\alpha)$: For all $k, k' \in \mathrm{var}(\alpha)$*

- *$k \sim_\lambda k'$ if and only if there are $j_1, j_2, \ldots, j_s \in \mathrm{var}(\alpha)$, $s \geq 1$, such that*

  1. $\Lambda_{j_1} \cap \Lambda_{j_2} \neq \emptyset$, $\Lambda_{j_2} \cap \Lambda_{j_3} \neq \emptyset$, $\ldots$, $\Lambda_{j_{s-1}} \cap \Lambda_{j_s} \neq \emptyset$ and

2. $k \in \Lambda_{j_1}$ and $k' \in \Lambda_{j_s}$.

- *$k \sim_\rho k'$ if and only if there are $j_1, j_2, \dots, j_t \in \text{var}(\alpha)$, $t \geq 1$, such that*

  1. $\mathrm{P}_{j_1} \cap \mathrm{P}_{j_2} \neq \emptyset$, $\mathrm{P}_{j_2} \cap \mathrm{P}_{j_3} \neq \emptyset$, ..., $\mathrm{P}_{j_{t-1}} \cap \mathrm{P}_{j_t} \neq \emptyset$ and

  2. $k \in \mathrm{P}_{j_1}$ and $k' \in \mathrm{P}_{j_t}$.

Then, given in arbitrary order, let $\Lambda_1^\sim, \Lambda_2^\sim, \dots, \Lambda_p^\sim \subseteq \text{var}(\alpha)$ be all equivalence classes resulting from $\sim_\lambda$ and $\mathrm{P}_1^\sim, \mathrm{P}_2^\sim, \dots, \mathrm{P}_q^\sim \subseteq \text{var}(\alpha)$ all equivalence classes resulting from $\sim_\rho$.

Consequently, in particular, two variables k, k' belong to the same equivalence class $\Lambda_i^\sim$ if they are in the same set Λ_j (i.e., in terms of Definition 5.5, $\Lambda_j = \Lambda_{j_1} = \Lambda_{j_s}$) or if they are in two different, but non-disjoint sets $\Lambda_j, \Lambda_{j'}$ (i.e. $s = 2$, $j = j_1$, $j' = j_2$ and $\Lambda_{j_1} \cap \Lambda_{j_2} \neq \emptyset$). Obviously, the same holds for the sets P_j and the equivalence classes $\mathrm{P}_i^\sim$. Note that it can be verified easily that $\sim_\lambda$ and $\sim_\rho$ indeed are equivalence relations. For our needs, however, their obvious transitivity is most important.

Since all $\Lambda_i^\sim$ (as well as all $\mathrm{P}_i^\sim$) are pairwise disjoint we can avoid those problems described in Example 5.4. Thus, we can use Definition 5.5 for introducing a well-defined morphism $\sigma_{\text{un},\alpha}$ which depends on the structure of a pattern α and implies that each $\Lambda_i^\sim$ is s-heterogeneous and each $\mathrm{P}_i^\sim$ is p-heterogeneous with respect to $\sigma_{\text{un},\alpha}$:

Definition 5.6 (Morphism $\sigma_{\text{un},\alpha}$). *Let Σ be an alphabet, $\{\mathtt{a}, \mathtt{b}\} \subseteq \Sigma$, and let $\alpha \in \mathbb{N}^+$. Let $\Lambda_1^\sim, \Lambda_2^\sim, \dots, \Lambda_p^\sim$ and $\mathrm{P}_1^\sim, \mathrm{P}_2^\sim, \dots, \mathrm{P}_q^\sim$ be the equivalence classes on $\text{var}(\alpha)$ given in Definition 5.5. Then, for $i \in \{1, 2, \dots, p\}$, $i' \in \{1, 2, \dots, q\}$ and for every $k \in \mathbb{N}$, the morphism $\sigma_{\text{un},\alpha} : \mathbb{N}^* \longrightarrow \Sigma^*$ is defined by*

$$\sigma_{\text{un},\alpha}(k) := \begin{cases} \mathtt{a}\,\mathtt{b}^{3k}\,\mathtt{a}\ \mathtt{a}\,\mathtt{b}^{3k+1}\,\mathtt{a}\ \mathtt{a}\,\mathtt{b}^{3k+2}\,\mathtt{a}\,, & \nexists i : k = \min \Lambda_i^\sim \wedge \nexists i' : k = \min \mathrm{P}_{i'}^\sim, \\ \mathtt{b}\,\mathtt{a}^{3k}\,\mathtt{b}\ \mathtt{a}\,\mathtt{b}^{3k+1}\,\mathtt{a}\ \mathtt{a}\,\mathtt{b}^{3k+2}\,\mathtt{a}\,, & \nexists i : k = \min \Lambda_i^\sim \wedge \exists i' : k = \min \mathrm{P}_{i'}^\sim, \\ \mathtt{a}\,\mathtt{b}^{3k}\,\mathtt{a}\ \mathtt{a}\,\mathtt{b}^{3k+1}\,\mathtt{a}\ \mathtt{b}\,\mathtt{a}^{3k+2}\,\mathtt{b}\,, & \exists i : k = \min \Lambda_i^\sim \wedge \nexists i' : k = \min \mathrm{P}_{i'}^\sim, \\ \mathtt{b}\,\mathtt{a}^{3k}\,\mathtt{b}\ \mathtt{a}\,\mathtt{b}^{3k+1}\,\mathtt{a}\ \mathtt{b}\,\mathtt{a}^{3k+2}\,\mathtt{b}\,, & \exists i : k = \min \Lambda_i^\sim \wedge \exists i' : k = \min \mathrm{P}_{i'}^\sim. \end{cases}$$

Obviously, for every $\alpha \in \mathbb{N}^+$, $\sigma_{\text{un},\alpha}$ is injective. More precisely, for every $\alpha \in \mathbb{N}^+$ and every $i \in \text{var}(\alpha)$, the morphism $\sigma_{\text{un},\alpha}$ maps i onto three distinct segments such that $\sigma_{\text{un},\alpha}$ belongs to the set 3-SEG introduced in Definition 5.1. This aspect is absolutely crucial for the proof of the main lemma of the present chapter (given in Lemma 5.10).

Furthermore, note that, in the conditions in the definition of $\sigma_{\text{un},\alpha}$, we choose the *minimum* variables in the $\Lambda_i^\sim$ and $\mathrm{P}_i^\sim$ merely for the sake of convenience: Actually, the claim for s-heterogeneity of the $\Lambda_i^\sim$ and p-heterogeneity of the $\mathrm{P}_i^\sim$ only requires that in each equivalence class with more than one element there is at least one variable k_1 matching the first case (i.e. $\sigma_{\text{un},\alpha}(k_1) = \mathtt{a}[\dots]\mathtt{a}$) and at least one variable k_2 (our definition changes this to *exactly* one variable k_2, namely the minimum one) matching an appropriate different case (i.e. $\sigma_{\text{un},\alpha}(k_2) \neq \mathtt{a}[\dots]\mathtt{a}$). Of course, such statements on morphic heterogeneity are based on the assumption that each $\Lambda_i^\sim$ and $\mathrm{P}_i^\sim$ contains at least two distinct variables. As to be revealed by the proof of our crucial Lemma 5.10, the morphic primitivity of the patterns under considerations entails that this really holds for all "standard" $\Lambda_i^\sim$ and $\mathrm{P}_i^\sim$. With regard to those equivalence classes $\Lambda_i^\sim$ and $\mathrm{P}_i^\sim$ satisfying $|\Lambda_i^\sim| = 1$ or $|\,\mathrm{P}_i^\sim\,| = 1$, we shall demonstrate that their inevitable morphic homogeneity does not foil our concepts.

We now consider an example pattern, so as to illustrate Definition 5.4, Definition 5.5 and Definition 5.6:

Example 5.6. Let $\alpha := 1\cdot 2\cdot 3\cdot 1\cdot 2\cdot 4\cdot 3\cdot 1\cdot 2\cdot 3\cdot 2\cdot 4$. Note that α is morphically primitive (cf. Theorem 5.2). Then $\Lambda_1 = \{3\}$, $\Lambda_2 = \{1,3\}$, $\Lambda_3 = \{2,4\}$, $\Lambda_4 = \{2\}$ and $\mathrm{P}_1 = \{2\}$, $\mathrm{P}_2 = \{3,4\}$, $\mathrm{P}_3 = \{1,2\}$, $\mathrm{P}_4 = \{3\}$ (see Definition 5.4). According to Definition 5.5, this leads to $\Lambda_1^{\sim} = \{1,3\}$, $\Lambda_2^{\sim} = \{2,4\}$ and $\mathrm{P}_1^{\sim} = \{1,2\}$, $\mathrm{P}_2^{\sim} = \{3,4\}$, and, thus, by Definition 5.6,

$$\begin{aligned}
\sigma_{\mathrm{un},\alpha}(1) &= \mathtt{b\,a^3\,b\;a\,b^4\,a\;b\,a^5\,b}, \\
\sigma_{\mathrm{un},\alpha}(2) &= \mathtt{a\,b^6\,a\;a\,b^7\,a\;b\,a^8\,b}, \\
\sigma_{\mathrm{un},\alpha}(3) &= \mathtt{b\,a^9\,b\;a\,b^{10}\,a\;a\,b^{11}\,a} \text{ and} \\
\sigma_{\mathrm{un},\alpha}(4) &= \mathtt{a\,b^{12}\,a\;a\,b^{13}\,a\;a\,b^{14}\,a}\,.
\end{aligned}$$

Hence, $\Lambda_1^{\sim}$, $\Lambda_2^{\sim}$ are s-heterogeneous and $\Lambda_1^{\sim}$, $\Lambda_2^{\sim}$ are p-heterogeneous with respect to $\sigma_{\mathrm{un},\alpha}$. ◇

It is evident, that $\sigma_{\mathrm{un},\alpha}$ is much more complicated than it is needed for assuring heterogeneity since this property solely depends on the first and the last letter of the morphic images of the variables. In this regard, the subsequent examples explain that, in general, mere heterogeneity of the equivalence classes introduced in Definition 5.5 is not sufficient for guaranteeing unambiguity:

Example 5.7. Let $\alpha := 1\cdot 1\cdot 2\cdot 2\cdot 3\cdot 3$. This pattern is morphically primitive (cf. Theorem 5.2), and $\Lambda_1^{\sim} = \mathrm{var}(\alpha)$, $\mathrm{P}_1^{\sim} = \mathrm{var}(\alpha)$ (cf. Definition 5.5). Then, for $\Lambda_1^{\sim}$ and $\mathrm{P}_1^{\sim}$, the non-injective morphism σ given by $\sigma(1) := \mathtt{b}$ and $\sigma(i) := \mathtt{a}$, $i \in \mathrm{var}(\alpha)\setminus\{1\}$ leads to the desired heterogeneity. Nevertheless, there is a morphism τ with $\tau(\alpha) = \sigma(\alpha)$ and $\tau \neq \sigma$, namely $\tau(1) := \mathtt{b}$, $\tau(2) := \varepsilon$, and $\tau(3) := \mathtt{a\,a}$.

Contrary to what might be suggested by the pattern $1\cdot 1\cdot 2\cdot 2\cdot 3\cdot 3$, our standard morphically primitive example $\alpha = 1\cdot 2\cdot 3\cdot 4\cdot 1\cdot 4\cdot 3\cdot 2$ shows that there are square-free patterns with the same property. With regard to α, $\mathrm{P}_1^{\sim} = \Lambda_1^{\sim} = \{1,3\}$ and $\mathrm{P}_2^{\sim} = \Lambda_2^{\sim} = \{2,4\}$. Thus, the morphism σ with $\sigma(1) := \sigma(2) := \mathtt{a}$ and $\sigma(3) := \sigma(4) := \mathtt{b}$ entails heterogeneity, but, with the morphism $\tau(1) := \tau(2) := \mathtt{a}$, $\tau(3) := \varepsilon$, $\tau(4) := \mathtt{b\,b}$, $\tau(\alpha) = \sigma(\alpha)$. ◇

Example 5.7 reveals that there is a second reason for ambiguity not covered by our considerations on the pattern $\alpha = j\cdot k\cdot j\cdot k'\cdot j'\cdot k\cdot j'\cdot k'$ introduced in the proof of Proposition 5.4 and discussed above Definition 5.4. For the patterns given in Example 5.7, the ambiguity of the regarded morphisms σ results from the fact that there are variables i, i' in these patterns, the images of which under the respective σ commute, i.e. $\sigma(i\cdot i') = \sigma(i'\cdot i)$. It is a classical insight in combinatorics on words that such a phenomenon can be avoided by choosing a morphism where the *primitive root* of $\sigma(i)$ does not equal the primitive root of $\sigma(i')$ (cf. Lothaire [48]), i.e., for some $m, n \in \mathbb{N}$, if $\rho(\sigma(i))$ is the shortest word such that $\rho(\sigma(i))^m = \sigma(i)$ and $\rho(\sigma(i'))$ is the shortest word such that $\rho(\sigma(i'))^n = \sigma(i')$ then $\rho(\sigma(i)) \neq \rho(\sigma(i'))$. Note that the unambiguous image $\sigma(\alpha) = \mathtt{aabbbaababab}$ of $\alpha = 1\cdot 2\cdot 3\cdot 4\cdot 1\cdot 4\cdot 3\cdot 2$ given in Chapter 1 is exactly based on these two requirements: the corresponding morphism σ leads to heterogeneity of all $\Lambda_i^{\sim}$ and $\mathrm{P}_i^{\sim}$, and, for each pair of variables $i, i' \in \mathrm{var}(\alpha)$ with $i \neq i'$, $\rho(\sigma(i)) \neq \rho(\sigma(i'))$.

As shown by this morphism σ and suggested by some simple additional considerations, there are morphisms with much shorter images than $\sigma_{\mathrm{un},\alpha}$ which guarantee heterogeneity of the equivalence classes introduced in Definition 5.5 and non-commutativity of the morphic images of variables (or, more general, of subpatterns). In fact, there is one particular reason for choosing $\sigma_{\mathrm{un},\alpha}$ in 3-SEG so that it maps every variable $j \in \mathbb{N}$ onto three distinct segments: because of this fact we can apply Lemma 5.8, which says that every $\sigma_{\text{3-seg}} \in$ 3-SEG is moderately ambiguous with respect to every morphically primitive α, i. e. the ambiguity of every $\sigma_{\text{3-seg}} \in$ 3-SEG with respect to every morphically primitive α is strongly restricted. Utilising this insight, our next statement, which is the decisive lemma of the present chapter, says that the ambiguity of $\sigma_{\mathrm{un},\alpha} \in$ 3-SEG is even maximally limited with respect to α:

Lemma 5.10. *Let $\alpha \in \mathbb{N}^+$ be morphically primitive. Then, for every morphism τ : $\mathbb{N}^* \longrightarrow \{\mathtt{a}, \mathtt{b}\}^*$ with $\tau(\alpha) = \sigma_{\mathrm{un},\alpha}(\alpha)$ and for every $i \in \mathrm{var}(\alpha)$, $\tau(i) = \sigma_{\mathrm{un},\alpha}(i)$.*

Proof. If $|\mathrm{var}(\alpha)| = 1$ then every morphic image of α is unambiguous, and therefore, in this case, Lemma 5.10 holds trivially. Hence, let $|\mathrm{var}(\alpha)| \geq 2$. Additionally note that, since $\sigma_{\mathrm{un},\alpha} \in$ 3-SEG (cf. Definitions 5.1 and 5.6) and according to Lemma 5.8 and Theorem 5.3, the morphic primitivity of α implies $\sigma_{\mathrm{un},\alpha}$ being moderately ambiguous with respect to α. More precisely, for every $i \in \mathrm{var}(\alpha)$, necessarily $\tau(i) = \ldots \, \mathtt{a}\,\mathtt{b}^{3i+1}\,\mathtt{a} \ldots$.

Now assume to the contrary that there is a morphism τ with $\tau(\alpha) = \sigma_{\mathrm{un},\alpha}(\alpha)$ such that, for every $i \in \mathrm{var}(\alpha)$, $\tau(i) = \ldots \, \mathtt{a}\,\mathtt{b}^{3i+1}\,\mathtt{a} \ldots$ and, for some $i' \in \mathrm{var}(\alpha)$, $\tau(i') \neq \sigma_{\mathrm{un},\alpha}(i')$. W. l. o. g., we may assume that $|\tau(i')| \neq |\sigma_{\mathrm{un},\alpha}(i')|$ since the assumption that, for all $i \in \mathrm{var}(\alpha)$, $|\tau(i)| = |\sigma_{\mathrm{un},\alpha}(i)|$ would imply, again for all $i \in \mathrm{var}(\alpha)$, $\tau(i) = \sigma_{\mathrm{un},\alpha}(i)$. Consequently, because of $|\tau(\alpha)| = |\sigma_{\mathrm{un},\alpha}(\alpha)|$, there is a $j \in \mathrm{var}(\alpha)$ such that $|\tau(j)| > |\sigma_{\mathrm{un},\alpha}(j)|$ and, thus, due to $\tau(j) = \ldots \, \mathtt{a}\,\mathtt{b}^{3j+1}\,\mathtt{a} \ldots$, for some $\mathtt{A}, \mathtt{B}, \mathtt{C} \in \{\mathtt{a}, \mathtt{b}\}$, $\mathtt{A} \neq \mathtt{B}$,

(a) $\tau(j) = \ldots \, \mathtt{C}\ \mathtt{A}\,\mathtt{B}^{3j}\,\mathtt{A}\ \mathtt{a}\,\mathtt{b}^{3j+1}\,\mathtt{a} \ldots$ <u>or</u>

(b) $\tau(j) = \ldots \, \mathtt{a}\,\mathtt{b}^{3j+1}\,\mathtt{a}\ \mathtt{A}\,\mathtt{B}^{3j+2}\,\mathtt{A}\ \mathtt{C} \ldots$.

We restrict the following reasoning to case (a) as an analogous argumentation can be applied to case (b) (using P_j and $\sim_\rho$ instead of Λ_j and $\sim_\lambda$): Note that, due to $|\mathrm{var}(\alpha)| \geq 2$ and the morphic primitivity of α, we have $|\alpha|_j \geq 2$, and therefore the set Λ_j (given in Definition 5.6) of left neighbours of j in α is not empty. With regard to Λ_j we can conclude immediately that, for no $k \in \Lambda_j$, $\tau(k)$ completely contains the left segment of $\sigma_{\mathrm{un},\alpha}(k)$:

Claim 1. For every $k \in \Lambda_j$ and any $\mathtt{C}, \mathtt{D} \in \{\mathtt{a}, \mathtt{b}\}$, $\mathtt{C} \neq \mathtt{D}$, $\tau(k) \neq \ldots \, \mathtt{C}\,\mathtt{D}^{3k+2}\,\mathtt{C} \ldots$.

Proof (Claim 1). As a result of the conditions $\tau(i) = \ldots \, \mathtt{a}\,\mathtt{b}^{3i+1}\,\mathtt{a} \ldots$, $i \in \mathrm{var}(\alpha)$, and $\tau(\alpha) = \sigma_{\mathrm{un},\alpha}(\alpha)$ we can apply Lemma 5.7 which says that, for some $\mathtt{A}, \mathtt{B}, \mathtt{C}, \mathtt{D} \in \{\mathtt{a}, \mathtt{b}\}$, $\mathtt{A} \neq \mathtt{B}$, $\mathtt{C} \neq \mathtt{D}$,

$$\tau(k \cdot j) = \ldots \, \mathtt{a}\ \mathtt{b}^{3k+1}\,\mathtt{a}\ \mathtt{C}\,\mathtt{D}^{3k+2}\boxed{\mathtt{C}}\ \mathtt{A}\,\mathtt{B}^{3j}\,\mathtt{A}\ \mathtt{a}\,\mathtt{b}^{3j+1}\,\mathtt{a} \ldots .$$

Since $\tau(j) = \ldots \boxed{\mathtt{C}}\,\mathtt{A}\,\mathtt{B}^{3j}\,\mathtt{A}\ \mathtt{a}\,\mathtt{b}^{3j+1}\,\mathtt{a} \ldots$ (cf. case (a)) and, because of Lemma 5.8, $\tau(k) = \ldots \, \mathtt{a}\,\mathtt{b}^{3k+1}\,\mathtt{a} \ldots$ it follows immediately from Lemma 5.4 that $\tau(k) \neq \ldots \, \mathtt{C}\,\mathtt{D}^{3k+2}\,\mathtt{C} \ldots$.

□ (Claim 1)

Concerning j and any $k \in \Lambda_j$ we now examine the following cases:

Case 1: $\alpha = j \ldots$.

Then we can directly conclude from Lemma 5.4: $\sigma_{\text{un},\alpha}(\alpha) = \texttt{C}\,\texttt{D}^{3j}\,\texttt{C} \ldots \neq \tau(\alpha) = \ldots \texttt{E}\ \texttt{A}\,\texttt{B}^{3j}\,\texttt{A} \ldots$, with $\texttt{A},\texttt{B},\texttt{C},\texttt{D},\texttt{E} \in \{\texttt{a},\texttt{b}\}$, $\texttt{A} \neq \texttt{B}$, $\texttt{C} \neq \texttt{D}$. This contradicts the condition $\sigma_{\text{un},\alpha}(\alpha) = \tau(\alpha)$.

Case 2: $\alpha = \ldots k$.

Since, from Claim 1 and for $\texttt{A},\texttt{B} \in \{\texttt{a},\texttt{b}\}$, $\texttt{A} \neq \texttt{B}$, $\tau(k) \neq \ldots \texttt{A}\,\texttt{B}^{3k+2}\,\texttt{A}$ we may conclude using Lemma 5.4 that $\tau(\alpha) \neq \ldots \texttt{A}\,\texttt{B}^{3k+2}\,\texttt{A} = \sigma_{\text{un},\alpha}(\alpha)$. This again contradicts the condition $\sigma_{\text{un},\alpha}(\alpha) = \tau(\alpha)$.

Case 3: $\alpha \neq j \ldots$ and $\alpha \neq \ldots k$.

For the equivalence classes $\Lambda_1^\sim, \Lambda_2^\sim, \ldots, \Lambda_p^\sim$ on $\text{var}(\alpha)$ – which are explained in Definition 5.5 – let $\iota \in \{1, 2, \ldots, p\}$ with $\Lambda_j \subseteq \Lambda_\iota^\sim$. Then, since all $\Lambda_1^\sim, \Lambda_2^\sim, \ldots, \Lambda_p^\sim$ are pairwise disjoint, this ι is unique. Now we can collect a number of facts that facilitate the argumentation in Case 3. The first holds as α is morphically primitive:

Claim 2. If $\alpha \neq j \ldots$ and $\alpha \neq \ldots k$ then $|\Lambda_\iota^\sim| \geq 2$.

Proof (Claim 2). If $|\Lambda_j| \geq 2$ then Claim 2 holds by definition. Hence, let $\Lambda_j = \{k\}$. Then, for every occurrence of j in α, the conditions $\alpha \neq j \ldots$ and $|\alpha|_j \geq 2$ lead to $\alpha = \ldots k \cdot j \ldots$. Thus, due to the morphic primitivity of α, there are some $j_1, j_2, \ldots, j_m \in \text{var}(\alpha)$, $m \geq 2$, with $\alpha = \ldots k \cdot j_r \ldots$, $1 \leq r \leq m$, since otherwise we could give a decomposition $\alpha = \beta_0 \gamma_1 \beta_1 \gamma_2 \beta_2 [\ldots] \beta_{|\alpha|_k - 1} \gamma_{|\alpha|_k} \beta_{|\alpha|_k}$ satisfying the conditions of Theorem 5.2 with, for every h, $1 \leq h \leq |\alpha|_k$, $\gamma_h = k \cdot j$. Additionally, because of $\alpha \neq \ldots k$, there must be an $s \in \{1, 2, \ldots, m\}$ and a $\bar{k} \in \text{var}(\alpha)$, $\bar{k} \neq k$, with $\alpha = \ldots \bar{k} \cdot j_s \ldots$, since, otherwise, α would either be morphically imprimitive – by a decomposition regarding each subpattern $k \cdot j_r$ as a γ_h – or start with a j_r, $r \in \{1, 2, \ldots, m\}$, leading to the same argumentation as in Case 1. Consequently, $\Lambda_{j_s} \supseteq \{k, \bar{k}\}$ and therefore $\Lambda_j \subset \{k, \bar{k}\} \subseteq \Lambda_{j_s} \subseteq \Lambda_\iota^\sim$. □ (Claim 2)

Now, for every $\Lambda^\sim$ among $\Lambda_1^\sim, \Lambda_2^\sim, \ldots, \Lambda_p^\sim$, the next fact follows by definition since these equivalence classes are composed by union of non-disjoint sets (cf. Definition 5.6):

Claim 3. If $|\Lambda^\sim| \geq 2$ then, for every $\hat{k} \in \Lambda^\sim$, there is an $\Lambda_{\hat{\jmath}} \subseteq \Lambda^\sim$ with $|\Lambda_{\hat{\jmath}}| \geq 2$ and $\hat{k} \in \Lambda_{\hat{\jmath}}$.

We conclude the list of preliminary claims with the following one, that deals with a crucial phenomenon which is reflected in the transitivity of $\sim_\lambda$, namely the insight that, for no $\hat{k} \in \Lambda_\iota^\sim$, $\tau(\hat{k})$ completely contains the left segment of $\sigma_{\text{un},\alpha}(\hat{k})$. Thus, the subsequent claim extends Claim 1:

Claim 4. For every $\hat{k} \in \Lambda_\iota^\sim$ and any $\texttt{C},\texttt{D} \in \{\texttt{a},\texttt{b}\}$, $\texttt{C} \neq \texttt{D}$, $\tau(\hat{k}) \neq \ldots \texttt{C}\,\texttt{D}^{3\hat{k}+2}\,\texttt{C} \ldots$.

Proof (Claim 4). We stepwise expand the argumentation on Claim 1 to all variables in $\Lambda_\iota^\sim$: With regard to any $\hat{k}' \in \Lambda_j \subseteq \Lambda_\iota^\sim$, Claim 4 directly holds according to Claim 1. We now regard all $\hat{k}'' \in \Lambda_\iota^\sim$ for which there is an $\Lambda_{j'}$ with $\hat{k}', \hat{k}'' \in$

$\Lambda_{j'}$ (recall that $\hat{k}' \in \Lambda_j$). Then – since Claim 4 is satisfied for $\hat{k}'$, i.e. $\tau(\hat{k}') \neq \ldots \mathtt{C}\,\mathtt{D}^{3\hat{k}+2}\,\mathtt{C} \ldots$, and on the other hand, for some $\mathtt{A}, \mathtt{B}, \mathtt{C}, \mathtt{D} \in \{\mathtt{a}, \mathtt{b}\}$, $\mathtt{A} \neq \mathtt{B}$, $\mathtt{C} \neq \mathtt{D}$,

$$\tau(\hat{k}' \cdot j') = \ldots\ \mathtt{a}\ \mathtt{b}^{3\hat{k}'+1}\ \mathtt{a}\ \mathtt{C}\,\mathtt{D}^{3\hat{k}'+2} \boxed{\mathtt{C}}\ \mathtt{A}\,\mathtt{B}^{3j'}\ \mathtt{A}\ \mathtt{a}\,\mathtt{b}^{3j'+1}\ \mathtt{a} \ldots$$

while we have $\tau(i) = \ldots \mathtt{a}\,\mathtt{b}^{3i+1}\,\mathtt{a} \ldots$, $i \in \mathrm{var}(\alpha)$ – Lemma 5.4 allows the conclusion that $\tau(j') = \ldots \boxed{\mathtt{C}}\,\mathtt{A}\,\mathtt{B}^{3j'}\,\mathtt{A}\,\mathtt{a}\,\mathtt{b}^{3j'+1}\,\mathtt{a} \ldots$. Thus, with regard to the variables $\hat{k}'' \in \Lambda_{j'}$ and all subpatterns $\hat{k}'' \cdot j'$ of α, our argumentation is verbatim the same as the reasoning on the variables $k \in \Lambda_j$ and the subpatterns $k \cdot j$ given in Claim 1. Therefore Claim 4 holds for each $\hat{k}'' \in \Lambda_{j'}$ as well. Now we proceed to all $\hat{k}''' \in \Lambda^{\sim}_{\iota}$ for which there is an $\Lambda_{j''}$ with $\hat{k}'', \hat{k}''' \in \Lambda_{j''}$ (recall that $\hat{k}'' \in \Lambda_{j'}$). Then, as Claim 4 is satisfied for $\hat{k}''$, Claim 4 holds for all $\hat{k}'''$ and so on. Consequently, according to the construction of $\Lambda^{\sim}_{\iota}$ (cf. Definition 5.6, which indirectly composes the equivalence classes $\Lambda^{\sim}_1, \Lambda^{\sim}_2, \ldots, \Lambda^{\sim}_p$ by a union operation on non-disjoint sets) Claim 4 holds for every $\hat{k} \in \Lambda^{\sim}_{\iota}$. □ (Claim 4)

We now can conclude our argumentation on Case 3: According to Claim 2, $|\Lambda^{\sim}_{\iota}| \geq 2$. Let $k_{\sharp} := \min \Lambda^{\sim}_{\iota}$; then, due to Claim 3, there is a $j_{\sharp} \in \mathrm{var}(\alpha)$ with $k_{\sharp} \in \Lambda_{j_{\sharp}}$ and $|\Lambda_{j_{\sharp}}| \geq 2$. Consequently, let $\bar{k}_{\sharp} \in \mathrm{var}(\alpha)$ with $k_{\sharp} \neq \bar{k}_{\sharp}$ and $\{k_{\sharp}, \bar{k}_{\sharp}\} \subseteq \Lambda_{j_{\sharp}}$. Then, because of $k_{\sharp} = \min \Lambda^{\sim}_{\iota}$ and $\bar{k}_{\sharp} \in \Lambda^{\sim}_{\iota}$, $\sigma_{\mathrm{un},\alpha}(k_{\sharp}) = \ldots \mathtt{b}$ and $\sigma_{\mathrm{un},\alpha}(\bar{k}_{\sharp}) = \ldots \mathtt{a}$ (cf. Definition 5.6). Referring to $\tau(i) = \ldots \mathtt{a}\,\mathtt{b}^{3i+1}\,\mathtt{a} \ldots$, $i \in \mathrm{var}(\alpha)$, as derived from Lemma 5.8, to Lemma 5.7 and to Claim 4, we can see that $\tau(j_{\sharp})$ contains the left segment of $\sigma_{\mathrm{un},\alpha}(j_{\sharp})$ and additionally, immediately to the left of this segment, the last letter of $\sigma_{\mathrm{un},\alpha}(k_{\sharp})$ and $\sigma_{\mathrm{un},\alpha}(\bar{k}_{\sharp})$. Therefore, the different rightmost letters of $\sigma_{\mathrm{un},\alpha}(k_{\sharp})$ and $\sigma_{\mathrm{un},\alpha}(\bar{k}_{\sharp})$ imply

$$\ldots \boxed{\mathtt{b}}\ \mathtt{A}\,\mathtt{B}^{3j_{\sharp}}\ \mathtt{A}\ \mathtt{a}\,\mathtt{b}^{3j_{\sharp}+1}\ \mathtt{a} \ldots = \tau(j_{\sharp}) = \ldots \boxed{\mathtt{a}}\ \mathtt{A}\,\mathtt{B}^{3j_{\sharp}}\ \mathtt{A}\ \mathtt{a}\,\mathtt{b}^{3j_{\sharp}+1}\ \mathtt{a} \ldots,$$

for some $\mathtt{A}, \mathtt{B} \in \{\mathtt{a}, \mathtt{b}\}$, $\mathtt{A} \neq \mathtt{B}$. This contradicts $\mathtt{a} \neq \mathtt{b}$.

Hence, in each of these three cases we can demonstrate that our assumption – according to which there is a morphism τ with $\tau(\alpha) = \sigma_{\mathrm{un},\alpha}(\alpha)$ such that, for every $i \in \mathrm{var}(\alpha)$, $\tau(i) = \ldots \mathtt{a}\,\mathtt{b}^{3i+1}\,\mathtt{a} \ldots$ and, for some $i' \in \mathrm{var}(\alpha)$, $\tau(i') \neq \sigma_{\mathrm{un},\alpha}(i')$ – is incorrect. As the above cases are exhaustive, this proves the lemma. □

Consequently, we can immediately give the main result of the present chapter:

Theorem 5.7. *Let $\alpha \in \mathbb{N}^+$ be morphically primitive. Then $\sigma_{\mathrm{un},\alpha}$ is unambiguous with respect to α.*

Proof. Directly from Lemma 5.10. □

We wish to emphasise that the morphisms $\sigma_{\mathrm{un},\alpha}$ are mainly meant to prove the *existence* of unambiguous morphisms for morphically primitive patterns. Thus, in a sense, they are "sufficient" for unambiguity, but, in general, they are not "necessary", since there are numerous patterns for which there are much shorter unambiguous morphic images. A large number of related examples can be found in the previous chapters (cf., e. g., Chapter 4). Additionally, in the end of the present chapter, we demonstrate that the morphism σ_{c} given by $\sigma_{\mathrm{c}}(j) := \mathtt{a}\,\mathtt{b}^j$, $j \in \mathbb{N}$, is unambiguous with respect to each pattern in some nontrivial set. Moreover, Freydenberger [20] provides several major sets of patterns with respect to which there are simpler unambiguous morphisms. In particular, Freydenberger characterises those patterns for which the simplified morphism $\sigma \in$ 3-SEG, introduced in Example 5.2 and given by $\sigma(j) := \mathtt{a}\,\mathtt{b}^{3j}\,\mathtt{a}\ \mathtt{a}\,\mathtt{b}^{3j+1}\,\mathtt{a}\ \mathtt{a}\,\mathtt{b}^{3j+2}\,\mathtt{a}$, $j \in \mathbb{N}$, leads to an unambiguous word. Thus, roughly speaking, Freydenberger distinguishes the patterns for which unambiguity can be achieved without heterogeneity of the equivalence classes $\Lambda_i^{\sim}$ and $\mathrm{P}_i^{\sim}$ from those where heterogeneity is necessary. Evidently, for instance due to our favourite example pattern $\alpha := 1 \cdot 2 \cdot 3 \cdot 4 \cdot 1 \cdot 4 \cdot 3 \cdot 2$, the former set is a proper subset of that of all morphically primitive patterns.

In spite of these options to partially improve our approach, we consider our method of constructing the $\sigma_{\mathrm{un},\alpha}$ to be of lasting value as it points out the need of morphic heterogeneity for selected patterns and, with regard to the equivalence relations $\sim_\lambda$ and $\sim_\rho$, it provides a relevant tool for analysing the structure of a pattern. Furthermore, it is unknown whether for every morphically primitive pattern α there exists an unambiguous morphism σ_α such that $|\sigma_\alpha| \ll |\sigma_{\mathrm{un},\alpha}|$, and therefore examples might exist for which $\sigma_{\mathrm{un},\alpha}$, in this sense, is optimal. We further discuss the question of whether our morphisms can be simplified below.

With regard to Question 4.3, we consider it noteworthy that Theorem 5.7 provides a weaker answer for unambiguity than Theorem 5.3 does for moderate ambiguity: while any morphism $\sigma_{\text{3-seg}} \in$ 3-SEG is moderately ambiguous with respect to every morphically primitive pattern, we can merely conclude from our studies in the present chapter that any morphism $\sigma_{\mathrm{un},\alpha}$ is unambiguous with respect to the patterns in any set Π satisfying, for all $\alpha, \beta \in \Pi$, $\sigma_{\mathrm{un},\alpha} = \sigma_{\mathrm{un},\beta}$. Due to the fact that there does not exist a single morphism σ which is unambiguous with respect to every morphically primitive pattern (cf. Proposition 5.4,), it is evident that every such set Π is a proper subset of the set of all morphically primitive patterns.

As briefly mentioned in Chapter 5.3, due to the direct relation between unambiguity and weak unambiguity, we can immediately interpret Theorem 5.7 as a statement on weak unambiguity:

Corollary 5.11. *Let $\alpha \in \mathbb{N}^+$ be morphically primitive. Then $\sigma_{\mathrm{un},\alpha}$ is weakly unambiguous with respect to α.*

Proof. Directly from Proposition 4.1 and Theorem 5.7. □

In addition to this we wish to give two alternative interpretations of Theorem 5.7 which result from analogies between the field of unambiguous morphisms and other prominent research areas in discrete mathematics described above. The first of these statements describes an elementary property of terminal-free E-pattern languages and answers a question posed by Reidenbach [76, 81]:

Corollary 5.12. *For every alphabet Σ, $|\Sigma| \geq 2$, and for every terminal-free E-pattern language $L \subseteq \Sigma^*$, there exists a word $w \in L$ such that w is unambiguous with respect to every succinct pattern $\alpha \in \mathrm{Pat}_{\mathrm{tf}}$ satisfying $L_{\mathrm{E},\Sigma}(\alpha) = L$.*

Proof. Directly from Corollary 5.3 and Theorem 5.7. □

While the connection between properties of pattern languages and the ambiguity of morphisms are widely discussed in the present thesis, our second (and last) alternative version of Theorem 5.7 is based on the relation between unambiguous morphisms and the Post Correspondence Problem briefly described in Chapter 3.1.1:

Corollary 5.13. *Let Σ be an alphabet, $|\Sigma| \geq 2$, and let $\alpha \in \mathbb{N}^+$ be a morphically primitive pattern. Then there exists a morphism $\sigma : \mathbb{N}^* \longrightarrow \Sigma^*$ such that σ is not a solution to the PCP for σ and any morphism τ satisfying $\tau(j) \neq \sigma(j)$ for some $j \in \mathrm{var}(\alpha)$.*

Proof. Directly from Theorem 5.7. □

Returning to the actual subject of our thesis, Theorem 5.7 allows for a terminal answer to Question 4.1:

Corollary 5.14. *Let Σ be an alphabet, $|\Sigma| \geq 2$, and let $\alpha \in \mathbb{N}^+$. Then α is morphically primitive if and only if there exists an injective morphism $\sigma : \mathbb{N}^* \longrightarrow \Sigma^*$ such that $\sigma(\alpha)$ is unambiguous.*

Proof. Directly from Corollary 5.10 and Theorem 5.7. □

Note that the analogous attempt to characterise morphic primitivity by the existence of *weakly* unambiguous injective morphisms fails. This is proven by, e. g., the morphically imprimitive pattern $\alpha := 1 \cdot 2 \cdot 2$, with respect to which the injective morphism σ_c given by $\sigma_c(j) := \mathtt{a}\,\mathtt{b}^j$, $j \in \mathbb{N}$, is weakly unambiguous.

Finally, by Corollary 5.14, we can add another characteristic property of morphically primitive patterns to our list noted in Corollary 5.7:

Corollary 5.15. *Let Σ be an alphabet, $|\Sigma| \geq 2$. Let $\alpha \in \mathbb{N}^+$. Then the following statements are equivalent:*

- *α is morphically primitive.*
- *α is not a fixed point of any nontrivial morphism $\phi : \mathrm{var}(\alpha)^* \longrightarrow \mathrm{var}(\alpha)^*$.*
- *α is succinct on Σ.*
- *Every morphism in 3-SEG is moderately ambiguous with respect to α.*
- *There exists a nonerasing morphism $\sigma : \mathbb{N}^* \longrightarrow \Sigma^*$ which is unambiguous with respect to α.*

Proof. Directly from Corollary 5.7 and Corollary 5.14. □

Corollary 5.14 concludes our main results on the unambiguity of common morphisms. For the remainder of the present chapter, we further discuss some potential directions for giving unambiguous morphisms that are more convenient (i. e., less complex) than the morphisms $\sigma_{\mathrm{un},\alpha}$.

As explained above, from our previous considerations we can merely conclude that the particular shape of any $\sigma_{\mathrm{un},\alpha} \in$ 3-SEG is vital for our *proof technique* validating Lemma 5.10, and hence it is *sufficient* for proving the *only if* part of Corollary 5.14. But, as long as we restrict ourselves to injective morphisms based on a number of segments (and we have no idea whether the proof for Lemma 5.10 and, in particular, Lemma 5.8

can be conducted for a different type of morphisms) then three segments can be considered a true necessity for defining an unambigous morphism. More precisely – just as a morphism mapping the variables onto two distinct segments generally is not moderately ambiguous and cannot be used for defining a telltale for a terminal-free E-pattern language (cf. Propositions 5.1 and Proposition 5.2) – if we omit the inner segment of $\sigma_{\mathrm{un},\alpha}$ then the resulting morphism is not unambiguous. This is shown by the following proposition, the proof of which again requires a more complicated example pattern than those of Propositions 5.1 and 5.2 since, for many patterns, numerous morphisms $\sigma \notin$ 3-SEG leading to heterogeneous equivalence classes $\Lambda_i^{\sim}$ and $\mathrm{P}_i^{\sim}$ (cf. Definition 5.5) are unambiguous:

Proposition 5.5. *For any pattern $\beta \in \mathbb{N}^+$, let $\Lambda_1^{\sim}, \Lambda_2^{\sim}, \ldots, \Lambda_p^{\sim}$ and $\mathrm{P}_1^{\sim}, \mathrm{P}_2^{\sim}, \ldots, \mathrm{P}_q^{\sim}$ be the equivalence classes on $\mathrm{var}(\beta)$ given in Definition 5.5. For $i \in \{1, 2, \ldots, p\}$, $i' \in \{1, 2, \ldots, q\}$ and for every $k \in \mathbb{N}$, let the morphism $\sigma_{\neg\mathrm{un},\beta} : \mathbb{N}^* \longrightarrow \{\mathtt{a}, \mathtt{b}\}^*$ be given by*

$$\sigma_{\neg\mathrm{un},\beta}(k) := \begin{cases} \mathtt{a}\,\mathtt{b}^{2k+1}\,\mathtt{a}\;\;\mathtt{a}\,\mathtt{b}^{2k+2}\,\mathtt{a}\,, & \nexists i : k = \min \Lambda_i^{\sim} \wedge \nexists i' : k = \min \mathrm{P}_{i'}^{\sim}, \\ \mathtt{b}\,\mathtt{a}^{2k+1}\,\mathtt{b}\;\;\mathtt{a}\,\mathtt{b}^{2k+2}\,\mathtt{a}\,, & \nexists i : k = \min \Lambda_i^{\sim} \wedge \exists i' : k = \min \mathrm{P}_{i'}^{\sim}, \\ \mathtt{a}\,\mathtt{b}^{2k+1}\,\mathtt{a}\;\;\mathtt{b}\,\mathtt{a}^{2k+2}\,\mathtt{b}\,, & \exists i : k = \min \Lambda_i^{\sim} \wedge \nexists i' : k = \min \mathrm{P}_{i'}^{\sim}, \\ \mathtt{b}\,\mathtt{a}^{2k+1}\,\mathtt{b}\;\;\mathtt{b}\,\mathtt{a}^{2k+2}\,\mathtt{b}\,, & \exists i : k = \min \Lambda_i^{\sim} \wedge \exists i' : k = \min \mathrm{P}_{i'}^{\sim}. \end{cases}$$

Then there exists a morphically primitive pattern $\alpha \in \mathbb{N}^+$ such that $\sigma_{\neg\mathrm{un},\alpha}$ is not unambiguous with respect to α.

Proof. Let the pattern $\alpha \in \mathbb{N}^+$ be given by

$$\begin{aligned} \alpha \;:=\; & 1\cdot2\cdot3\cdot3\cdot4\cdot1\cdot2\cdot3\cdot3\cdot4\cdot5\cdot2\cdot6\cdot5\cdot7\cdot8\cdot6\cdot8\cdot6\cdot9\cdot7\cdot9\cdot7\cdot \\ & 10\cdot4\cdot11\cdot4\cdot10\cdot12\cdot11\cdot12\cdot3\cdot13\cdot3\cdot13\cdot14\cdot3\cdot2\cdot15\cdot14\cdot3\cdot2\cdot15. \end{aligned}$$

According to Definition 5.5, $\mathrm{P}_1^{\sim} = \Lambda_1^{\sim} = \mathrm{var}(\alpha)$, so that $\sigma_{\neg\mathrm{un},\alpha}(1) = \mathtt{b}\,\mathtt{a}^3\,\mathtt{b}\;\;\mathtt{b}\,\mathtt{a}^4\,\mathtt{b}$ and, for every $k \in \mathbb{N} \setminus \{1\}$, $\sigma_{\neg\mathrm{un},\alpha}(k) = \mathtt{a}\,\mathtt{b}^{2k+1}\,\mathtt{a}\;\;\mathtt{a}\,\mathtt{b}^{2k+2}\,\mathtt{a}$. Furthermore, by Theorem 5.2, the pattern α is morphically primitive. Let now the morphism τ be given by

$$\begin{aligned}
\tau(1) &:= \mathtt{b}\,\mathtt{a}^3\,\mathtt{b}\;\mathtt{b}\,\mathtt{a}^4\,\mathtt{b}\;\mathtt{a}\,\mathtt{b}^5\,\mathtt{a}\;\mathtt{a}\,\mathtt{b}^6\,\mathtt{a}\;\mathtt{a}\,\mathtt{b}^7\,\mathtt{a}\;\mathtt{a}\,\mathtt{b}^8\,\mathtt{a}\;\mathtt{a}\,\mathtt{b}^7\,\mathtt{a}\;\mathtt{a}\,\mathtt{b}^5, \\
\tau(2) &:= \mathtt{b}^3\,\mathtt{a}\;\mathtt{a}\,\mathtt{b}^5, \\
\tau(3) &:= \varepsilon, \\
\tau(4) &:= \mathtt{b}^4\,\mathtt{a}\;\mathtt{a}\,\mathtt{b}^{10}\,\mathtt{a}, \\
\tau(5) &:= \mathtt{a}\,\mathtt{b}^{11}\,\mathtt{a}\;\mathtt{a}\,\mathtt{b}^{12}\,\mathtt{a}\;\mathtt{a}\,\mathtt{b}^2, \\
\tau(6) &:= \mathtt{b}\,\mathtt{a}\;\mathtt{a}\,\mathtt{b}^{13}\,\mathtt{a}\;\mathtt{a}\,\mathtt{b}^{14}\,\mathtt{a}, \\
\tau(7) &:= \mathtt{b}^{13}\,\mathtt{a}\;\mathtt{a}\,\mathtt{b}^{16}\,\mathtt{a}, \\
\tau(8) &:= \mathtt{a}\,\mathtt{b}^{17}\,\mathtt{a}\;\mathtt{a}\,\mathtt{b}^{17}, \\
\tau(9) &:= \mathtt{a}\,\mathtt{b}^{19}\,\mathtt{a}\;\mathtt{a}\,\mathtt{b}^{20}\,\mathtt{a}\;\mathtt{a}\,\mathtt{b}^2, \\
\tau(10) &:= \mathtt{a}\,\mathtt{b}^{21}\,\mathtt{a}\;\mathtt{a}\,\mathtt{b}^{22}\,\mathtt{a}\;\mathtt{a}\,\mathtt{b}^5, \\
\tau(11) &:= \mathtt{a}\,\mathtt{b}^{23}\,\mathtt{a}\;\mathtt{a}\,\mathtt{b}^{24}\,\mathtt{a}\;\mathtt{a}\,\mathtt{b}^5, \\
\tau(12) &:= \mathtt{b}^{20}\,\mathtt{a}\;\mathtt{a}\,\mathtt{b}^{26}\,\mathtt{a}, \\
\tau(13) &:= \mathtt{a}\,\mathtt{b}^7\,\mathtt{a}\;\mathtt{a}\,\mathtt{b}^8\,\mathtt{a}\;\mathtt{a}\,\mathtt{b}^{27}\,\mathtt{a}\;\mathtt{a}\,\mathtt{b}^{28}\,\mathtt{a}, \\
\tau(14) &:= \mathtt{a}\,\mathtt{b}^{29}\,\mathtt{a}\;\mathtt{a}\,\mathtt{b}^{27}, \\
\tau(15) &:= \mathtt{b}^2\,\mathtt{a}\;\mathtt{a}\,\mathtt{b}^8\,\mathtt{a}\;\mathtt{a}\,\mathtt{b}^5\,\mathtt{a}\;\mathtt{a}\,\mathtt{b}^6\,\mathtt{a}\;\mathtt{a}\,\mathtt{b}^{31}\,\mathtt{a}\;\mathtt{a}\,\mathtt{b}^{32}\,\mathtt{a}.
\end{aligned}$$

Then $\tau \neq \sigma_{\neg\mathrm{un},\alpha}$, but nevertheless $\tau(\alpha) = \sigma_{\neg\mathrm{un},\alpha}(\alpha)$:

$$\overbrace{\mathtt{ba^3b\,ba^4b}}^{\sigma_{\neg\mathrm{un},\alpha}(1)}\,\overbrace{\mathtt{ab^5a\,ab^6a}}^{\sigma_{\neg\mathrm{un},\alpha}(2)}\,\overbrace{\mathtt{ab^7a\,ab^8a}}^{\sigma_{\neg\mathrm{un},\alpha}(3)}\,\overbrace{\mathtt{ab^7a\,ab^5\,b^3a}}^{\sigma_{\neg\mathrm{un},\alpha}(3)}\,\overbrace{\mathtt{ab^5\,b^4a\,ab^{10}a}}^{\sigma_{\neg\mathrm{un},\alpha}(4)}$$
$$\underbrace{\mathtt{ba^3b\,ba^4b\,ab^5a\,ab^6a\,ab^7a\,ab^8a\,ab^7a\,ab^5}}_{\tau(1)}\,\underbrace{\mathtt{b^3a\,ab^5}}_{\tau(2)}\,\underbrace{\mathtt{b^4a\,ab^{10}a}}_{\tau(4)}$$

$$\overbrace{\mathtt{ba^3b\,ba^4b}}^{\sigma_{\neg\mathrm{un},\alpha}(1)}\,\overbrace{\mathtt{ab^5a\,ab^6a}}^{\sigma_{\neg\mathrm{un},\alpha}(2)}\,\overbrace{\mathtt{ab^7a\,ab^8a}}^{\sigma_{\neg\mathrm{un},\alpha}(3)}\,\overbrace{\mathtt{ab^7a\,ab^5\,b^3a}}^{\sigma_{\neg\mathrm{un},\alpha}(3)}\,\overbrace{\mathtt{ab^5\,b^4a\,ab^{10}a}}^{\sigma_{\neg\mathrm{un},\alpha}(4)}$$
$$\underbrace{\mathtt{ba^3b\,ba^4b\,ab^5a\,ab^6a\,ab^7a\,ab^8a\,ab^7a\,ab^5}}_{\tau(1)}\,\underbrace{\mathtt{b^3a\,ab^5}}_{\tau(2)}\,\underbrace{\mathtt{b^4a\,ab^{10}a}}_{\tau(4)}$$

$$\overbrace{\mathtt{ab^{11}a\,ab^{12}a}}^{\sigma_{\neg\mathrm{un},\alpha}(5)}\,\overbrace{\mathtt{ab^2\,b^3a\,ab^5\,ba}}^{\sigma_{\neg\mathrm{un},\alpha}(2)}\,\overbrace{\mathtt{ab^{13}a\,ab^{14}a}}^{\sigma_{\neg\mathrm{un},\alpha}(6)}\,\overbrace{\mathtt{ab^{11}a\,ab^{12}a}}^{\sigma_{\neg\mathrm{un},\alpha}(5)}\,\overbrace{\mathtt{ab^2\,b^{13}a\,ab^{16}a}}^{\sigma_{\neg\mathrm{un},\alpha}(7)}$$
$$\underbrace{\mathtt{ab^{11}a\,ab^{12}a\,ab^2}}_{\tau(5)}\,\underbrace{\mathtt{b^3a\,ab^5}}_{\tau(2)}\,\underbrace{\mathtt{ba\,ab^{13}a\,ab^{14}a}}_{\tau(6)}\,\underbrace{\mathtt{ab^{11}a\,ab^{12}a\,ab^2}}_{\tau(5)}\,\underbrace{\mathtt{b^{13}a\,ab^{16}a}}_{\tau(7)}$$

$$\overbrace{\mathtt{ab^{17}a\,ab^{17}\,ba}}^{\sigma_{\neg\mathrm{un},\alpha}(8)}\,\overbrace{\mathtt{ab^{13}a\,ab^{14}a}}^{\sigma_{\neg\mathrm{un},\alpha}(6)}\,\overbrace{\mathtt{ab^{17}a\,ab^{17}\,ba}}^{\sigma_{\neg\mathrm{un},\alpha}(8)}\,\overbrace{\mathtt{ab^{13}a\,ab^{14}a}}^{\sigma_{\neg\mathrm{un},\alpha}(6)}$$
$$\underbrace{\mathtt{ab^{17}a\,ab^{17}}}_{\tau(8)}\,\underbrace{\mathtt{ba\,ab^{13}a\,ab^{14}a}}_{\tau(6)}\,\underbrace{\mathtt{ab^{17}a\,ab^{17}}}_{\tau(8)}\,\underbrace{\mathtt{ba\,ab^{13}a\,ab^{14}a}}_{\tau(6)}$$

$$\overbrace{\mathtt{ab^{19}a\,ab^{20}a}}^{\sigma_{\neg\mathrm{un},\alpha}(9)}\,\overbrace{\mathtt{ab^2\,b^{13}a\,ab^{16}a}}^{\sigma_{\neg\mathrm{un},\alpha}(7)}\,\overbrace{\mathtt{ab^{19}a\,ab^{20}a}}^{\sigma_{\neg\mathrm{un},\alpha}(9)}\,\overbrace{\mathtt{ab^2\,b^{13}a\,ab^{16}a}}^{\sigma_{\neg\mathrm{un},\alpha}(7)}$$
$$\underbrace{\mathtt{ab^{19}a\,ab^{20}a\,ab^2}}_{\tau(9)}\,\underbrace{\mathtt{b^{13}a\,ab^{16}a}}_{\tau(7)}\,\underbrace{\mathtt{ab^{19}a\,ab^{20}a\,ab^2}}_{\tau(9)}\,\underbrace{\mathtt{b^{13}a\,ab^{16}a}}_{\tau(7)}$$

$$\overbrace{\mathtt{ab^{21}a\,ab^{22}a}}^{\sigma_{\neg\mathrm{un},\alpha}(10)}\,\overbrace{\mathtt{ab^5\,b^4a\,ab^{10}a}}^{\sigma_{\neg\mathrm{un},\alpha}(4)}\,\overbrace{\mathtt{ab^{23}a\,ab^{24}a}}^{\sigma_{\neg\mathrm{un},\alpha}(11)}\,\overbrace{\mathtt{ab^5\,b^4a\,ab^{10}a}}^{\sigma_{\neg\mathrm{un},\alpha}(4)}$$
$$\underbrace{\mathtt{ab^{21}a\,ab^{22}a\,ab^5}}_{\tau(10)}\,\underbrace{\mathtt{b^4a\,ab^{10}a}}_{\tau(4)}\,\underbrace{\mathtt{ab^{23}a\,ab^{24}a\,ab^5}}_{\tau(11)}\,\underbrace{\mathtt{b^4a\,ab^{10}a}}_{\tau(4)}$$

$$\overbrace{\mathtt{ab^{21}a\,ab^{22}a}}^{\sigma_{\neg\mathrm{un},\alpha}(10)}\,\overbrace{\mathtt{ab^5\,b^{20}a\,ab^{26}a}}^{\sigma_{\neg\mathrm{un},\alpha}(12)}\,\overbrace{\mathtt{ab^{23}a\,ab^{24}a}}^{\sigma_{\neg\mathrm{un},\alpha}(11)}\,\overbrace{\mathtt{ab^5\,b^{20}a\,ab^{26}a}}^{\sigma_{\neg\mathrm{un},\alpha}(12)}$$
$$\underbrace{\mathtt{ab^{21}a\,ab^{22}a\,ab^5}}_{\tau(10)}\,\underbrace{\mathtt{b^{20}a\,ab^{26}a}}_{\tau(12)}\,\underbrace{\mathtt{ab^{23}a\,ab^{24}a\,ab^5}}_{\tau(11)}\,\underbrace{\mathtt{b^{20}a\,ab^{26}a}}_{\tau(12)}$$

$$\overbrace{\mathtt{ab^7a\,ab^8a}}^{\sigma_{\neg\mathrm{un},\alpha}(3)}\,\overbrace{\mathtt{ab^{27}a\,ab^{28}a}}^{\sigma_{\neg\mathrm{un},\alpha}(13)}\,\overbrace{\mathtt{ab^7a\,ab^8a}}^{\sigma_{\neg\mathrm{un},\alpha}(3)}\,\overbrace{\mathtt{ab^{27}a\,ab^{28}a}}^{\sigma_{\neg\mathrm{un},\alpha}(13)}$$
$$\underbrace{\mathtt{ab^7a\,ab^8a\,ab^{27}a\,ab^{28}a}}_{\tau(13)}\,\underbrace{\mathtt{ab^7a\,ab^8a\,ab^{27}a\,ab^{28}a}}_{\tau(13)}$$

$$\overbrace{\mathtt{ab^{29}a\,ab^{27}\,b^3a}}^{\sigma_{\neg\mathrm{un},\alpha}(14)}\,\overbrace{\mathtt{ab^5\,b^2a\,ab^8a}}^{\sigma_{\neg\mathrm{un},\alpha}(3)}\,\overbrace{\mathtt{ab^5a\,ab^6a}}^{\sigma_{\neg\mathrm{un},\alpha}(2)}\,\overbrace{\mathtt{ab^{31}a\,ab^{32}a}}^{\sigma_{\neg\mathrm{un},\alpha}(15)}$$
$$\underbrace{\mathtt{ab^{29}a\,ab^{27}}}_{\tau(14)}\,\underbrace{\mathtt{b^3a\,ab^5}}_{\tau(2)}\,\underbrace{\mathtt{b^2a\,ab^8a\,ab^5a\,ab^6a\,ab^{31}a\,ab^{32}a}}_{\tau(15)}$$

$$\overbrace{\mathtt{ab^{29}a\,ab^{27}\,b^3a}}^{\sigma_{\neg\mathrm{un},\alpha}(14)}\,\overbrace{\mathtt{ab^5\,b^2a\,ab^8a}}^{\sigma_{\neg\mathrm{un},\alpha}(3)}\,\overbrace{\mathtt{ab^5a\,ab^6a}}^{\sigma_{\neg\mathrm{un},\alpha}(2)}\,\overbrace{\mathtt{ab^{31}a\,ab^{32}a}}^{\sigma_{\neg\mathrm{un},\alpha}(15)}$$
$$\underbrace{\mathtt{ab^{29}a\,ab^{27}}}_{\tau(14)}\,\underbrace{\mathtt{b^3a\,ab^5}}_{\tau(2)}\,\underbrace{\mathtt{b^2a\,ab^8a\,ab^5a\,ab^6a\,ab^{31}a\,ab^{32}a}}_{\tau(15)}$$

Thus, $\sigma_{\neg\mathrm{un},\alpha}$ is not unambiguous with respect to α. □

We do not give a comprehensive explanation on the (admittedly sophisticated) structure of the pattern α introduced in the above proof since Proposition 5.5, within the scope of the present thesis, is just a minor result. Nevertheless, on behalf of the reader who intends to study this pattern in more detail, we wish to give a hint of where to start the analysis. Among the variables in the pattern, the role of the variable 2, concerning the two morphisms $\sigma_{\neg\mathrm{un},\alpha}$ and τ, is the most remarkable one. From our current knowledge, it seems inevitable that any pattern α which can prove Proposition 5.5 must contain such a variable j for which τ can satisfy $\tau(j) = \mathtt{b}^s\,\mathtt{a}\,\mathtt{a}\,\mathtt{b}^t$, $s, t \in \mathbb{N}_0$, (note that, with regard to the variable 2, it is $s = 3$ and $t = 5$) and which has one occurrence in α (in our example it is the third one) where this $\tau(j)$ corresponds to an (inner) subword of an occurrence of $\sigma_{\neg\text{un},\alpha}(j)$ in $\sigma_{\neg\text{un},\alpha}(\alpha)$, whereas, for some other occurrence in α (namely, in our example, the first and the second one), $\tau(j)$ does not generate any subword of an occurrence of $\sigma_{\neg\text{un},\alpha}(j)$. The need for such a variable can be indirectly derived from the proof for Lemma 5.8 and, a bit more precisely, from the relation examined therein between, on the one hand, morphic imprimitivity of patterns and, on the other hand, the existence of variables j for which $\tau(j)$ *of at least one segment of* $\sigma_{3\text{-seg}}(j)$ (and hence, since $\sigma_{\text{un},\alpha} \in$ 3-SEG, of $\sigma_{\text{un},\alpha}(j)$) *does not contain a single letter* (cf. the formal and informal definition of the set X_2 of variables in that proof). If we now examine the variable 2 in the pattern α and the morphism τ as defined in the above proof then we can see that the question of whether $\tau(2)$ shows this property cannot be uniformly solved for all occurrences of 2 in α as the third occurrence suggests an answer in the negative, but, e. g., the second occurrence answers it in the affirmative. Contrary to this, if we consider a morphism mapping the variables onto words that consist of three segments (as done by $\sigma_{\text{un},\alpha}$) then we can unequivocally tell by τ whether $\tau(j)$, $j \in \text{var}(\alpha)$, of at least one segment of $\sigma_{\text{un},\alpha}(j)$ does not contain a single letter. In particular, with regard to any variable j satisfying $\tau(j) = \mathtt{b}^s\,\mathtt{a}\,\mathtt{a}\,\mathtt{b}^t$, $s, t \in \mathbb{N}_0$, it is obvious that there always exists at least one segment of $\sigma_{\text{un},\alpha}(\alpha)$, namely the left one or the right one, which, for any occurrence of j in α, cannot be partially covered by $\tau(j)$. Since the proof for Lemma 5.8 demonstrates that a variable j is redundant in a pattern α if there exists a morphism τ such that, for *any* (and hence, due to the above explanations, for *every*) occurrence of j in α, $\tau(j)$ of one segment of $\sigma_{\text{un},\alpha}(j)$ does not contain a single letter, this major difference between $\sigma_{\text{un},\alpha} \in$ 3-SEG and $\sigma_{\neg\text{un},\alpha}$ is one of the main reasons for the existence of a *morphically primitive* pattern proving Proposition 5.5. As demonstrated by this first step towards an explanation of the structure of α, the comprehension of this pattern requires a deep understanding of the probably most difficult proof of the present thesis, namely that of Lemma 5.8. With regard to a second step, which, e. g., might describe the fact that the moderate ambiguity of morphisms is a necessary condition for the effectiveness of morphic heterogeneity of the equivalence classes $\mathrm{P}_i^{\sim}$ and $\Lambda_i^{\sim}$, we do not give any further details, but we leave it to the interested reader.

Returning to the concrete shape of α, the prefix $1\cdot2\cdot3\cdot3\cdot4\cdot1\cdot2\cdot3\cdot3\cdot4\cdot5\cdot2\cdot6$ therefore is decisive for our argumentation, and the rest of the pattern more or less is canonically (but by no means trivially) constructed with the objective of guaranteeing morphic primitivity and $\mathrm{P}_1^{\sim} = \Lambda_1^{\sim} = \text{var}(\alpha)$. We wish to emphasise that a similar structure (in a less intricate manner) can be found in the patterns verifying Proposition 5.1 and Proposition 5.2. As the three patterns used for the proofs of Propositions 5.1, 5.2 and 5.5 do not result from an exhaustive computer-implemented search[†], but are "handmade", we do not know

[†]Note that, unfortunately, today's (and probably also tomorrow's) computers and the current state of knowledge on algorithms for morphisms do not allow for an extensive use of computers when seeking for such patterns. This is largely caused by the fact that the length of the patterns and words involved in such a construction conflicts, first, with the NP-completeness of the membership problem for E-pattern languages (cf. Proposition 3.8), second, the insight that the number of terminal-free patterns of length $n \in \mathbb{N}$ in canonical form equals the nth Bell number (cf. Proposition 2.5) and, third, the time complexity of the known decision procedures for morphic primitivity, which – similar to the membership test for E-pattern languages – check the existence of morphisms mapping two patterns onto each other.

whether there are shorter patterns which could also verify these propositions, and we do not dare to give any corresponding conjecture. Furthermore, and due to the same reason, we do not know whether, in spite of Proposition 5.5, there exists an unambiguous word w with respect to the pattern α given in the above proof which satisfies $|w| < |\sigma_{\text{un},\alpha}(\alpha)|$.

As stated above, we consider Proposition 5.5 a strong hint that there might be patterns α, with respect to which there is no unambiguous morphic image that is significantly shorter than $\sigma_{\text{un},\alpha}$. Contrary to this, if we do not wish to find an unambiguous morphism for every morphically primitive pattern in $\mathbb{N}^+$, but if we merely consider specific sets $\Pi \subset \mathbb{N}^+$ then, subject to the shape of Π, we can possibly give simpler morphisms which are unambiguous for each pattern in Π (cf. our notes below Theorem 5.7). We now regard an example for this approach; hence, we introduce a set Π for which we can answer Question 4.3 in the affirmative. Since selected unambiguous (or at least moderately ambiguous) words can be used as meaningful parts of a telltale (which from a practical point of view should consist of a preferably small number of preferably short words; see our notes in Chapter 5.2.1) for a terminal-free E-pattern language, such an analysis is particularly worthwile for the application of our results to algorithmic problems for pattern languages.

The set of patterns which we exemplarily wish to examine, is derived by Reidenbach [76] from a set introduced and named by Shinohara [98]):

Definition 5.7 (Terminal-free non-cross pattern). *A pattern $\alpha \in \mathbb{N}^+$ is a* terminal-free non-cross pattern *if, for some $n \in \mathbb{N}$ and $r_1, r_2, \ldots, r_n \geq 2$, it satisfies*

$$\alpha = 1^{r_1} \cdot 2^{r_2} \cdot [\ldots] \cdot n^{r_n}.$$

We designate the set of all terminal-free non-cross patterns as $\Pi_{\text{nc,tf}}$.

Reidenbach [76] claims that the morphism σ_{c} given by $\sigma_{\text{c}}(j) := \mathtt{a}\,\mathtt{b}^j$, $j \in \mathbb{N}$, is unambiguous with respect to every terminal-free non-cross pattern α. Moreover it is shown in [76] that, as a consequence of this, the unary set $T_\alpha := \{\sigma_{\text{c}}(\alpha)\}$ is a telltale for $L_{\text{E},\Sigma}(\alpha)$ with respect to the class of all E-pattern languages over any alphabet Σ, $\Sigma \supseteq \{\mathtt{a}, \mathtt{b}\}$, generated by terminal-free non-cross patterns. However, the proof for the unambiguity of σ_{c} in [76] is incomplete. We therefore now give the full proof:

Theorem 5.8. *Let Σ be an alphabet, $\{\mathtt{a}, \mathtt{b}\} \subseteq \Sigma$, and let $\alpha \in \Pi_{\text{nc,tf}}$. Then the morphism $\sigma_{\text{c}} : \mathbb{N}^* \longrightarrow \Sigma^*$ given by $\sigma_{\text{c}}(j) := \mathtt{a}\,\mathtt{b}^j$, $j \in \mathbb{N}$, is unambiguous with respect to α.*

Proof. According to Definition 5.7, $\alpha = 1^{r_1} \cdot 2^{r_2} \cdot [\ldots] \cdot n^{r_n}$ for an $n \geq 1$ and $r_1, r_2, \ldots, r_n \geq 2$. Then obviously $\sigma_{\text{c}}(\alpha) = (\mathtt{a}\,\mathtt{b})^{r_1}\,(\mathtt{a}\,\mathtt{b}^2)^{r_2}\,[\ldots]\,(\mathtt{a}\,\mathtt{b}^n)^{r_n}$ and, thus, the sequence of the lengths of the uniform subwords over $\mathtt{b}$ in $\sigma_{\text{c}}(\alpha)$ (from the left to the right) is monotonic increasing. We show that $\tau(j) = \sigma_{\text{c}}(j)$ necessarily holds true for every morphism τ with $\tau(\alpha) = \sigma_{\text{c}}(\alpha)$ and for all $j \in \text{var}(\alpha)$.

As a prerequisite of our main line of reasoning we list a number of major restrictions which the conditions $\alpha \in \Pi_{\text{nc,tf}}$ and $\tau(j) = \sigma_{\text{c}}(j)$, for every $j \in \text{var}(\alpha)$, impose on $\tau(j)$:

Claim 1. For every morphism τ with $\tau(\alpha) = \sigma_{\text{c}}(\alpha)$, there does not exist any $j \in \text{var}(\alpha)$ such that $\tau(j)$ satisfies one of the following equations:

$$\tau(j) = \mathtt{a}\,\mathtt{b}^p\ u_1\ \mathtt{a}\,\mathtt{b}^q\ u_2 \tag{5.1}$$

with $p, q \geq 0$, $p \neq q$, $u_1 = \varepsilon$ *or* $u_1 = \mathtt{a}\ v_1$, $u_2 = \varepsilon$ *or* $u_2 = \mathtt{a}\ v_2$, $v_1, v_2 \in \{\mathtt{a}, \mathtt{b}\}^*$, *or*

$$\tau(j) = u_1\ \mathtt{b}^p\,\mathtt{a}\ u_2\ \mathtt{b}^q\,\mathtt{a} \tag{5.2}$$

with $p, q \geq 0$, $p \neq q$, $u_1 = \varepsilon$ *or* $u_1 = v_1\ \mathtt{a}$, $u_2 = \varepsilon$ *or* $u_2 = v_2\ \mathtt{a}$, $v_1, v_2 \in \{\mathtt{a}, \mathtt{b}\}^*$, *or*

$$\tau(j) = \mathtt{b}^p\ u_1\ \mathtt{a}\,\mathtt{b}^j\,\mathtt{a}\ u_2\ \mathtt{b}^q \tag{5.3}$$

with $p, q \geq 0$, $p \neq q$, $u_1 = \varepsilon$ *or* $u_1 = \mathtt{a}\ v_1$, $u_2 = \varepsilon$ *or* $u_2 = v_2\ \mathtt{a}$, $v_1, v_2 \in \{\mathtt{a}, \mathtt{b}\}^*$.

Proof (Claim 1). We regard the equations (5.1) and (5.2) first: Assume to the contrary that there exists a morphism τ with $\tau(\alpha) = \sigma_{\mathrm{c}}(\alpha)$ and τ satisfying (5.1) or (5.2) for a $j' \in \mathrm{var}(\alpha)$. Since $\alpha \in \Pi_{\mathrm{nc,tf}}$, we may conclude that $r_{j'} \geq 2$. Consequently, for some $w_1, w_2, w_3, w_4, w_5 \in \{\mathtt{a}, \mathtt{b}\}^*$, $\tau(\alpha)$ satisfies

$$\tau(\alpha) = w_1\ \mathtt{a}\,\mathtt{b}^p\ w_2\ \mathtt{a}\,\mathtt{b}^q\ w_3\ \mathtt{a}\,\mathtt{b}^p\ w_4\ \mathtt{a}\,\mathtt{b}^q\ w_5$$

or

$$\tau(\alpha) = w_1\ \mathtt{b}^p\,\mathtt{a}\ w_2\ \mathtt{b}^q\,\mathtt{a}\ w_3\ \mathtt{b}^p\,\mathtt{a}\ w_4\ \mathtt{b}^q\,\mathtt{a}\ w_5,$$

where, by definition, the subwords $\mathtt{b}^p$ and $\mathtt{b}^q$ are empty or uniform (the latter holds due to the conditions on u_1, u_2 in equations (5.1) and (5.2)). This implies that $p \neq 0$ and $q \neq 0$ since $\sigma_{\mathrm{c}}(\alpha)$ does not contain any subword $\mathtt{a}\,\mathtt{a}$. As $p \neq q$, $\tau(\alpha)$ contains some uniform subwords over $\mathtt{b}$ such that the sequence of their lengths is not monotonic increasing. This contradicts the shape of $\sigma_{\mathrm{c}}(\alpha)$ described above. Thus, for every τ with $\tau(\alpha) = \sigma_{\mathrm{c}}(\alpha)$ there does not exist any $j \in \mathrm{var}(\alpha)$ such that $\tau(j)$ satisfies (5.1) or (5.2).

Concerning equation (5.3) we argue as follows: Assume to the contrary that there exists a morphism τ with $\tau(\alpha) = \sigma_{\mathrm{c}}(\alpha)$ and τ satisfying (5.3) for a $j' \in \mathrm{var}(\alpha)$. Then, for some $w_1, w_2, w_3, w_4, w_5, w_6 \in \{\mathtt{a}, \mathtt{b}\}^*$, we have

$$\tau(\alpha) = w_1\ \mathtt{b}^p\ w_2\ \mathtt{a}\,\mathtt{b}^{j'}\,\mathtt{a}\ w_3\ \mathtt{b}^q\,\mathtt{b}^p\ w_4\ \mathtt{a}\,\mathtt{b}^{j'}\,\mathtt{a}\ w_5\ \mathtt{b}^q\ w_6,$$

where the subword $\mathtt{b}^q\,\mathtt{b}^p$ is empty or uniform (the latter holds due to the conditions on u_1, u_2 in equation (5.3)). Obviously, $p+q \neq 0$ since $\sigma_{\mathrm{c}}(\alpha)$ does not contain any subword $\mathtt{a}\,\mathtt{a}$. Furthermore, if $p+q \neq j'$ then there are some uniform subwords over $\mathtt{b}$ in $\tau(\alpha)$ such that the sequence of their lengths is not monotonic increasing. Thus, $p+q$ must equal j'. However, with (5.3), $\tau(\alpha)$ contains at least $2r_{j'} - 1$ uniform subwords $\mathtt{b}^{j'}$, whereas there are exactly $r_{j'}$ occurrences of this uniform subword in $\sigma_{\mathrm{c}}(\alpha)$. This contradicts the assumption. □ (Claim 1)

We now proceed with the main part of our proof; for the respective argumentation, recall that $\alpha \in \Pi_{\mathrm{nc,tf}}$ and therefore, for *every* morphism $\sigma : \mathbb{N}^* \longrightarrow \Sigma^*$ and for every $j \in \mathrm{var}(\alpha)$, $\sigma(\alpha) = u_1\ \sigma(j)\sigma(j)\ u_2$, $u_1, u_2 \in \Sigma^*$. Assume to the contrary that there exists a leftmost variable j' with $\tau(j') \neq \mathtt{a}\,\mathtt{b}^{j'}$. Clearly, $\tau(j')$ must not begin with the letter $\mathtt{b}$ and therefore – as $\sigma_{\mathrm{c}}(\alpha)$ does not contain any subword $\mathtt{a}\,\mathtt{a}$ – it must not end with the letter $\mathtt{a}$. Furthermore,

- $\tau(j') \neq (\mathtt{a}\,\mathtt{b}^{j'})^{r'}\ u$ with $r' \geq 2$, $u = \varepsilon$ or $u = \mathtt{a}\ v$, $v \in \{\mathtt{a}, \mathtt{b}\}^*$, and
- $\tau(j') \neq \mathtt{a}\,\mathtt{b}^{j''}$ with $j'' \neq j'$,

as these morphisms would cause uniform subwords over b with the wrong number or with the wrong length with respect to $\sigma_c(\alpha)$. Hence, and due to the assumption and Claim 1, $\tau(j')$ must equal the empty word. Consequently, since the number of uniform subwords over b in $\sigma_c(\alpha)$ equals the length of α, there must be some variable $j'' \in \text{var}(\alpha)$ such that $\tau(j'')$ contains at least two uniform subwords over b with different length. Moreover, we even may assume w. l. o. g. that one of these uniform subwords must be of length j'' as a simple combinatorial consideration reveals that the existence of j'' implies the existence of a variable with such a feature. Since, obviously, $\tau(j'') \neq \mathtt{a}\ u\ \mathtt{a}$ for any $u \in \{\mathtt{a}, \mathtt{b}\}^*$, it must satisfy equation (5.1), (5.2), or (5.3). This contradicts Claim 1. Thus, the assumption is incorrect. □

As mentioned above, several analogous examples of sets $\Pi \subset \mathbb{N}^+$ for which Question 4.3 has an answer in the affirmative are presented by Freydenberger [20] and also by Hechler [29].

Chapter 6

The ambiguity of terminal-preserving morphisms

In the present chapter, we deal with the ambiguity of terminal-preserving morphisms $\sigma : (\mathbb{N} \cup \Sigma)^* \longrightarrow \Sigma^*$ for arbitrary alphabets Σ (though we focus on small alphabet sizes). Hence, we intensively examine Question 4.5 and Question 4.6.

Of course there are various other ways to modify our initial setting, such as, e. g., restricting the domain to finite sets of variables and analysing the ambiguity of morphisms in the adapted context. Our decision to deal with general patterns in $(\mathbb{N} \cup \Sigma)^+$ – which implies an extension of the domain that, in a sense, leads to a restriction imposed on the morphisms (see our corresponding notes in Chapter 2.2) – is particularly motivated by the concept of pattern languages (which in general is based on substitutions, i. e., terminal-preserving morphisms). Thus, in addition to the intrinsic interest involved in the questions examined in the present chapter, we also strive for *useful* insights that allow for drawing conclusions on general E-pattern languages, and our subsequent results shall justify our choice. More precisely, we shall demonstrate that – similar to the relation between the ambiguity of common morphisms and the inferrability of ePAT_{tf} (cf. Chapter 5.2.1) – the ambiguity of terminal-preserving morphisms has a strong impact on the inferrability of and the inclusion and equivalence problem for classes of general E-pattern languages. Still, as to be explained below, our related results are not as comprehensive as those on common morphisms, and many of the stronger statements remain restricted to small terminal alphabets Σ with at most four distinct letters. At first glance, this might be a bit surprising, as the step from terminal-free to general patterns merely adds some symbols to the patterns that rest unchanged by any morphism. So it seems as if the most challenging problems on the ambiguity of morphisms are addressed (and solved) in the previous chapter. However, our studies shall soon reveal that the combinatorial properties of terminal-preserving morphisms show a number of very complex and counter-intuitive traits, and this explains why numerous basic properties of pattern languages, in spite of decades of research, are still unresolved.

As we largely examine the extensibility of our results on the ambiguity of common morphisms $\sigma : \mathbb{N}^* \longrightarrow \Sigma^*$ gained in Chapter 5 (which completely answer our elementary questions on this subject) to terminal-preserving morphisms, we mainly deal with terminal-comprising ambiguity (cf. Definition 4.5) and its *avoidability* (i. e. the possibility of giving, for a fixed pattern $\alpha \in (\mathbb{N} \cup \Sigma)^+$, a substitution $\sigma : (\mathbb{N} \cup \Sigma)^+ \longrightarrow \Sigma^+$ such that σ is not terminal-comprisingly ambiguous with respect to α) in the present chapter.

This focus is justified by the following, immediate observation:

Lemma 6.1. *Let Σ be an alphabet. For an $n \in \mathbb{N}$, let $\alpha := \beta_0 w_1 \beta_1 w_2 \beta_2 [\ldots] \beta_{n-1} w_n \beta_n \in (\mathbb{N} \cup \Sigma)^+ \setminus \mathbb{N}^+$ with $w_j \in \Sigma^+$, $1 \leq j \leq n$, $\beta_0, \beta_n \in \mathbb{N}^*$ and $\beta_k \in \mathbb{N}^+$, $1 \leq k \leq n-1$. Let $\sigma : (\mathbb{N} \cup \Sigma)^* \longrightarrow \Sigma^*$ be a substitution. If*

1. *σ is not unambiguous with respect to α and*
2. *there is no morphism τ such that, for every k, $0 \leq k \leq n$, $\tau(\beta_k) = \sigma(\beta_k)$ and, for some $i \in \mathrm{var}(\alpha)$, $\tau(i) \neq \sigma(i)$*

then σ is terminal-comprisingly ambiguous with respect to α.

Proof. If σ is not unambiguous with respect to α then there exists a morphism τ with $\tau(\alpha) = \sigma(\alpha)$ and, for some $i \in \mathrm{var}(\alpha)$, $\tau(i) \neq \sigma(i)$. For this τ, according to the second condition of the proposition, there is a k', $0 \leq k' \leq n$, such that $\tau(\beta_{k'}) \neq \sigma(\beta_{k'})$. Let k'' be the smallest index satisfying this inequality. Then, since $\tau(\alpha) = \sigma(\alpha)$, $\tau(\beta_{k''})$ is a prefix of $\sigma(\beta_{k''})$ or $\sigma(\beta_{k''})$ is a prefix of $\tau(\beta_{k''})$. As $\tau(\beta_{k''}) \neq \sigma(\beta_{k''})$ this implies that $|\tau(\beta_{k''})| \neq |\sigma(\beta_{k''})|$. Thus, by definition, σ is terminal-comprisingly ambiguous with respect to α. □

Hence, if we can, conversely, give substitutions which are not terminal-comprisingly ambiguous then we can examine their unambiguity by canonically adapting our insights into the ambiguity of common morphisms given in Chapter 5. Additional basic properties of terminal-comprising ambiguity are presented by Propositions 4.5 and 4.6.

With regard to the application of Lemma 6.1 it is important to notice that, roughly speaking, the step from common to terminal-preserving morphisms, in the given technical setting, can only lead to *more* ambiguity, not less. More precisely, if, for any morphism $\sigma : \mathbb{N}^* \longrightarrow \Sigma^*$ and terminal-free patterns $\beta_0, \beta_1, \ldots, \beta_n$, $n \in \mathbb{N}$, there exist exactly $m \in \mathbb{N}_0$ morphisms $\tau : \mathbb{N}^* \longrightarrow \Sigma^*$ satisfying, for every i, $1 \leq i \leq n$, $\tau(\beta_i) = \sigma(\beta_i)$ and, for some $j \in \mathrm{var}(\beta_0) \cup \mathrm{var}(\beta_1) \cup \ldots \cup \mathrm{var}(\beta_n)$, $\tau(j) \neq \sigma(j)$ then, concerning the unique substitutions $\sigma' : (\mathbb{N} \cup \Sigma)^* \longrightarrow \Sigma^*$ with $\sigma'(j) = \sigma(j)$ and $\tau' : (\mathbb{N} \cup \Sigma)^* \longrightarrow \Sigma^*$ with $\tau'(j) = \tau(j)$, $j \in \mathbb{N}$, and any pattern $\alpha \in \mathrm{Pat}_\Sigma$ with, for some $w_1, w_2, \ldots, w_n$, $\alpha := \beta_0\, w_1\, \beta_1\, w_2\, \beta_2 \,[\ldots]\, \beta_{n-1}\, w_n\, \beta_n$, it is necessarily $\tau'(\alpha) = \sigma'(\alpha)$. Hence, the number of substitutions τ' satisfying $\tau'(\alpha) = \sigma'(\alpha)$ can*not* be strictly smaller than m.

In particular, the comprehensive negative result on morphically imprimitive patterns presented in Theorem 5.1 also holds for terminal-preserving morphisms (regardless of their terminal-comprising ambiguity):

Theorem 6.1. *Let Σ be an alphabet. Let $\alpha \in (\mathbb{N} \cup \Sigma)^+$ be morphically imprimitive. Then every nonerasing substitution $\sigma : (\mathbb{N} \cup \Sigma)^* \longrightarrow \Sigma^*$ is substantially ambiguous with respect to α.*

Proof. Since α is morphically imprimitive, according to Proposition 2.1 there exists a variable $j_\sharp \in \mathrm{var}(\alpha)$ such that $j_\sharp$ is redundant. Hence there exist residual-preserving morphisms $\phi, \psi : (\mathbb{N} \cup \Sigma)^* \longrightarrow (\mathbb{N} \cup \Sigma)^*$ such that $\psi(\phi(\alpha)) = \alpha$ and $\psi(\phi(j_\sharp)) = \varepsilon$. We consider any morphism $\sigma : (\mathbb{N} \cup \Sigma)^+ \longrightarrow \Sigma^+$, and we define a morphism $\tau : (\mathbb{N} \cup \Sigma)^* \longrightarrow \Sigma^*$ by $\tau(j) := \sigma(\psi(\phi(j)))$, $j \in \mathrm{var}(\alpha)$. Then $\tau(\alpha) = \sigma(\alpha)$, but $\tau(j_\sharp) = \varepsilon \neq \sigma(j_\sharp)$. Hence, σ is substantially ambiguous for $j_\sharp$ in α. □

Therefore we can largely restrict our subsequent considerations to morphically primitive patterns and ask for the extensibility of the positive results on *morphically primitive* terminal-free patterns presented in Chapter 5 to morphically primitive general patterns.

It can be easily observed that our overall technical goal of avoiding terminal-comprising ambiguity, which is mainly motivated by Lemma 6.1, for certain patterns can be achieved effortlessly. For instance, with respect to the alphabet $\Sigma := \{\mathtt{a}, \mathtt{b}, \mathtt{c}\}$ and the pattern $\alpha := 1 \cdot \mathtt{a} \cdot 2 \cdot \mathtt{b} \cdot 3 \in (\mathbb{N} \cup \Sigma)^+$, we can simply choose a morphic image in $\{\mathtt{c}\}$ for each variable in α. Thus, in general, terminal-comprising ambiguity can be easily avoided if, e. g., for every variable j in a pattern β (for instance, the variable $j := 2$ in our example pattern $\beta := \alpha$), there exists a letter $\mathtt{A}$ (hence, in our example, $\mathtt{A} := \mathtt{c}$) in the alphabet under consideration such that $\mathtt{A}$ differs from all terminal symbols which, for any occurrence j_k of j, are the neighbours of j_k in β (in α, these two neighbours of the only occurrence of 2 are the letters $\mathtt{a}$ and $\mathtt{b}$). Our subsequent analysis of the ambiguity of terminal-preserving morphisms is largely divided into two parts: a first part where we examine the consequences of such a simple avoidability of terminal-comprising ambiguity (cf. Chapter 6.1) and a second part where we deal with arbitrary patterns, thus comprising those patterns where such an option does not exist (cf. Chapters 6.2 and 6.3).

More precisely, the present chapter is organised as follows: After some additional formal explanations to be given below, we introduce in Chapter 6.1 the so-called *quasi-terminal-free* patterns, for which terminal-comprising ambiguity can be effortlessly avoided, and we shall explain that most (but, surprisingly, not all) of our results on common morphisms in Chapter 5 can be canonically extended to terminal-preserving morphisms when restricting their domain to quasi-terminal-free patterns. As most of the corresponding considerations are straightforward and boring, we do not formally extend all the possible previous results to quasi-terminal-free patterns, but we largely focus on statements dealing with moderately ambiguous morphisms. Our choice of moderate ambiguity instead of unambiguity as the main subject of our corresponding considerations is largely for the sake of convenience since there exists a single morphism which is moderately ambiguous with respect to every morphically primitive terminal-free pattern (as demonstrated by Theorem 5.3), whereas any unambiguous morphism must be tailor-made for the pattern under consideration (cf. Proposition 5.4). So, our explanations on moderate ambiguity of terminal-preserving morphisms provide all crucial elements of an extension of insights for terminal-free patterns to quasi-terminal-free patterns, and, furthermore, they require fewer technical details than the corresponding analysis of unambiguity. In Chapter 6.1.1, we apply a particular result for quasi-terminal-free patterns to inductive inference of E-pattern languages. This application shall yield a partial answer to the longstanding unresolved Question 3.2, which nicely contrasts with our learnability result on ePAT_{tf} presented in Chapter 5.2.1. In Chapter 6.2, we turn our attention to general patterns, and we demonstrate that morphically primitive example patterns can be constructed for which our main results on terminal-free patterns do not hold any longer, i. e. there is no nonerasing substitution which is moderately ambiguous with respect to these patterns (and, thus, there does not exist a corresponding unambiguous substitution, either). Chapter 6.2.1 describes the impact of this insight on the equivalence problem for E-pattern languages by disproving the prevailing conjecture on this topic to be found in literature. To this end, however, we have to restrict ourselves to terminal alphabets with at most four letters. Finally, in Chapter 6.3, we take a closer look at the ambiguity of terminal-preserving morphisms when applied to particular morphically *im*primitive patterns, and thereby, in spite of Theorem 6.1, we can observe a counter-intuitive phenomenon on the substantial ambiguity of morphisms for non-redundant variables. This observation results in a stunning insight into the expressive power of injective substitutions, the unexpected implications of which for a prominent proof technique on the inclusion and equivalence problem for classes of E-pattern languages is discussed in Chapter 6.3.1.

Before we can start our analysis of the ambiguity of terminal-preserving morphisms, we have to explain two terms used above and widely discussed in the subsequent chapters. First of all, we have to define the *moderate ambiguity* of terminal-preserving morphisms since, for the sake of convenience, in Chapter 4 only a definition for common morphisms is given (cf. Definition 4.3). Our terminal-preserving counterpart of this definition reads as follows:

Definition 6.1 (Moderate ambiguity of substitutions). *Let Σ be an alphabet, and, for some $n \in \mathbb{N}$, let $\alpha := \beta_0\, w_1\, \beta_1\, w_2\, \beta_2\, [\dots]\, \beta_{n-1}\, w_n\, \beta_n \in (\mathbb{N} \cup \Sigma)^+ \setminus \mathbb{N}^+$ with $w_j \in \Sigma^+$, $1 \leq j \leq n$, $\beta_0, \beta_n \in \mathbb{N}^*$ and $\beta_k \in \mathbb{N}^+$, $1 \leq k \leq n-1$. Let $\mathrm{tf}(\alpha) = i_1 \cdot i_2 \cdot [\dots] \cdot i_m$, $m \in \mathbb{N}$, $i_1, i_2, \dots, i_m \in \mathbb{N}$, and let $w \in \Sigma^+$. Then w is called* moderately ambiguous (with respect to α) *provided that*

1. *w is not terminal-comprisingly ambiguous with respect to α and*
2. *there exist $l_2, l_3, \dots, l_n \in \mathbb{N}$ and $r_1, r_2, \dots, r_{n-1} \in \mathbb{N}$ with*

$$1 \leq r_1 < l_2 \leq r_2 < l_3 \leq r_3 < l_4 \leq \dots \leq r_{n-2} < l_{n-1} \leq r_{n-1} < l_n \leq |w|$$

such that there is at least one substitution $\sigma : (\mathbb{N} \cup \Sigma)^* \longrightarrow \Sigma^*$ with $\sigma(\alpha) = w$ and, for every such substitution σ and for every k, $1 \leq k \leq n-1$,

- $|\,\sigma(i_1 \cdot i_2 \cdot [\dots] \cdot i_k)\,| < l_{k+1}$ *and*
- $|\,\sigma(i_1 \cdot i_2 \cdot [\dots] \cdot i_k)\,| \geq r_k$.

Additionally, we designate each substitution σ with $\sigma(\alpha) = w$ as moderately ambiguous (with respect to α) *as well.*

Consequently, a substitution is said to be moderately ambiguous with respect to a pattern α provided that it is sort of moderately ambiguous with respect to $\mathrm{tf}(\alpha)$ and not terminal-comprisingly ambiguous. Referring to our notes on Propositions 4.2 and 4.8 it can be easily seen that the relation of substantial ambiguity and moderate ambiguity of terminal-preserving morphisms equals that of common morphisms. Hence we can immediately extend Proposition 4.8:

Proposition 6.1. *Let Σ be an alphabet. Let $\alpha \in (\mathbb{N} \cup \Sigma)^+$, and let $\sigma : (\mathbb{N} \cup \Sigma)^* \longrightarrow \Sigma^*$ be a nonerasing substitution. Then $\sigma(\alpha)$ is not moderately ambiguous if it is substantially ambiguous. The converse of this statement does not hold true.*

Analogously, we can adapt Proposition 4.2 which describes the connection between moderately ambiguous and nonerasing morphisms:

Proposition 6.2. *Let Σ be an alphabet, and let $\alpha \in (\mathbb{N} \cup \Sigma)^+$. Let $\sigma : (\mathbb{N} \cup \Sigma)^* \longrightarrow \Sigma^*$ be a substitution with, for some $i \in \text{var}(\alpha)$, $\sigma(i) = \varepsilon$. Then neither σ nor $\sigma(\alpha)$ are moderately ambiguous with respect to α.*

Finally, the equivalent to Proposition 4.3 – explaining that unambiguity is a sufficient condition for moderate ambiguity – also holds for terminal-preserving morphisms:

Proposition 6.3. *Let Σ be an alphabet. Let $\alpha \in (\mathbb{N} \cup \Sigma)^+$, and let $\sigma : (\mathbb{N} \cup \Sigma)^* \longrightarrow \Sigma^*$ be a nonerasing substitution. Then $\sigma(\alpha)$ is moderately ambiguous if it is unambiguous. The converse of this statement does not hold true.*

The second precondition for our subsequent considerations is a formal explanation of the term of an *injective substitution.* Since injectivity is a vital property of many moderately ambiguous and unambiguous morphisms discussed in the previous chapter (and in literature), our quest for an extension of the corresponding results to terminal-preserving morphisms obviously implies the need of giving a proper definition of injectivity. However, it is inappropriate to call a substitution $\sigma : (\mathbb{N} \cup \Sigma)^* \longrightarrow \Sigma^*$ injective if and only if it is injective on $(\mathbb{N} \cup \Sigma)^+$ (cf. our definition of injectivity as presented in Chapter 2.1) as it can be easily observed that *no* substitution σ is injective on this set. This is caused by the fact that, for every $\alpha \in (\mathbb{N} \cup \Sigma)^+ \setminus \Sigma^+$, necessarily $\alpha \neq \sigma(\alpha)$, but $\sigma(\alpha) = \sigma(\sigma(\alpha))$; consequently, there are two different "patterns", namely α and $\sigma(\alpha)$, which σ maps onto the same word – a property which, by definition, implies the non-injectivity of σ on each set comprising α and $\sigma(\alpha)$ (such as $(\mathbb{N} \cup \Sigma)^+$). Therefore we shall henceforth designate a *substitution* σ as *injective* whenever $\sigma : (\mathbb{N} \cup \Sigma)^* \longrightarrow \Sigma^*$ is injective on $\mathbb{N}^+$. As to be demonstrated by the subsequent chapters (and, in particular, by Chapter 6.3), this at first glance questionable redefinition allows for numerous meaningful statements as the properties of terminal-preserving morphisms which are injective on the set of all terminal-free patterns in part strongly differ from those which are not injective on this set.

6.1 The ambiguity of morphisms on quasi-terminal-free patterns

In the present chapter, we examine a particular type of patterns, which goes back to the research by Ohlebusch and Ukkonen [69] on the equivalence problem for E-pattern languages (cf. our corresponding explanations presented in Chapter 6.1.1):

Definition 6.2 (Quasi-terminal-free pattern). *Let Σ be an alphabet, $|\Sigma| \geq 3$. Then a pattern $\alpha \in (\mathbb{N} \cup \Sigma)^+$ is said to be* quasi-terminal-free (on Σ) *provided that $|\Sigma| - |\text{term}(\alpha)| \geq 2$.*

Hence, for each quasi-terminal-free pattern α on an alphabet Σ, there exist at least two distinct letters in Σ which do not occur in α.

Our name for these patterns follows from the fact that we can easily give a multitude of nonerasing substitutions for each of them which are not terminal-comprisingly ambiguous:

Lemma 6.2. *Let Σ, Σ' be alphabets with $\Sigma' \subset \Sigma$ and $|\Sigma| - |\Sigma'| \geq 2$, and let $\alpha \in (\mathbb{N} \cup \Sigma')^+$. Then a substitution $\sigma : (\mathbb{N} \cup \Sigma)^* \longrightarrow \Sigma^*$ is not terminal-comprisingly ambiguous with respect to α if, for every $i \in \text{var}(\alpha)$, it satisfies $\sigma(i) \in (\Sigma \setminus \Sigma')^*$.*

Proof. For some $n \in \mathbb{N}$, let $\alpha := \beta_0 w_1 \beta_1 w_2 \beta_2 [\ldots] \beta_{n-1} w_n \beta_n$ with $w_j \in \Sigma'^+$, $1 \leq j \leq n$, $\beta_0, \beta_n \in \mathbb{N}^*$ and $\beta_k \in \mathbb{N}^+$, $1 \leq k \leq n-1$. Now assume to the contrary that σ is terminal-comprisingly ambiguous with respect to α. Then there exists a morphism τ with $\tau(\alpha) = \sigma(\alpha)$ and, for some k', $|\tau(\beta_{k'})| \neq |\sigma(\beta_{k'})|$. A straightforward combinatorial consideration shows that, among all k' satisfying this inequality, there is at least one index k'' such that $\tau(\beta_{k''}) \in \Sigma^+ \setminus (\Sigma \setminus \Sigma')^+$. Consequently, the number of occurrences of letters $\mathtt{A} \in \Sigma'$ in $\tau(\alpha)$ exceeds that of such letters in $\sigma(\alpha)$. This contradicts $\tau(\alpha) = \sigma(\alpha)$. □

Note that Lemma 6.2 can analogously be given for $|\Sigma| - |\Sigma'| = 1$, but in this case there is only a one letter alphabet left for defining the images of the variables in the pattern α under consideration. As a unary alphabet in general cannot be used to define an unambiguous morphism we prefer to restrict ourselves to the version given above, which exactly corresponds to the quasi-terminal-free patterns. As most of the morphisms examined in Chapter 5 map the patterns onto words over a binary alphabet it seems promising to analyse to which extent their properties are preserved if they are interpreted as substitutions and applied to quasi-terminal-free patterns. To this end, of course, it is necessary that they avoid terminal-comprising ambiguity, and therefore they must map the variables onto words over two suitable letters not occurring in the pattern under considerations (just as suggested by Lemma 6.2). We now introduce a substitution potentially satisfying this requirement; it assigns three distinct segments (as extensively discussed in Chapter 5) to each variable:

Definition 6.3 (Substitution $\sigma_{\text{3-seg},\mathtt{A},\mathtt{B}}$). *Let Σ be an alphabet, $|\Sigma| \geq 2$. Let $\mathtt{A}, \mathtt{B} \in \Sigma$, $\mathtt{A} \neq \mathtt{B}$. Then the substitution $\sigma_{\text{3-seg},\mathtt{A},\mathtt{B}} : (\mathbb{N} \cup \Sigma)^* \longrightarrow \Sigma^*$ is given by, for every $j \in \mathbb{N}$,*

$$\sigma_{\text{3-seg},\mathtt{A},\mathtt{B}}(j) = \mathtt{A}\,\mathtt{B}^{3j}\,\mathtt{A}\ \mathtt{A}\,\mathtt{B}^{3j+1}\,\mathtt{A}\ \mathtt{A}\,\mathtt{B}^{3j+2}\,\mathtt{A}\,.$$

Hence, $\sigma_{\text{3-seg},\mathtt{A},\mathtt{B}}$ is derived from Definition 5.1, but we only consider one particular subtype of the morphisms in 3-SEG, namely those where all three segments have the same left letters and the same right letters. Thus, on the one hand, Definition 6.3 is more restrictive than Definition 5.1, but, on the other hand, we allow to choose arbitrary letters $\mathtt{A}, \mathtt{B} \in \Sigma$ instead of the letters $\mathtt{a}, \mathtt{b}$ fixed by Definition 5.1, so that not each substitution $\sigma_{\text{3-seg},\mathtt{A},\mathtt{B}}$ – when interpreted as a morphism on $\mathbb{N}^*$ – is contained in 3-SEG. While the latter modification is necessary for avoiding terminal-comprising ambiguity (cf. Lemma 6.2), the former is just for the sake of convenience – we claim that our subsequent results can be easily extended such that they cover arbitrary substitutions following the structure proposed by Definition 5.1.

Since Chapter 5.2 provides comprehensive insights into moderate ambiguity and Lemma 6.2 demonstrates how to easily avoid terminal-comprising ambiguity we now can complement Theorem 5.3 by the analogous result for terminal-preserving morphisms on quasi-terminal-free patterns. More precisely, Theorem 5.3 states that, due to the fact that each variable is mapped by any $\sigma \in$ 3-SEG onto a word which consists of three distinct segments, σ is moderately ambiguous with respect to every morphically primitive terminal-free pattern α. We shall demonstrate that the same holds for the substitution

$\sigma_{3\text{-seg},\mathtt{A},\mathtt{B}}$ when applied to arbitrary morphically primitive quasi-terminal-free patterns, provided that suitable letters are chosen for $\mathtt{A}$ and $\mathtt{B}$ so that terminal-comprising ambiguity is avoided:

Theorem 6.2. *Let Σ be an alphabet, $|\Sigma| \geq 3$. Let $\alpha \in (\mathbb{N} \cup \Sigma)^+$ be morphically primitive and quasi-terminal-free on Σ. Let $\mathtt{A}, \mathtt{B} \in \Sigma$ with $\mathtt{A} \neq \mathtt{B}$ and $\mathtt{A}, \mathtt{B} \notin \mathrm{term}(\alpha)$. Then the substitution $\sigma_{3\text{-seg},\mathtt{A},\mathtt{B}}$ is moderately ambiguous with respect to α.*

Proof. Assume to the contrary that $\sigma_{3\text{-seg},\mathtt{A},\mathtt{B}}$ is not moderately ambiguous with respect to α. We show that then there exists a morphically primitive terminal-free pattern γ and a morphism $\sigma_{3\text{-seg}} \in$ 3-SEG (cf. Definition 5.1) such that $\sigma_{3\text{-seg}}$ is not moderately ambiguous with respect to γ. This conclusion evidently contradicts Theorem 5.3.

For some $n \in \mathbb{N}$, let $\alpha := \beta_0 w_1 \beta_1 w_2 \beta_2 [\ldots] \beta_{n-1} w_n \beta_n \in (\mathbb{N} \cup \Sigma)^+ \setminus \mathbb{N}^+$ with $w_j \in \Sigma^+$, $1 \leq j \leq n$, $\beta_0, \beta_n \in \mathbb{N}^*$ and $\beta_k \in \mathbb{N}^+$, $1 \leq k \leq n-1$. Let $\mathrm{tf}(\alpha) = i_1 \cdot i_2 \cdot [\ldots] \cdot i_m$, $m \in \mathbb{N}$, $i_1, i_2, \ldots, i_m \in \mathbb{N}$. According to Lemma 6.2, $\sigma_{3\text{-seg},\mathtt{A},\mathtt{B}}$ is not terminal-comprisingly ambiguous with respect to α. Consequently, by definition, if $\sigma_{3\text{-seg},\mathtt{A},\mathtt{B}}$ is not moderately ambiguous with respect to α then there exists an s, $1 \leq s \leq m-1$, which can*not*, at a time, satisfy the following two requirements:

- there are r_s, l_{s+1}, r_{s+1} with $1 \leq r_s < l_{s+1} \leq r_{s+1} \leq |\,\sigma_{3\text{-seg},\mathtt{A},\mathtt{B}}(\alpha)|$ and
- for each substitution σ with $\sigma(\alpha) = \sigma_{3\text{-seg},\mathtt{A},\mathtt{B}}(\alpha)$, $|\,\sigma(i_1 \cdot i_2 \cdot [\ldots] \cdot i_s)\,| < l_{s+1}$, $|\,\sigma(i_1 \cdot i_2 \cdot [\ldots] \cdot i_s)\,| \geq r_s$ and $|\,\sigma(i_1 \cdot i_2 \cdot [\ldots] \cdot i_{s+1})\,| \geq r_{s+1}$.

Consequently, there are substitutions τ_1, τ_2 with $\tau_1(\alpha) = \tau_2(\alpha) = \sigma_{3\text{-seg},\mathtt{A},\mathtt{B}}(\alpha)$ that do *not* satisfy

$$r_s \leq |\,\tau_1(i_1 \cdot i_2 \cdot [\ldots] \cdot i_s)\,| < l_{s+1} \leq r_{s+1} \leq |\,\tau_2(i_1 \cdot i_2 \cdot [\ldots] \cdot i_{s+1})\,|$$

and therefore

$$|\,\tau_1(i_1 \cdot i_2 \cdot [\ldots] \cdot i_s)\,| \geq |\,\tau_2(i_1 \cdot i_2 \cdot [\ldots] \cdot i_{s+1})\,|.$$

Furthermore – as $\sigma_{3\text{-seg},\mathtt{A},\mathtt{B}}$ is not terminal-comprisingly ambiguous with respect to α – $\tau_1(\mathrm{tf}(\alpha)) = \tau_2(\mathrm{tf}(\alpha)) = \sigma_{3\text{-seg},\mathtt{A},\mathtt{B}}(\mathrm{tf}(\alpha))$. Thus, because of the existence of two different substitutions generating $\sigma_{3\text{-seg},\mathtt{A},\mathtt{B}}(\mathrm{tf}(\alpha))$ which conflict with the requirements of moderate ambiguity for terminal-free patterns as determined by Lemma 4.1, we may immediately conclude that $\sigma_{3\text{-seg},\mathtt{A},\mathtt{B}}$ is not moderately ambiguous with respect to $\mathrm{tf}(\alpha)$, either. However, $\mathrm{tf}(\alpha)$ does not need to be morphically primitive so that this insight does not necessarily contradict Theorem 5.3.

Let $i_\sharp$ be the maximum variable in $\mathrm{var}(\alpha)$. Then we define the pattern $\gamma \in \mathbb{N}^+$ by

$$\begin{aligned}\gamma \;:=\; & \beta_0 \cdot (i_\sharp + 1)^2 \cdot \beta_1 \cdot (i_\sharp + 2)^2 \cdot \beta_2 \cdot [\ldots] \cdot (i_\sharp + n)^2 \cdot \beta_n \cdot (i_\sharp + n + 1)^2 \cdot \\ & \beta_0 \cdot (i_\sharp + n + 2)^2 \cdot \beta_1 \cdot (i_\sharp + n + 3)^2 \cdot \beta_2 \cdot [\ldots] \cdot (i_\sharp + 2n + 1)^2 \cdot \beta_n\end{aligned}$$

where the superscript ^{2} does not refer to the multiplication, but to the concatenation, i. e., for instance, $(i_\sharp + 1)^2$ designates the string $(i_\sharp + 1) \cdot (i_\sharp + 1)$. Referring to Theorem 5.2 and the fact that α is morphically primitive, it can be verified that γ is morphically primitive as well.

We now consider the morphism $\sigma_{3\text{-seg}}$ given by, for every $j \in \mathbb{N}$,

$$\sigma_{3\text{-seg}}(j) := \mathtt{a}\,\mathtt{b}^{3j}\,\mathtt{a}\;\mathtt{a}\,\mathtt{b}^{3j+1}\,\mathtt{a}\;\mathtt{a}\,\mathtt{b}^{3j+2}\,\mathtt{a}$$

and the morphism $\phi : \Sigma^* \longrightarrow \{\mathtt{a},\mathtt{b}\}^*$ given by $\phi(\mathtt{A}) := \mathtt{a}$ and $\phi(\mathtt{B}) := \mathtt{b}$. Evidently, $\sigma_{\text{3-seg}} = \phi \circ \sigma_{\text{3-seg},\mathtt{A},\mathtt{B}}$. Additionally, we define two more morphisms τ_1' and τ_2' by

$$\tau_1'(j) := \begin{cases} \phi(\tau_1(j)) & , \quad j \in \text{var}(\alpha)\,, \\ \sigma_{\text{3-seg}}(j) & , \quad \text{else}\,, \end{cases}$$

and

$$\tau_2'(j) := \begin{cases} \phi(\tau_2(j)) & , \quad j \in \text{var}(\alpha)\,, \\ \sigma_{\text{3-seg}}(j) & , \quad \text{else}\,. \end{cases}$$

Consequently, due to $\tau_1(\text{tf}(\alpha)) = \tau_2(\text{tf}(\alpha)) = \sigma_{\text{3-seg},\mathtt{A},\mathtt{B}}(\text{tf}(\alpha))$, we have $\tau_1'(\gamma) = \tau_2'(\gamma) = \sigma_{\text{3-seg}}(\gamma)$. Additionally – because of the fact that $\sigma_{\text{3-seg},\mathtt{A},\mathtt{B}}$ is not terminal-comprisingly ambiguous with respect to α – for every k, $0 \leq k \leq n$, it is $|\tau_1(\beta_k)| = |\tau_2(\beta_k)|$ so that $|\tau_1'(\beta_k)| = |\tau_2'(\beta_k)|$ and furthermore, for every $i \notin \text{var}(\alpha)$, $|\tau_1'(i)| = |\tau_2'(i)|$. We now write $\gamma =: i_1' \cdot i_2' \cdot [\ldots] \cdot i_{m'}'$ and identify the occurrence $i_{s'}'$ of a variable $i \in \text{var}(\alpha) \cap \text{var}(\gamma)$ which satisfies $i_{s'}' = i_s = i$ and $|\, i_1 \cdot i_2 \cdot [\ldots] \cdot i_{s-1}\,|_i = |\, i_1' \cdot i_2' \cdot [\ldots] \cdot i_{s'-1}'\,|_i$. Then the properties of τ_1' and τ_2' stated above imply that the behaviour of τ_1' and τ_2' on the prefix of γ up to $i_{s'}'$ is equivalent to the behaviour of τ_1 and τ_2 on the prefix of $\text{tf}(\alpha)$ up to i_s, i. e., more precisely,

$$|\, \tau_1'(i_1' \cdot i_2' \cdot [\ldots] \cdot i_{s'}')\,| \geq |\, \tau_2'(i_1' \cdot i_2' \cdot [\ldots] \cdot i_{s'+1}')\,|.$$

Thus, by Lemma 4.1, we can conclude that $\sigma_{\text{3-seg}}$ is not moderately ambiguous with respect to the morphically primitive pattern $\gamma \in \mathbb{N}^+$, a statement which contradicts Theorem 5.3. □

Obviously, the only difficulty in the proof for Theorem 6.2 is caused by the fact that the knowledge on the (un-)ambiguity of a morphism σ with respect to a terminal-free pattern $\alpha := \alpha_1 \cdot \alpha_2 \cdot [\ldots] \cdot \alpha_n$ does not allow for reliable conclusions on the ambiguity of σ with respect to any α_i, $1 \leq i \leq n$ (and vice versa) – a phenomenon which largely results from Corollary 5.1. Therefore, we expect that an even more straightforward proof can be given if one extends our definitions on (un-)ambiguity to *sets* of terminal-free patterns, i. e. if one calls a morphism σ e. g. unambiguous with respect to a set $\Pi := \{\alpha_1, \alpha_2, \ldots, \alpha_n\}$ of terminal-free patterns provided that there exists no morphism τ such that, for every i, $1 \leq i \leq n$, $\tau(\alpha_i) = \sigma(\alpha_i)$, and for some $i' \leq n$ and some $j \in \text{var}(\alpha_{i'})$, $\tau(j) \neq \sigma(j)$. Referring to such a definition (in the present case, of course, dealing with moderate ambiguity), we could shorten the proof for Theorem 6.2 by utilising that if a substitution is not terminal-comprisingly ambiguous with respect to $\alpha := \beta_0\, w_1\, \beta_1\, w_2\, \beta_2\, [\ldots]\, \beta_{n-1}\, w_n\, \beta_n$ then its (un-)ambiguity on $\Pi_\alpha := \{\beta_0, \beta_1, \ldots, \beta_n\}$ can be easily extended to α. With regard to the focus of the present thesis, however, we do not consider such a view sufficiently important, and therefore we do not adapt our definitions as indicated above.

We expect that – after some potential canonical modifications of several definitions – all major results obtained in Chapters 5.2, 5.3 and 5.4 can be extended to quasi-terminal-free patterns. With regard to the negative statements, this follows immediately, and concerning the positive ones, we refer (where necessary) to the method presented in the proof for Theorem 6.2, which is essentially based on the applicability of Lemma 6.2.

As we do not consider these topics overly interesting, we omit the details and proceed with cases where terminal-comprising ambiguity cannot be avoided that easily, and we shall demonstrate that such a less restricted focus immediately leads to severe new difficulties and surprisingly different results. To this end, we first keep our focus on quasi-terminal-free patterns, but we examine an important different type of morphisms which is introduced in Chapter 5 and which has not yet been considered in the present chapter: The "telltale morphism" $\sigma_{\text{tt},\text{3-seg},i}$ introduced in Definition 5.3 and, more general, the telltale candidates given in Definition 5.2. Since the latter so far is only defined as a common morphism, we now introduce its terminal-preserving counterpart:

Definition 6.4 (Telltale candidate substitution). *Let Σ be an alphabet, $|\Sigma| \geq 2$. Then a substitution $\sigma : (\mathbb{N} \cup \Sigma)^* \longrightarrow \Sigma^*$ is said to be a* telltale candidate for a variable $j \in \mathbb{N}$ *(in a pattern $\alpha \in (\mathbb{N} \cup \Sigma)^+$) provided that there exists an $\mathtt{A} \in \Sigma$ such that $|\sigma(j)|_{\mathtt{A}} = 1$ and $|\sigma(\text{tf}(\alpha))|_{\mathtt{A}} = |\alpha|_j$. Additionally, we call σ a* telltale candidate (with respect to α) *if there exists a $j \in \text{var}(\alpha)$ such that σ is a telltale candidate for j.*

We now demonstrate that, when applied to quasi-terminal-free patterns, the avoidability of terminal-comprising ambiguity for telltale candidate substitutions does not have the same friendly properties as that for the other substitutions considered above and derived from our considerations on common morphisms in Chapter 5 (i. e., in particular, the substitution $\sigma_{\text{3-seg},\mathtt{A},\mathtt{B}}$ introduced in Definition 6.3). To this end, rather complex example patterns are required:

Definition 6.5 (Patterns $\alpha^{\text{qtf}}_{\mathtt{abc}}$ and $\alpha^{\text{qtf}}_{\mathtt{abcd}}$). *The patterns $\alpha^{\text{qtf}}_{\mathtt{abc}}$ and $\alpha^{\text{qtf}}_{\mathtt{abcd}}$ are given by*

$$\begin{aligned}\alpha^{\text{qtf}}_{\mathtt{abc}} &:= 1 \cdot \mathtt{a} \cdot 2 \cdot 3^2 \cdot 4^2 \cdot 5^2 \cdot 6^2 \cdot \mathtt{a} \cdot 7 \cdot \mathtt{a} \cdot 2 \cdot 8^2 \cdot 4^2 \cdot 5^2 \cdot 6^2\,,\\ \alpha^{\text{qtf}}_{\mathtt{abcd}} &:= 1 \cdot \mathtt{a} \cdot 2 \cdot 3^2 \cdot 4^2 \cdot 5^2 \cdot 6^2 \cdot 7^2 \cdot 8 \cdot \mathtt{b} \cdot 9 \cdot \mathtt{a} \cdot 2 \cdot 10^2 \cdot 4^2 \cdot 5^2 \cdot 6^2 \cdot 11^2 \cdot 8 \cdot \mathtt{b} \cdot 12\,,\end{aligned}$$

where the superscript ^{2} refers to the concatenation.

By definition, $\alpha^{\text{qtf}}_{\mathtt{abc}}$ is quasi-terminal-free on any alphabet Σ satisfying $|\Sigma| \geq 3$ and $\mathtt{a} \in \Sigma$, and $\alpha^{\text{qtf}}_{\mathtt{abcd}}$ is quasi-terminal-free on any alphabet Σ with $|\Sigma| \geq 4$ and $\mathtt{a}, \mathtt{b} \in \Sigma$.

As we wish to show that the ambiguity of telltale candidates on quasi-terminal-free patterns differs from that of other significant morphisms on these patterns, we first have to state that the patterns $\alpha^{\text{qtf}}_{\mathtt{abc}}$ and $\alpha^{\text{qtf}}_{\mathtt{abcd}}$ introduced in Definition 6.5 are not morphically imprimitive:

Proposition 6.4. *The patterns $\alpha^{\text{qtf}}_{\mathtt{abc}}$ and $\alpha^{\text{qtf}}_{\mathtt{abcd}}$ are morphically primitive.*

Proof. Our argumentation is based on the characterisation of morphic primitivity given in Proposition 2.1, according to which a pattern is morphically primitive if and only if it does not contain a redundant variable.

We write $\alpha^{\text{qtf}}_{\mathtt{abc}} =: 1 \cdot \mathtt{a} \cdot 2 \cdot \beta_1 \cdot \mathtt{a} \cdot 7 \cdot \mathtt{a} \cdot 2 \cdot \beta_2$. The variables $1, 7$ are not redundant in $\alpha^{\text{qtf}}_{\mathtt{abc}}$ because they do not have any neighbouring variables, and 2 is not redundant as in both β_1 and β_2 each variable occurs twice. Finally, assume to the contrary that a variable i in β_1 or β_2 is redundant. Then, for the morphism ϕ given by $\phi(i) := \varepsilon$ and $\phi(j) := j$, $j \in (\text{var}(\beta_1) \cup \text{var}(\beta_2)) \setminus \{i\}$, there exists a morphism ψ such that $\psi(\phi(2 \cdot \beta_1)) = 2 \cdot \beta_1$ and $\psi(\phi(2 \cdot \beta_2)) = 2 \cdot \beta_2$. Hence, according to Lemma 5.1, there must be an anchor variable in β_1 and in β_2, or 2 must be an anchor variable with respect to ϕ and ψ. The latter assumption implies that $\psi(2)$ has to satisfy both $\psi(\phi(2)) = 2 \cdot 3 \ldots$ and $\psi(\phi(2)) = 2 \cdot 8 \ldots$, which contradicts the fact that ϕ and ψ are morphisms. Thus β_1 and β_2 must contain an anchor variable each. However, since all the variables in these subpatterns have squared occurrences, this implies that, e. g., $\psi(\phi(\beta_1)) = \ldots j' \ldots j'' \ldots j' \ldots j'' \ldots$ with $j' \neq j''$. As β_1 does not have a subpattern like this, we may conclude that $\psi(\phi(\beta_1)) \neq \beta_1$. This is a contradiction. Consequently, $\alpha^{\text{qtf}}_{\mathtt{abc}}$ does not contain any redundant variable, and therefore, according to Proposition 2.1, it is morphically primitive.

The analogous statement on $\alpha^{\text{qtf}}_{\mathtt{abcd}}$ follows from a straightforward extension of our argumentation on $\alpha^{\text{qtf}}_{\mathtt{abc}}$. □

We now present our crucial argument on $\alpha^{\text{qtf}}_{\mathtt{abc}}$ and $\alpha^{\text{qtf}}_{\mathtt{abcd}}$, which says that these patterns contain a particular variable (namely $6 \in \text{var}(\alpha^{\text{qtf}}_{\mathtt{abc}}) \cap \text{var}(\alpha^{\text{qtf}}_{\mathtt{abcd}})$) for which every corresponding telltale candidate is substantially ambiguous (concerning substantial ambiguity, see Definition 4.6 and the related remarks). We first informally explain this phenomenon with respect to $\alpha^{\text{qtf}}_{\mathtt{abc}}$, so as to illustrate our formal reasoning below: The core of $\alpha^{\text{qtf}}_{\mathtt{abc}}$ is the pattern $\alpha_0 := 4 \cdot 4 \cdot 5 \cdot 5 \cdot 6 \cdot 6$, that is a renaming of the pattern $\alpha^{\text{tf}}_{\mathtt{ab}} = 1 \cdot 1 \cdot 2 \cdot 2 \cdot 3 \cdot 3$ introduced by Reidenbach [76, 81] for verifying the non-learnability of the class of terminal-free E-pattern languages over binary alphabets (cf. Theorem 3.15). Concerning α_0, every telltale candidate σ_1 for $6 \in \text{var}(\alpha_0)$ over a unary or binary alphabet (i. e., $|\,\text{term}(\sigma_1(\alpha_0))| \leq 2$) is substantially ambiguous for 6;[†] this can be easily verified by noting that in this case it is $|\,\text{term}(\sigma_1(4 \cdot 4 \cdot 5 \cdot 5)| \leq 1$, and therefore we can define a morphism τ_1 satisfying $\tau_1(\alpha_0) = \sigma_1(\alpha_0)$ by $\tau_1(4) := \sigma_1(4 \cdot 5)$, $\tau_1(5) := \sigma_1(6)$ and $\tau_1(6) := \varepsilon$. Hence, we only have to examine the case that any substitution σ_2 is a telltale candidate for 6 which, furthermore, satisfies $|\,\text{term}(\sigma_2(\alpha_0))| = 3$. We now turn our attention to the pattern $\alpha_1 := 1 \cdot \mathtt{a} \cdot 2 \cdot 3 \cdot 3 \cdot \alpha_0$, which evidently contains α_0 and, moreover, is a prefix of $\alpha^{\text{qtf}}_{\mathtt{abc}}$. As σ_2 maps α_0 onto a word over the three letters in Σ, $\sigma_2(\alpha_0)$ includes the letter a. However, this implies that σ_2 is terminal-comprisingly ambiguous with respect to α_1, i. e., more precisely, we can define a substitution τ_2 by $\tau_2(\alpha_0) := \varepsilon$, $\tau_2(1) := \sigma_2(1 \cdot \mathtt{a} \cdot 2 \cdot 3 \cdot 3)[\sigma_2(\alpha_0)/\,\mathtt{a}]$ and $\tau_2(2) := [\mathtt{a}\,\backslash \sigma_2(\alpha_0)]$, which implies $\tau_2(\alpha_1) = \sigma_2(\alpha_1)$ and $\tau_2(6) = \varepsilon$. Thus, we can conclude that each telltale candidate for 6 in α_1 is substantially ambiguous for 6 in α_1. This, in turn, is by no means surprising since, due to the single occurrence of the variable 2, α_1 is morphically imprimitive and 6 is a redundant variable in this pattern. Consequently, we have to add some suffix to α_1 which leads to a morphically primitive pattern – i. e. it must at least contain a second occurrence of the variable 2 – and nevertheless preserves the terminal-comprising ambiguity of σ_2 (this of course is the most challenging part of the construction of $\alpha^{\text{qtf}}_{\mathtt{abc}}$ and can only be done because of the special shape of α_1). This suffix reads $\alpha_1' := 7 \cdot \mathtt{a} \cdot 2 \cdot 8 \cdot 8 \cdot \alpha_0$ and, consequently, $\alpha^{\text{qtf}}_{\mathtt{abc}} := \alpha_1 \cdot \mathtt{a} \cdot \alpha_1'$. Note that the inner letter a separating α_1 and α_1' is not required for our argumentation, but it is merely meant to keep the structure of $\alpha^{\text{qtf}}_{\mathtt{abc}}$ as clearly arranged as possible. Our above proof for Proposition 6.4 demonstrates that $\alpha^{\text{qtf}}_{\mathtt{abc}}$ now indeed is morphically primitive, and our subsequent formal reasoning explains why any telltale candidate σ_2 mapping α_0 onto a word which contains the letter a still is terminal-comprisingly (and, additionally, substantially) ambiguous.

In fact our above explanations, due to presentational reasons, slightly differ from the formal reasoning given below as the former is mainly meant to describe the structure of $\alpha^{\text{qtf}}_{\mathtt{abc}}$, whereas the latter wishes to emphasise the importance of terminal-comprising

[†]Note that this fact is a decisive aspect for the reasoning on Theorem 3.15 as presented by Reidenbach [76, 81].

ambiguity. More precisely, with respect to any telltale candidate σ for 6, whenever σ allows both types of unambiguity, i. e. $\sigma(\alpha_0)$ contains the letter $\mathtt{a}$ *and* σ maps α_0 onto a word which contains at most two distinct letters, then we choose a morphism τ with $\tau(\alpha^{\mathrm{qtf}}_{\mathtt{abc}}) = \sigma(\alpha^{\mathrm{qtf}}_{\mathtt{abc}})$ which utilises the terminal-comprising ambiguity of σ; hence, τ follows the above definition of τ_2 rather than that of τ_1. Consequently, our main statement and reasoning on the substantial ambiguity of telltale candidates with respect to $\alpha^{\mathrm{qtf}}_{\mathtt{abc}}$ formally read as follows:

Lemma 6.3. *Let* $\Sigma := \{\mathtt{a}, \mathtt{b}, \mathtt{c}\}$ *be an alphabet. If a substitution* $\sigma : (\mathbb{N} \cup \Sigma)^* \longrightarrow \Sigma^*$ *is a telltale candidate for the variable* 6 *in* $\alpha^{\mathrm{qtf}}_{\mathtt{abc}}$ *then* σ *is substantially ambiguous for* 6 *in* $\alpha^{\mathrm{qtf}}_{\mathtt{abc}}$.

Proof. If σ is a telltale candidate for 6 then, by Definition 6.4, there exists an $\mathtt{A} \in \Sigma$ such that $|\sigma(6)|_{\mathtt{A}} = 1$ and, for each $j \in \mathrm{var}(\alpha^{\mathrm{qtf}}_{\mathtt{abc}}) \setminus \{6\}$, $|\sigma(j)|_{\mathtt{A}} = 0$. We now show that this implies the existence of a substitution $\tau : (\mathbb{N} \cup \Sigma)^* \longrightarrow \Sigma^*$ with $\tau(\alpha^{\mathrm{qtf}}_{\mathtt{abc}}) = \sigma(\alpha^{\mathrm{qtf}}_{\mathtt{abc}})$ and $\tau(6) = \varepsilon$. To this end, we consider the following cases:

Case 1: $\mathtt{A} = \mathtt{a}$.

Define
$$\begin{aligned}
\tau(1) &:= \sigma(1 \cdot \mathtt{a} \cdot 2 \cdot 3^2 \cdot 4^2 \cdot 5^2)\, [\sigma(6^2)/\, \mathtt{a}]\,,\\
\tau(2) &:= [\mathtt{a} \backslash \sigma(6^2)]\,,\\
\tau(7) &:= \sigma(7 \cdot \mathtt{a} \cdot 2 \cdot 8^2 \cdot 4^2 \cdot 5^2)\, [\sigma(6^2)/\, \mathtt{a}]\,,\\
\tau(j) &:= \varepsilon,\ j \in \{3, 4, 5, 6, 8\}\,.
\end{aligned}$$

Case 2: $\mathtt{A} = \mathtt{b}$ and therefore, due to the conditions on σ, $\sigma(4^2 \cdot 5^2) \in \{\mathtt{a}, \mathtt{c}\}^*$.

Case 2.1: $\sigma(4^2 \cdot 5^2) \in \{\mathtt{a}\}^* \cup \{\mathtt{c}\}^*$.

Define
$$\begin{aligned}
\tau(4) &:= \sigma(4 \cdot 5)\,,\\
\tau(5) &:= \sigma_i(6)\,,\\
\tau(6) &:= \varepsilon\,,\\
\tau(j) &:= \sigma(j),\ j \in \{1, 2, 3, 7, 8\}\,.
\end{aligned}$$

Case 2.2: $\sigma(4^2 \cdot 5^2) \in \{\mathtt{a}, \mathtt{c}\}^+ \setminus (\{\mathtt{a}\}^+ \cup \{\mathtt{c}\}^+)$.

Define
$$\begin{aligned}
\tau(1) &:= \sigma(1 \cdot \mathtt{a} \cdot 2 \cdot 3^2)\, [\sigma(4^2 \cdot 5^2)/\, \mathtt{a}]\,,\\
\tau(2) &:= [\mathtt{a} \backslash \sigma(4^2 \cdot 5^2 \cdot 6^2)]\,,\\
\tau(7) &:= \sigma(7 \cdot \mathtt{a} \cdot 2 \cdot 8^2)\, [\sigma(4^2 \cdot 5^2)/\, \mathtt{a}]\,,\\
\tau(j) &:= \varepsilon,\ j \in \{3, 4, 5, 6, 8\}\,.
\end{aligned}$$

Case 3: $\mathtt{A} = \mathtt{c}$ and therefore, due to the conditions on σ, $\sigma(4^2 \cdot 5^2) \in \{\mathtt{a}, \mathtt{b}\}^*$.

Case 3.1: $\sigma(4^2 \cdot 5^2) \in \{\mathtt{a}\}^* \cup \{\mathtt{b}\}^*$.

The definition of τ is identical to that in Case 2.1.

Case 3.2: $\sigma(4^2 \cdot 5^2) \in \{\mathtt{a}, \mathtt{b}\}^+ \setminus (\{\mathtt{a}\}^+ \cup \{\mathtt{b}\}^+)$.

The definition of τ is identical to that in Case 2.2.

Hence, with regard to $|\Sigma| = 3$ and $\alpha^{\mathrm{qtf}}_{\mathtt{abc}}$, if σ is a telltale candidate for 6 then we can give a substitution τ with $\tau(6) = \varepsilon$ and $\tau(\alpha^{\mathrm{qtf}}_{\mathtt{abc}}) = \sigma(\alpha^{\mathrm{qtf}}_{\mathtt{abc}})$. The verification of the equality of $\tau(\alpha^{\mathrm{qtf}}_{\mathtt{abc}})$ and $\sigma(\alpha^{\mathrm{qtf}}_{\mathtt{abc}})$ is straightforward. □

Note that, in the above proof, the Cases 1, 2.2 and 3.2 are based on the terminal-comprising ambiguity of the telltale candidate σ.

We now prove the analogous result for $|\Sigma| = 4$. To this end, the corresponding example pattern $\alpha^{\mathrm{qtf}}_{\mathtt{abcd}}$ is a canonical extension of $\alpha^{\mathrm{qtf}}_{\mathtt{abc}}$ (i. e., in particular, $\alpha^{\mathrm{qtf}}_{\mathtt{abcd}}$ also contains the subpattern α_0 introduced above in our explanation on the proof of Lemma 6.3), so that any morphism σ is terminal-comprisingly ambiguous with respect to $\alpha^{\mathrm{qtf}}_{\mathtt{abcd}}$ whenever $\sigma(\alpha_0)$ contains the letter a *or* the letter b. This is implemented by adding the suffix $7 \cdot 7 \cdot 8 \cdot \mathtt{b} \cdot 9$ to α_1, leading to the prefix $\alpha_2 := \alpha_1 \cdot 7 \cdot 7 \cdot 8 \cdot \mathtt{b} \cdot 9$ of $\alpha^{\mathrm{qtf}}_{\mathtt{abcd}}$. Due to the need of giving a morphically primitive pattern, the structure of α_2 must occur twice in $\alpha^{\mathrm{qtf}}_{\mathtt{abcd}}$ – after a renaming of certain variables. Therefore we introduce the pattern $\alpha_2' := 9 \cdot \mathtt{a} \cdot 2 \cdot 10 \cdot 10 \cdot \alpha_0 \cdot 11 \cdot 11 \cdot 8 \cdot \mathtt{b} \cdot 12$, which is the suffix of $\alpha^{\mathrm{qtf}}_{\mathtt{abcd}}$. As only a single occurrence of the variable 9 is appropriate for guaranteeing terminal-comprising ambiguity, the prefix α_2 and the suffix α_2' of $\alpha^{\mathrm{qtf}}_{\mathtt{abc}}$ are overlapping by 9. Apart from these modifications, the underlying ideas of $\alpha^{\mathrm{qtf}}_{\mathtt{abcd}}$ largely equal that of $\alpha^{\mathrm{qtf}}_{\mathtt{abc}}$.

Hence, concerning alphabet size 4, the ambiguity of telltale candidates with respect to $\alpha^{\mathrm{qtf}}_{\mathtt{abcd}}$ is equivalent to that of telltale candidates with respect to $\alpha^{\mathrm{qtf}}_{\mathtt{abc}}$ (over an alphabet containing three letters):

Lemma 6.4. *Let $\Sigma := \{\mathtt{a}, \mathtt{b}, \mathtt{c}, \mathtt{d}\}$ be an alphabet. If a substitution $\sigma : (\mathbb{N} \cup \Sigma)^* \longrightarrow \Sigma^*$ is a telltale candidate for the variable 6 in $\alpha^{\mathrm{qtf}}_{\mathtt{abcd}}$ then σ is substantially ambiguous for 6 in $\alpha^{\mathrm{qtf}}_{\mathtt{abcd}}$.*

Proof. We adapt our argumentation on $\alpha^{\mathrm{qtf}}_{\mathtt{abc}}$ as given in the proof for Lemma 6.3 to the shape of $\alpha^{\mathrm{qtf}}_{\mathtt{abcd}}$. Hence, let σ be any telltale candidate for 6 which means that, by Definition 6.4, there exists an $\mathtt{A} \in \Sigma$ such that $|\sigma(6)|_{\mathtt{A}} = 1$ and $|\sigma(j)|_{\mathtt{A}} = 0$, $j \in \mathrm{var}(\alpha^{\mathrm{qtf}}_{\mathtt{abc}}) \setminus \{6\}$. We use this condition for constructing a morphism τ with $\tau(\alpha^{\mathrm{qtf}}_{\mathtt{abcd}}) = \sigma(\alpha^{\mathrm{qtf}}_{\mathtt{abcd}})$ and $\tau(6) = \varepsilon$. To this end, we consider the following cases:

<u>Case 1:</u> $\mathtt{A} = \mathtt{a}$.

Define
$$\begin{aligned}
\tau(1) &:= \sigma(1 \cdot \mathtt{a} \cdot 2 \cdot 3^2 \cdot 4^2 \cdot 5^2)\,[\sigma(6^2)/\,\mathtt{a}]\,,\\
\tau(2) &:= [\mathtt{a} \setminus \sigma(6^2)]\,,\\
\tau(9) &:= \sigma(9 \cdot \mathtt{a} \cdot 2 \cdot 10^2 \cdot 4^2 \cdot 5^2)\,[\sigma(6^2)/\,\mathtt{a}]\,,\\
\tau(j) &:= \varepsilon,\ j \in \{3,4,5,6,10\}\,,\\
\tau(j) &:= \sigma(j),\ j \in \{7,8,11,12\}\,.
\end{aligned}$$

<u>Case 2:</u> $\mathtt{A} = \mathtt{b}$.

Define
$$\begin{aligned}
\tau(8) &:= \sigma(4^2 \cdot 5^2)\,[\sigma(6^2)/\,\mathtt{b}]\,,\\
\tau(9) &:= [\mathtt{b} \setminus \sigma(6^2 \cdot 7^2 \cdot 8 \cdot \mathtt{b} \cdot 9)]\,,\\
\tau(12) &:= [\mathtt{b} \setminus \sigma(6^2 \cdot 11^2 \cdot 8 \cdot \mathtt{b} \cdot 12)]\,,\\
\tau(j) &:= \varepsilon,\ j \in \{4,5,6,7,11\}\,,\\
\tau(j) &:= \sigma(j),\ j \in \{1,2,3,10\}\,.
\end{aligned}$$

Case 3: $\mathtt{A} = \mathtt{c}$ and therefore, due to the conditions on σ, $\sigma(4^2 \cdot 5^2) \in \{\mathtt{a}, \mathtt{b}, \mathtt{d}\}^*$.

Case 3.1: $\sigma(4^2 \cdot 5^2) \in \{\mathtt{a}\}^* \cup \{\mathtt{b}\}^* \cup \{\mathtt{d}\}^*$.

$$\begin{aligned} \text{Define} \quad \tau(4) &:= \sigma(4 \cdot 5)\,, \\ \tau(5) &:= \sigma(6)\,, \\ \tau(6) &:= \varepsilon\,, \\ \tau(j) &:= \sigma(j),\ j \in \{1,2,3,7,8,9,10,11,12\}\,. \end{aligned}$$

Case 3.2: $\sigma(4^2 \cdot 5^2) \in \{\mathtt{a}, \mathtt{d}\}^+ \setminus (\{\mathtt{a}\}^+ \cup \{\mathtt{d}\}^+)$.

$$\begin{aligned} \text{Define} \quad \tau(1) &:= \sigma(1 \cdot \mathtt{a} \cdot 2 \cdot 3^2)\,[\sigma(4^2 \cdot 5^2)/\,\mathtt{a}]\,, \\ \tau(2) &:= [\mathtt{a} \backslash \sigma(4^2 \cdot 5^2 \cdot 6^2)]\,, \\ \tau(9) &:= \sigma(9 \cdot \mathtt{a} \cdot 2 \cdot 10^2)\,[\sigma(4^2 \cdot 5^2)/\,\mathtt{a}]\,, \\ \tau(j) &:= \varepsilon,\ j \in \{3,4,5,6,10\}\,, \\ \tau(j) &:= \sigma(j),\ j \in \{7,8,11,12\}\,. \end{aligned}$$

Case 3.3: $\sigma(4^2 \cdot 5^2) \in \{\mathtt{a}, \mathtt{b}, \mathtt{d}\}^+ \setminus (\{\mathtt{a}\}^+ \cup \{\mathtt{b}\}^+ \cup \{\mathtt{d}\}^+ \cup \{\mathtt{a}, \mathtt{d}\}^+)$.

$$\begin{aligned} \text{Define} \quad \tau(8) &:= [\sigma(4^2 \cdot 5^2)/\,\mathtt{b})\,, \\ \tau(9) &:= [\mathtt{b} \backslash \sigma(4^2 \cdot 5^2 \cdot 6^2 \cdot 7^2 \cdot 8 \cdot \mathtt{b} \cdot 9)]\,, \\ \tau(12) &:= [\mathtt{b} \backslash \sigma(4^2 \cdot 5^2 \cdot 6^2 \cdot 11^2 \cdot 8 \cdot \mathtt{b} \cdot 12)]\,, \\ \tau(j) &:= \varepsilon,\ j \in \{4,5,6,7,11\}\,, \\ \tau(j) &:= \sigma(j),\ j \in \{1,2,3,10\}\,. \end{aligned}$$

Case 4: $\mathtt{A} = \mathtt{d}$ and therefore, due to the conditions on σ, $\sigma(4^2 \cdot 5^2) \in \{\mathtt{a}, \mathtt{b}, \mathtt{c}\}^*$.

Case 4.1: $\sigma(4^2 \cdot 5^2) \in \{\mathtt{a}\}^* \cup \{\mathtt{b}\}^* \cup \{\mathtt{c}\}^*$.
The definition of τ is identical to that in Case 3.1.

Case 4.2: $\sigma(4^2 \cdot 5^2) \in \{\mathtt{a}, \mathtt{c}\}^+ \setminus (\{\mathtt{a}\}^+ \cup \{\mathtt{c}\}^+)$.
The definition of τ is identical to that in Case 3.2.

Case 4.3: $\sigma(4^2 \cdot 5^2) \in \{\mathtt{a}, \mathtt{b}, \mathtt{c}\}^+ \setminus (\{\mathtt{a}\}^+ \cup \{\mathtt{b}\}^+ \cup \{\mathtt{c}\}^+ \cup \{\mathtt{a}, \mathtt{c}\}^+)$.
The definition of τ is identical to that in Case 3.3.

It can be verified easily that $\tau(\alpha^{\mathrm{qtf}}_{\mathtt{abcd}}) = \sigma(\alpha^{\mathrm{qtf}}_{\mathtt{abcd}})$ and $\tau(6) = \varepsilon$. This proves the lemma. □

From the above proof, namely from the Cases 1, 2, 3.2, 3.3, 4.2, 4.3, it can be immediately derived that the phenomenon discussed in Lemma 6.4 again essentially depends on the terminal-comprising ambiguity of a major class of telltale candidates for $\alpha^{\mathrm{qtf}}_{\mathtt{abcd}}$.

By Lemma 6.3 and Lemma 6.4 it is proven that the moderate ambiguity of telltale candidates with respect to certain morphically primitive patterns differs from that of other substitutions derived from the morphisms introduced in Chapter 5:

Theorem 6.3. *Let Σ be an alphabet, $|\Sigma| \in \{3,4\}$. Then there exists a morphically primitive quasi-terminal-free pattern α on Σ and an $i \in \mathrm{var}(\alpha)$ such that no telltale candidate $\sigma : (\mathbb{N} \cup \Sigma)^* \longrightarrow \Sigma^*$ for i is moderately ambiguous with respect to α.*

Proof. Evidently, $\alpha^{\text{qtf}}_{\text{abc}}$ is quasi-terminal-free on any suitable alphabet with three letters and $\alpha^{\text{qtf}}_{\text{abcd}}$ is quasi-terminal-free as soon as we have four letters. Furthermore, according to Proposition 6.4, the patterns $\alpha^{\text{qtf}}_{\text{abc}}$ and $\alpha^{\text{qtf}}_{\text{abcd}}$ are morphically primitive. In addition to this, Lemma 6.3 and Lemma 6.4 say that, with regard to $|\Sigma| = 3$, each telltale candidate $\sigma : (\mathbb{N} \cup \Sigma)^* \longrightarrow \Sigma^*$ is substantially ambiguous with respect to $\alpha^{\text{qtf}}_{\text{abc}}$ and, concerning $|\Sigma| = 4$, each telltale candidate $\sigma : (\mathbb{N} \cup \Sigma)^* \longrightarrow \Sigma^*$ is substantially ambiguous with respect to $\alpha^{\text{qtf}}_{\text{abcd}}$. Since, for any fixed pattern α, no substitution which is substantially ambiguous with respect to α can be moderately ambiguous with respect to α (cf. Proposition 6.1), Theorem 6.3 is correct. □

Consequently, the positive result on common telltale candidate morphisms $\sigma : \mathbb{N}^* \longrightarrow \Sigma^*$ presented in Lemma 5.9 in general cannot be extended to telltale candidate substitutions, and this even holds for the combinatorially rather simple quasi-terminal-free patterns. This additionally implies that Theorem 6.3 contrasts with the other results on quasi-terminal-free patterns, a fact which can be easily observed when comparing it to Theorem 6.2.

Our focus on moderate ambiguity in Theorem 6.3 is mainly caused by the fact that we wish to allow for an immediate comparison of its statement to those in Lemma 5.9 and Theorem 6.2. In addition to this, we can also give a slightly weaker version of it in terms of unambiguity:

Corollary 6.1. *Let Σ be an alphabet, $|\Sigma| \in \{3, 4\}$. Then there exists a morphically primitive quasi-terminal-free pattern α on Σ and an $i \in \text{var}(\alpha)$ such that no telltale candidate $\sigma : (\mathbb{N} \cup \Sigma)^* \longrightarrow \Sigma^*$ for i is unambiguous with respect to α.*

Proof. Directly from Theorem 6.3 and Proposition 6.3. □

Summarising our results in the present chapter we, thus, can conclude that, as long as we restrict ourselves to quasi-terminal-free patterns, many of the interesting and useful morphisms introduced so far can be adapted in such a way that they avoid terminal-comprising ambiguity. This is mainly due to the fact that these morphisms may map each variable in the pattern under consideration onto a word over the same two distinct letters. As soon as it is necessary to map a terminal-free subpattern of a quasi-terminal-free pattern onto a word which contains three distinct letters (as it is frequently necessary for morphisms generating telltales, cf. Lemma 5.9 and our remarks below Corollary 5.8), particular patterns can be constructed with respect to which numerous of such morphisms are terminal-comprisingly (and even substantially) ambiguous. In order to show this result, however, we have to restrict ourselves to alphabet sizes 3 or 4. We do not know, whether analogous examples to $\alpha^{\text{qtf}}_{\text{abc}}$ and $\alpha^{\text{qtf}}_{\text{abcd}}$ can be given for larger alphabets. At least, we do not see any possibility to straightforward extend these example patterns.

In the next chapter we apply our results on telltale candidate substitutions to the problem of the inferrability of the full class of E-pattern languages (cf. Question 3.2). Thereby, we show that our previous crucial insight into inductive inference – according to which the ambiguity of common morphisms decides on the learnability of terminal-free E-pattern languages (see the problem statement in Chapter 3.3 and its substantiation in Chapter 5.2.1) – also holds for the (in particular, terminal-comprising) ambiguity of terminal-preserving morphisms and the learnability of classes of more general E-pattern languages.

6.1.1 Application: Inductive inference of the class of quasi-terminal-free E-pattern languages

In the present chapter we demonstrate that the phenomenon described in Theorem 6.3 has a significant impact on inductive inference of E-pattern languages. More precisely, we shall prove that – because of the substantial ambiguity of telltale candidates for the patterns $\alpha_{\mathtt{abc}}^{\mathrm{qtf}}$ and $\alpha_{\mathtt{abcd}}^{\mathrm{qtf}}$ given in Definition 6.5 – the class of quasi-terminal E-pattern languages (and, as an immediate consequence thereof, the full class as well) is not inferrable from positive data. Hence, the different ambiguity of telltale candidates when comparing terminal-free (cf. Lemma 5.9) and quasi-terminal-free patterns implies the analogous difference in the learnability of the classes of E-pattern languages generated by these patterns. Due to the restriction on alphabet sizes 3 and 4 in the previous chapter, our corresponding statements on the learnability of course have the same constraint.

According to the focus of this chapter, we now consider quasi-terminal-free patterns as generators of languages:

Definition 6.6 (Quasi-terminal-free E-pattern language). *Let Σ be an alphabet, $|\Sigma| \geq 3$. When considered as generators of languages, the set of all quasi-terminal-free patterns on Σ is denoted by* $\mathrm{Pat}_{\text{q-tf},\Sigma}$. *An E-pattern language $L \subseteq \Sigma^*$ is called* quasi-terminal-free (on Σ) *if there exists a pattern $\alpha \in \mathrm{Pat}_{\text{q-tf},\Sigma}$ which satisfies $L_{\mathrm{E},\Sigma}(\alpha) = L$. The class of all quasi-terminal-free E-pattern languages on Σ is referred to by* $\mathrm{ePAT}_{\text{q-tf},\Sigma}$.

Quasi-terminal-free E-pattern languages have been introduced by Ohlebusch and Ukkonen [69].[†] Motivated by the research on the equivalence problem, these authors mainly deal with the inclusion problem for sets of similar patterns (which do not necessarily have to be quasi-terminal-free) – a focus that is justified by Theorem 3.9. We discuss this problem in Chapter 6.2.1.

In the context of the present chapter, it is vital that the inclusion of E-pattern languages generated by *similar* quasi-terminal-free patterns has the same characteristics as that of terminal-free E-pattern languages (cf. Theorem 3.6):

Theorem 6.4 (Ohlebusch, Ukkonen [69]). *Let Σ be an alphabet, $|\Sigma| \geq 3$, and let $\alpha, \beta \in \mathrm{Pat}_{\text{q-tf},\Sigma}$ be similar patterns. Then $L_{\mathrm{E},\Sigma}(\beta) \subseteq L_{\mathrm{E},\Sigma}(\alpha)$ if and only if there exists a residual-preserving morphism $\phi : (\mathbb{N} \cup \Sigma)^* \longrightarrow (\mathbb{N} \cup \Sigma)^*$ such that $\phi(\alpha) = \beta$.*

Thus, the inclusion of E-pattern languages generated by similar quasi-terminal-free patterns provides another example where the sufficient condition on the inclusion of arbitrary E-pattern languages given in Theorem 3.5 is characteristic. Note that the proof for Theorem 6.4 again is largely based on the fact that many substitutions on quasi-terminal-free patterns are not terminal-comprisingly ambiguous (due to Lemma 6.2). The decidability of the inclusion problem for each class of similar quasi-terminal-free E-pattern languages follows immediately from Theorem 6.4.

By definition, Theorem 6.4 does not only cover the inclusion problem, but also the equivalence problem for E-pattern languages generated by similar patterns. Hence (and due to Theorem 3.9), it can be used for characterising the succinct quasi-terminal-free patterns in a manner that is equivalent to the corresponding statement for terminal-free patterns given by Corollary 5.3:

[†]Note that Ohlebusch and Ukkonen neither for these languages nor for their generating patterns use the term "quasi-terminal-free".

Corollary 6.2. *Let Σ be an alphabet, $|\Sigma| \geq 3$, and let $\alpha \in \mathrm{Pat}_{\text{q-tf},\Sigma}$. Then α is succinct on Σ if and only if α is morphically primitive.*

Proof. Directly from Theorem 3.9 and Theorem 6.4. □

In addition to this, we claim that Theorem 6.4 implies that any quasi-terminal-free-pattern is succinct if and only if it is not a fixed point of a residual-preserving morphism. As shown by Schneider [93], these fixed points of residual-preserving morphisms are characterised by the fact that they cannot be decomposed in a particular manner which can be derived from the decomposition of morphically imprimitive terminal-free patterns (introduced by Theorem 5.2). Hence, all equivalences of properties of terminal-free patterns mentioned in Corollary 5.4 after minor and canonical modifications also hold for quasi-terminal-free patterns. Thus, we can summarise that the existence of significant nonerasing substitutions which avoid terminal-comprising ambiguity does not only imply major analogies between the existence of moderately ambiguous and unambiguous morphisms for terminal-free and quasi-terminal-free patterns (as discussed in Chapter 6.1), but also between crucial properties of the corresponding classes of E-pattern languages. Once again, as stated in the previous chapter, we do not consider this topic to be of sufficient interest for justifying extensive formal statements describing all facets of the subject. Instead of this, we focus on the use involved in Theorem 6.4 and Corollary 6.2 when analysing the impact of the remarkable difference between terminal-free and quasi-terminal-free patterns stated in Theorem 6.3 on inductive inference of quasi-terminal-free E-pattern languages.

Before we can do so, however, we first need to note that Lemma 5.3 also holds for residual-preserving morphisms:

Lemma 6.5. *Let Σ be an alphabet and let $\alpha, \beta \in (\mathbb{N} \cup \Sigma)^+$, α morphically primitive. Let $\phi, \psi : (\mathbb{N} \cup \Sigma)^* \longrightarrow (\mathbb{N} \cup \Sigma)^*$ be residual-preserving morphisms with $\phi(\alpha) = \beta$ and $\psi(\beta) = \alpha$. Then, for every $j \in \mathrm{var}(\alpha)$, $\psi(\phi(j)) = j$.*

Proof. As Lemmata 5.1 and 5.2 can be easily extended to residual-preserving morphisms, the same holds for the proof of Lemma 5.3. □

In spite of its simplicity, Lemma 6.5 shall be an important tool for the proof of the subsequent Lemma 6.6.

We now proceed with the main argument of the present chapter, namely a lemma which determines the (potentially disastrous) impact of substantial ambiguity of telltale candidates on the question of whether these telltale candidates can generate a meaningful word of a telltale – and, hence, on the question of whether any reliable information about the pattern under consideration is preserved by a telltale candidate:

Lemma 6.6. *Let Σ be an alphabet, $|\Sigma| \geq 3$, and, for some $m \in \mathbb{N}$, let $\{\sigma_i : (\mathbb{N} \cup \Sigma)^* \longrightarrow \Sigma^* \mid 1 \leq i \leq m\}$ be a set of substitutions. Let $\alpha \in \mathrm{Pat}_{\text{q-tf},\Sigma}$ be succinct. If there exists a $j_\sharp \in \mathrm{var}(\alpha)$ such that, for every i, $1 \leq i \leq m$,*

- *σ_i is substantially ambiguous for $j_\sharp$ or*
- *σ_i is not a telltale candidate for $j_\sharp$*

then the set $\{\sigma_i(\alpha) \mid 1 \leq i \leq m\}$ is not a telltale for $L_{\mathrm{E},\Sigma}(\alpha)$ with respect to $\mathrm{ePAT}_{\text{q-tf},\Sigma}$.

Proof. We write

$$\alpha := j_1 \cdot j_2 \cdot [\dots] \cdot j_{p_1} \cdot v_1 \cdot j_{p_1+1} \cdot j_{p_1+2} \cdot [\dots] \cdot j_{p_2} \cdot v_2 \cdot [\dots] \cdot v_n \cdot j_{p_n+1} \cdot j_{p_n+2} \cdot [\dots] \cdot j_{p_{n+1}}$$

with $v_k \in \Sigma^+$, $1 \leq k \leq n$, and $p_1, p_2, \dots p_{n+1} \in \mathbb{N}$, $p_1 < p_2 < \dots < p_{n+1}$, and $j_0, j_1, \dots, j_{p_{n+1}} \in \mathbb{N}$. For every i, $1 \leq i \leq m$, let $w_i := \sigma_i(\alpha)$. Furthermore, let $W := \{w_1, w_2, \dots, w_m\}$.

We now construct a passe-partout $\beta \in \mathrm{Pat}_{\text{q-tf},\Sigma}$ for α and W (cf. Definition 3.1):

For every variable $j_k \in \mathrm{var}(\alpha)$, $1 \leq k \leq p_{n+1}$, we introduce an initially empty string $\gamma_{0,k} \in \mathbb{N}^*$, i.e. $\gamma_{0,k} = \varepsilon$. These empty "patterns" (note that we actually consider a pattern to be nonempty) are to be modified depending on W so that they can serve as a part of the passe-partout β.

For some $r \geq 3$, let $\Sigma := \{\mathtt{A}_1, \mathtt{A}_2, \dots, \mathtt{A}_r\}$. For every i, $1 \leq i \leq m$ and every q, $1 \leq q \leq r$, we define an inverse substitution $\bar{\sigma}_i : \Sigma^* \longrightarrow \mathbb{N}^*$ by $\bar{\sigma}_i(\mathtt{A}_q) := r(i-1) + q$.

For every $i = 1, 2, \dots, m$, we now consider the following cases:

<u>Case 1:</u> σ_i is substantially ambiguous for $j_\sharp$.

Thus, by definition, there exists a substitution τ such that $\tau(\alpha) = \sigma_i(\alpha)$ and $\tau(j_\sharp) = \varepsilon$.

Then, for every k, $1 \leq k \leq p_{n+1}$, define $\gamma_{i,k} := \gamma_{i-1,k} \cdot \bar{\sigma}_i(\tau(k))$.

<u>Case 2:</u> σ_i is not substantially ambiguous for $j_\sharp$, and σ_i is not a telltale candidate for $j_\sharp$.

In this case, for every k, $1 \leq k \leq p_{n+1}$, simply define $\gamma_{i,k} := \gamma_{i-1,k} \cdot \bar{\sigma}_i(\sigma_i(k))$.

When this is accomplished for every i, $1 \leq i \leq m$, then we define

$$\begin{aligned}\beta \;:=\;& \gamma_{m,1} \cdot \gamma_{m,2} \cdot [\dots] \cdot \gamma_{m,p_1} \cdot v_1 \cdot \\ & \gamma_{m,p_1+1} \cdot \gamma_{m,p_1+2} \cdot [\dots] \cdot \gamma_{m,p_2} \cdot v_2 \cdot \\ & [\dots] \cdot \\ & v_n \cdot \gamma_{m,p_n+1} \cdot \gamma_{m,p_n+2} \cdot [\dots] \cdot \gamma_{m,p_{n+1}}\end{aligned}$$

with $v_1, v_2, \dots, v_n$ taken from α.

We now prove that β is a passe-partout for α and W. Thus, we have to show that $L_{\mathrm{E},\Sigma}(\beta) \subset L_{\mathrm{E},\Sigma}(\alpha)$ and $W \subseteq L_{\mathrm{E},\Sigma}(\beta)$. With regard to the latter aspect, we introduce substitutions $\sigma'_i : (\mathbb{N} \cup \Sigma)^* \longrightarrow \Sigma^*$, $1 \leq i \leq m$, by

$$\sigma'_i(j) := \begin{cases} \mathtt{A}_{j-r(i-1)} & , \quad 1 \leq j - r(i-1) \leq r\,, \\ \varepsilon & , \quad \text{else}\,, \end{cases}$$

$j \in \mathrm{var}(\beta)$. From the construction of β – which is based on an inverse substitution applied to the words in W – it follows immediately that, for every i, $1 \leq i \leq m$, $\sigma'_i(\beta) = w_i$. Thus, $W \subseteq L_{\mathrm{E},\Sigma}(\beta)$.

In order to conclude the proof on Lemma 6.6, we now show that $L_{\mathrm{E},\Sigma}(\beta)$ is a proper sublanguage of $L_{\mathrm{E},\Sigma}(\alpha)$. In this regard note that (again due to the construction of β) whenever, for some k, k' with $1 \leq k \leq p_{n+1}$ and $1 \leq k' \leq p_{n+1}$, it is $j_k = j_{k'}$ then $\gamma_{i,k} = \gamma_{i,k'}$. Consequently, for every $j \in \mathrm{var}(\alpha)$, let $\hat{k} \in \{1, 2, \dots, p_{n+1}\}$ be any index such that $j_{\hat{k}} = j$. Then we can define a residual-preserving morphism $\phi : (\mathbb{N} \cup \Sigma)^* \longrightarrow (\mathbb{N} \cup \Sigma)^*$ by

$\phi(j) := \gamma_{m,j_k}$. This implies $\phi(\alpha) = \beta$, and therefore – according to Theorem 3.5, which is applicable since each residual-preserving morphism by definition is terminal-preserving – $L_{\mathrm{E},\Sigma}(\beta) \subseteq L_{\mathrm{E},\Sigma}(\alpha)$.

Hence, we finally have to show that $L_{\mathrm{E},\Sigma}(\beta) \neq L_{\mathrm{E},\Sigma}(\alpha)$. To this end, we consider two cases: If, on the one hand, there exists a t, $0 \leq t \leq n$, such that $\gamma_{p_t+1} = \gamma_{p_t+2} = \ldots = \gamma_{p_{t+1}} = \varepsilon$ (note that, in order to cover all possible cases, we define $p_0 := 0$) then we refer to the substitution σ_0 given by $\sigma_0(j_{p_t+1}) := \mathtt{A}$ and $\sigma_0(j) := \varepsilon$, $j \in \mathbb{N} \setminus \{j_{p_t+1}\}$, where $\mathtt{A} \in \Sigma \setminus \mathrm{term}(\alpha)$. As α is quasi-terminal-free on Σ, such a letter $\mathtt{A}$ exists. It follows immediately from Lemma 6.2 that σ_0 is not terminal-comprisingly ambiguous with respect to α. If $t = 0$ then $\sigma_0(\alpha)$ begins with $\mathtt{A}$, whereas β begins with a letter $\mathtt{B} \in \Sigma$ satisfying $\mathtt{A} \neq \mathtt{B}$; in the case of $t = n$, $\sigma_0(\alpha)$ ends with $\mathtt{A}$ and β ends with such a $\mathtt{B}$; finally, for $1 \leq t \leq n-1$, the substrings w_t and w_{t+1} in $\sigma_0(\alpha)$ are separated by the letter $\mathtt{A} \notin \mathrm{term}(\alpha)$, but, in β, they are not separated by any variable. Hence, for every $t \in \{0, 1, \ldots, n\}$, we can conclude $\sigma_0(\alpha) \notin L_{\mathrm{E},\Sigma}(\beta)$. Note that the proof for Theorem 3.9 as presented in [32] uses a similar argument.

On the other hand, if α and β are similar then we cannot easily give a word separating $L_{\mathrm{E},\Sigma}(\alpha)$ and $L_{\mathrm{E},\Sigma}(\beta)$. Thus, we show that $L_{\mathrm{E},\Sigma}(\beta) \not\supseteq L_{\mathrm{E},\Sigma}(\alpha)$, which – in terms of Theorem 6.4 – is equivalent to the statement that there is no residual-preserving morphism ψ mapping β onto α. Assume to the contrary that there is such a morphism ψ. With regard to the construction of $\phi(j_\sharp)$ we know that it either is empty (as a consequence of Case 1), or it exclusively contains variables which occur more frequently in β than in α. Thus $\psi(\phi(j_\sharp)) \neq j_\sharp$. Since α is succinct and succinctness of quasi-terminal-free patterns is equivalent to morphic primitivity (cf. Corollary 6.2), this statement contradicts Lemma 6.5. Thus, there is no such morphism ψ, and therefore $L_{\mathrm{E},\Sigma}(\beta)$ is a proper sublanguage of $L_{\mathrm{E},\Sigma}(\alpha)$.

Consequently, $W \subseteq L_{\mathrm{E}}(\beta)$ and $L_{\mathrm{E},\Sigma}(\beta) \subset L_{\mathrm{E},\Sigma}(\alpha)$. Hence, β is a passe-partout for α and W. This statement proves Lemma 6.6. □

Note that the proof for Lemma 6.6 generalises the technique used by Reidenbach [76, 81] for showing the non-learnability of the class of terminal-free E-pattern languages over binary alphabets (cf. Theorem 3.15): with regard to the pattern α under consideration and a finite set $W \subset L_{\mathrm{E}}(\alpha)$, a passe-partout β is constructed by applying selected inverse substitutions $\bar{\sigma}_i$ to the words w_i in W and combining the resulting patterns in a particular manner. For a fixed variable $j_\sharp \in \mathrm{var}(\alpha)$ and each $w_i \in W$, the definition of $\bar{\sigma}_i$ is based on an arbitrary substitution τ with $\tau(\alpha) = w_i$ that is *not* a telltale candidate for $j_\sharp$. The existence of such a τ is guaranteed by the fact that whenever there is a telltale candidate σ_i for $j_\sharp$ satisfying $\sigma_i(\alpha) = w_i$ then σ_i is substantially ambiguous for $j_\sharp$ in α. From this construction it can be concluded that, first, $W \subseteq L_{\mathrm{E}}(\beta)$ and, second, there is no terminal-preserving morphism mapping β onto α.

By Lemma 6.6, we can immediately transform the combinatorial statement on some quasi-terminal-free example patterns that contain a variable for which every corresponding telltale candidate is substantially ambiguous (cf. Lemmata 6.3 and 6.4) into the learning theoretically relevant conclusion that the E-pattern languages generated by these patterns do not have a telltale with respect to $\mathrm{ePAT}_{\text{q-tf},\Sigma}$ (for an appropriate alphabet Σ each):

Lemma 6.7. *If* $\Sigma = \{\mathtt{a}, \mathtt{b}, \mathtt{c}\}$ *then* $L_\Sigma(\alpha^{\text{qtf}}_{\mathtt{abc}})$ *does not have a telltale with respect to* $\text{ePAT}_{\text{q-tf},\Sigma}$*, and if* $\Sigma = \{\mathtt{a}, \mathtt{b}, \mathtt{c}, \mathtt{d}\}$ *then* $L_\Sigma(\alpha^{\text{qtf}}_{\mathtt{abcd}})$ *does not have a telltale with respect to* $\text{ePAT}_{\text{q-tf},\Sigma}$.

Proof. We focus on $\alpha^{\text{qtf}}_{\mathtt{abc}}$. According to Proposition 6.4, $\alpha^{\text{qtf}}_{\mathtt{abc}} \in \text{Pat}_{\text{q-tf},\Sigma}$ is morphically primitive, which implies that it is succinct (cf. Corollary 6.2). Furthermore, due to Lemma 6.3, each telltale candidate $\sigma' : (\mathbb{N} \cup \Sigma)^* \longrightarrow \Sigma^*$ for the variable $6 \in \text{var}(\alpha^{\text{qtf}}_{\mathtt{abc}})$ is substantially ambiguous for 6 in $\alpha^{\text{qtf}}_{\mathtt{abc}}$. Hence, with respect to $\alpha^{\text{qtf}}_{\mathtt{abc}}$, each substitution $\sigma : (\mathbb{N} \cup \Sigma)^* \longrightarrow \Sigma^*$ is substantially ambiguous for 6 or it is not a telltale candidate for 6. Consequently, by $j_\sharp := 6$, the conditions of Lemma 6.6 are satisfied, which immediately implies that the statement on $\alpha^{\text{qtf}}_{\mathtt{abc}}$ in Lemma 6.7 is correct.

Referring to Lemma 6.4 instead of Lemma 6.3, the proof on $\alpha^{\text{qtf}}_{\mathtt{abcd}}$ is verbatim the same. □

Hence, for alphabets Σ of size 3 or 4, the class of quasi-terminal-free E-pattern languages is not identifiable in the limit:

Theorem 6.5. *Let* Σ *be an alphabet,* $|\Sigma| \in \{3, 4\}$*. Then* $\text{ePAT}_{\text{q-tf},\Sigma}$ *is not inferrable from positive data.*

Proof. Directly from Theorem 3.12 and Lemma 6.7. □

Actually, from the technical details presented in Lemmata 6.3, 6.4 and 6.7, we can derive an even stronger statement than Theorem 6.5: as, for any passe-partout β constructed by the procedure in the proof of Lemma 6.6, there exists a residual-preserving morphism ϕ mapping the pattern α under consideration onto β, it is evident that the class $\{L_{\text{E},\Sigma}(\beta) \mid \beta = \phi(\alpha^{\text{qtf}}_{\mathtt{abc}}), \phi \text{ residual-preserving}\}$ (and, alternatively, $\{L_{\text{E},\Sigma}(\beta) \mid \beta = \phi(\alpha^{\text{qtf}}_{\mathtt{abcd}}), \phi \text{ residual-preserving}\}$) – which is a proper subclass of $\text{ePAT}_{\text{q-tf},\Sigma}$ for the respective alphabet Σ – is not learnable, either.

In addition to this, we claim that the insights presented so far on the avoidability of terminal-comprising ambiguity for quasi-terminal-free patterns (cf. Chapter 6.1) and on telltales for terminal-free E-pattern languages over alphabets with at least three distinct letters (cf. Chapter 5.2.1) allow for a simple verification of the statement that, with respect to any alphabet Σ, $|\Sigma| \geq 4$, the class of all quasi-terminal-free E-pattern languages over Σ which are generated by patterns α satisfying $|\Sigma| - |\text{term}(\alpha)| = 3$ is inferrable from positive data. Since we do not consider this result overly interesting or surprising, we omit the corresponding details. Furthermore we expect that the tools provided in the present chapter can be used for giving a characterisation of telltales of quasi-terminal-free E-pattern languages (with respect to $\text{ePAT}_{\text{q-tf},\Sigma}$ for arbitrary alphabets Σ) which is largely equivalent to the analogous characterisation for terminal-free E-pattern languages presented by Theorem 3.16. Such a characterisation might contribute to an analysis of the learnability of $\text{ePAT}_{\text{q-tf},\Sigma}$ over alphabets Σ with five or more letters, but we expect the corresponding (presumably necessary) discussion of the avoidability of terminal-comprising ambiguity of telltale candidates over such alphabets when applied to the patterns in $\text{Pat}_{\text{q-tf},\Sigma}$ to be the overriding and more challenging problem.

By definition, every superclass of a non-inferrable class of languages is not learnable, either. Consequently, Theorem 6.5 provides the following partial answer to the prominent Question 3.2:

Corollary 6.3. *Let Σ be an alphabet, $|\Sigma| \in \{3, 4\}$. Then* ePAT_Σ *is not inferrable from positive data.*

Proof. Directly from Theorem 6.5. □

From a *learning theoretical* and *language theoretical* point of view, we consider this main result of the present Chapter 6.1.1 counter-intuitive. In particular, Corollary 6.3 demonstrates that the following seemingly auspicious algorithmic idea for learning the full class of E-pattern languages necessarily fails: first, stepwise present the text for the language to be inferred to a learner for the class of E-pattern languages generated by regular patterns (given by Shinohara [97]), second, using the hypothesis generated by this learner, extract from the words read so far the scattered subwords which are generated by the substitution of variables and, third, feed these words to an (adapted) learner for terminal-free E-pattern languages (which exists due to Theorem 5.5) for specifying the dependencies of variables. In addition to this, the non-learnability of $\mathrm{ePAT}_{\text{q-tf},\Sigma}$, $|\Sigma| \in \{3, 4\}$, conflicts with the positive learnability result for the class of terminal-free E-pattern languages over alphabets with at least three distinct letters (cf. Theorem 5.5) and the fact that the equivalence and inclusion of similar quasi-terminal-free patterns is characterised analogously to that of terminal-free patterns (a statement which can be verified by referring to Theorem 3.6 and Theorem 6.4). As the terminal-free E-pattern languages, hence, can be interpreted as a mere special case of the quasi-terminal-free E-pattern languages and as there is a manifest impact of the inclusion problem on learnability questions (cf. Theorem 3.12, Theorem 3.13 and our respective remarks), one might have expected that the learnability of $\mathrm{ePAT}_{\text{tf},\Sigma}$ equals that of $\mathrm{ePAT}_{\text{q-tf},\Sigma}$.

From a *combinatorial* point of view, however, Corollary 6.3 is not really surprising since it can be easily seen that the step from terminal-free patterns to those containing terminal symbols potentially adds a new type of ambiguity – i. e., more precisely, the terminal-comprising ambiguity – which, as demonstrated in the previous Chapter 6.1, can lead to the substantial ambiguity of certain telltale candidates. Substantial ambiguity, in turn, according to our informal information theoretical intuition is an undesirable property, and Lemma 6.6 confirms that it is indeed a major obstacle to inductive inference of E-pattern languages. Consequently, we feel that, unexpectedly, our combinatorial approach provides a deeper and, thus, more adequate intuition on inductive inference of E-pattern languages than the language theoretical or the original algorithmic understanding of the subject.

Incorporating the main result of the present chapter, we conclude our statements on inductive inference of E-pattern languages with a summary of those alphabet sizes for which the problem of the learnability of the full class of E-pattern languages is resolved now:

Corollary 6.4. *Let Σ be an alphabet. Then* ePAT_Σ *is inferrable from positive data if $|\Sigma| \in \{1, \infty\}$, and it is not inferrable from positive data if $|\Sigma| \in \{2, 3, 4\}$.*

Proof. Directly from Proposition 3.14, Corollary 3.5 and Corollary 6.3. □

If finite alphabets with five or more letters are considered, we have to leave Question 3.2 open.

Though the above considerations draw conclusions on the inferrability of the full class of E-pattern languages, it is important to note that the corresponding argumentation solely discusses the ambiguity of terminal-preserving morphisms when applied to quasi-terminal-free patterns. In the subsequent chapter we skip this restriction and, thus, finally turn our attention to *arbitrary* patterns in $(\mathbb{N} \cup \Sigma)^+$.

6.2 The ambiguity of morphisms on general patterns

In this chapter we mainly examine the extensibility of the positive results on quasi-terminal-free patterns given in Chapter 6.1 – i. e., in particular, the fact that, for every alphabet Σ, $|\Sigma| \geq 3$, and for every morphically primitive pattern $\alpha \in \mathrm{Pat}_{\mathtt{q\text{-}tf},\Sigma}$ there exists a substitution $\sigma_{\mathtt{3\text{-}seg,A,B}} : (\mathbb{N} \cup \Sigma)^* \longrightarrow \Sigma^*$ such that $\sigma_{\mathtt{3\text{-}seg,A,B}}$ is moderately ambiguous with respect to α (cf. Theorem 6.2) – to general patterns over some alphabet Σ. A first closer look at this task immediately suggests that the combinatorial properties of substitutions are more intricate if they are applied to arbitrary patterns α over some alphabet Σ. This is mainly due to the fact that – unlike the situation for quasi-terminal-free patterns – in general every variable in the pattern under consideration must be mapped onto a word which exclusively consists of letters additionally occurring in the pattern. Thus, the simple trick for avoiding terminal-comprising ambiguity presented by the proof for Lemma 6.2 is not applicable here, and therefore we now have to handle the full combinatorial richness involved in terminal-preserving morphisms.

While – among the important types of common morphisms introduced in Chapter 5 – only the ambiguity of telltale candidates necessarily changes when interpreting them as substitutions (a fact, the undesirable consequences of which on inductive inference of E-pattern languages are stated in Corollary 6.3), we now shall demonstrate that, as soon as we do not restrict ourselves to quasi-terminal-free patterns, morphically primitive example patterns can be given for which *no* moderately ambiguous morphism (and, thus, no unambiguous nonerasing morphism) exists. This again has a significant impact on a prominent open algorithmic problem for E-pattern languages, namely the decidability of the equivalence problem (see Open Problem 3.3), which shall be examined in Chapter 6.2.1.

Our above claim on the existence of particular general patterns in $(\mathbb{N} \cup \Sigma)^+$ with respect to which no nonerasing substitution is moderately ambiguous is a simple and well-known phenomenon for unary and binary alphabets Σ:

Proposition 6.5. *Let Σ be an alphabet, $|\Sigma| \in \{1, 2\}$. Then there exists a morphically primitive pattern $\alpha \in (\mathbb{N} \cup \Sigma)^+$ such that no nonerasing substitution $\sigma : (\mathbb{N} \cup \Sigma)^* \longrightarrow \Sigma^*$ is moderately ambiguous with respect to α.*

Proof. For $|\Sigma| = 1$ we define $\alpha := \alpha_{\mathtt{a}}$ and, for $|\Sigma| = 2$, $\alpha := \alpha_{\mathtt{ab}}$, where

$$\alpha_{\mathtt{a}} := 1 \cdot \mathtt{a} \cdot 2\,,$$
$$\alpha_{\mathtt{ab}} := 1 \cdot \mathtt{a} \cdot 2 \cdot \mathtt{b} \cdot 3\,.$$

These patterns are morphically primitive. Referring to $\alpha_{\mathtt{a}}$ and $\alpha_{\mathtt{ab}}$ and utilising terminal-comprising ambiguity, it can be easily verified that each $\sigma : (\mathbb{N} \cup \{\mathtt{a}\})^+ \longrightarrow \{\mathtt{a}\}^+$ is substantially ambiguous for the variable 2 in $\alpha_{\mathtt{a}}$ and each $\sigma : (\mathbb{N} \cup \{\mathtt{a}, \mathtt{b}\})^+ \longrightarrow \{\mathtt{a}, \mathtt{b}\}^+$ is substantially ambiguous for the variable 2 in $\alpha_{\mathtt{ab}}$. By Proposition 6.1, this implies that σ is not moderately ambiguous. □

If we restrict ourselves to unary alphabets then simple terminal-free patterns such as $\alpha_{\mathtt{a}}^{\mathrm{tf}} := 1 \cdot 1 \cdot 2 \cdot 2$ easily prove the analogous result for common nonerasing morphisms $\sigma : \mathbb{N}^* \longrightarrow \{\mathtt{a}\}^*$. For binary alphabets, however, Proposition 6.5 shows an interesting and informative difference to the moderate ambiguity of nonerasing morphisms when applied to terminal-free patterns (cf. Theorem 5.3) – and, in a sense, also of quasi-terminal-free patterns (cf. Theorem 6.2) though these patterns are only defined for $|\Sigma| \geq 3$. Nevertheless, this phenomenon is by no means surprising since, for the pattern $\alpha_{\mathtt{ab}}$ introduced in the above proof, the substantial ambiguity of every nonerasing substitution (and even every substitution mapping the variable 2 onto a nonempty word) simply results from the fact that the alphabet $\{\mathtt{a}, \mathtt{b}\}$ does not provide any terminal symbol which differs from the terminal symbols occurring to the left and to the right of 2 in $\alpha_{\mathtt{ab}}$, and this structural property of the pattern leads to the unavoidability of terminal-comprising ambiguity.

Obviously, this particular problem disappears as soon as there are at least three distinct letters in the alphabet Σ, which implies that for every pattern $\alpha \in (\mathbb{N} \cup \Sigma)^+$ a nonerasing substitution $\sigma : (\mathbb{N} \cup \Sigma)^* \longrightarrow \Sigma^*$ can be chosen such that, for each variable $j \in \mathrm{var}(\alpha)$ and at least one occurrence j_k of j in α, the word $\sigma(j)$ contains a letter which does not equal the terminal symbols to the left and to the right of j_k in α. From this observation, the natural question arises whether this basic combinatorial property of patterns over alphabets with strictly more than two letters is sufficient for avoiding terminal-comprising (and, thus, potentially substantial) ambiguity. As demonstrated in the previous chapter and extensively explained above, the quasi-terminal-free patterns easily allow for the construction of an appropriate substitution σ, but it seems that, for general patterns and for each variable $j \in \mathrm{var}(\alpha)$, the subalphabet $\Sigma_j \subset \Sigma$ with $\sigma(j) \in \Sigma_j^*$ must be carefully chosen – provided that this idea is appropriate for avoiding terminal-comprising ambiguity at all. At least, from previous results in literature it can be easily derived that some elementary properties of the equivalence of E-pattern languages and, hence, combinatorial properties of morphisms change as soon as the terminal alphabet contains at least three (cf. Theorem 3.9 and Corollary 5.8) or at least four (cf. Theorem 3.10) letters, and therefore it seems possible that the basic difference between patterns over alphabets with at most two letters and over alphabets with an least three letters explained above might result in the desirable non-extensibility of Proposition 6.5 to larger alphabets.

The main result of the present chapter, however, demonstrates that analogous patterns over alphabets of size 3 and 4 can be given, the relevant features of which are equivalent to $\alpha_{\mathtt{a}}$ and $\alpha_{\mathtt{ab}}$. Our corresponding examples read as follows:

Definition 6.7 (Patterns $\alpha_{\mathtt{abc}}$ and $\alpha_{\mathtt{abcd}}$; first version). *The patterns $\alpha_{\mathtt{abc}}$ and $\alpha_{\mathtt{abcd}}$ are given by*

$$\begin{aligned} \alpha_{\mathtt{abc}} &:= 1 \cdot \mathtt{a} \cdot 2 \cdot 3^2 \cdot 4^2 \cdot 5^2 \cdot 6^2 \cdot 7 \cdot \mathtt{b} \cdot 8 \cdot \mathtt{a} \cdot 2 \cdot 9^2 \cdot 4^2 \cdot 5^2 \cdot 10^2 \cdot 7 \cdot \mathtt{b} \cdot 11, \\ \alpha_{\mathtt{abcd}} &:= 1 \cdot \mathtt{a} \cdot 2 \cdot 3^2 \cdot 4^2 \cdot 5^2 \cdot 6^2 \cdot 7 \cdot \mathtt{b} \cdot 8 \cdot \mathtt{a} \cdot 2 \cdot 9^2 \cdot 4^2 \cdot 5^2 \cdot 10^2 \cdot 7 \cdot \mathtt{b} \cdot 11 \\ &\quad\; \cdot \mathtt{c} \cdot 12 \cdot 13^2 \cdot 14^2 \cdot 15^2 \cdot 16^2 \cdot 17 \cdot \mathtt{d} \cdot 18 \cdot \mathtt{c} \cdot 12 \cdot 19^2 \cdot 14^2 \cdot 15^2 \cdot 20^2 \cdot 17 \cdot \mathtt{d} \\ &\quad\; \cdot 21 \cdot 14^2 \cdot 15^2 \cdot 14^2 \cdot 15^2 \cdot 14^2 \cdot 15^2 \cdot 22 \cdot 4^2 \cdot 5^2 \cdot 4^2 \cdot 5^2 \cdot 4^2 \cdot 5^2, \end{aligned}$$

where the superscript ^{2} *refers to the concatenation.*

Note that, in the present chapter, $\alpha_{\mathtt{abc}}$ is considered to be a pattern over $\{\mathtt{a}, \mathtt{b}, \mathtt{c}\}$. Contrary to this, in Chapter 6.1, the almost identical pattern $\alpha^{\mathrm{qtf}}_{\mathtt{abcd}}$ – which is used for proving the existence of a pattern with respect to which every telltale candidate for a certain variable is substantially ambiguous (cf. Lemma 6.4) – is interpreted as a quasi-terminal-free pattern over $\{\mathtt{a}, \mathtt{b}, \mathtt{c}, \mathtt{d}\}$. In fact, $\alpha_{\mathtt{abc}}$ is simply the canonical form of the pattern $\phi(\alpha^{\mathrm{qtf}}_{\mathtt{abcd}})$, where the residual-preserving morphism $\phi : (\mathbb{N} \cup \Sigma)^* \longrightarrow (\mathbb{N} \cup \Sigma)^*$ is given by $\phi(6) := \varepsilon$ and $\phi(j) = j$, $j \in \mathbb{N} \setminus \{6\}$. Therefore it is not surprising that many of the subsequent combinatorial considerations on $\alpha_{\mathtt{abc}}$ are very similar to those on $\alpha^{\mathrm{qtf}}_{\mathtt{abcd}}$ presented in the proof of Lemma 6.4.

Since $\alpha_{\mathtt{abc}}$ and $\alpha_{\mathtt{abcd}}$ are rather sophisticated, we now give a second version of the definition, so as to reveal their structure:

Definition 6.7 (Patterns $\alpha_{\mathtt{abc}}$ and $\alpha_{\mathtt{abcd}}$; second version). *Consider the patterns*

$$\begin{aligned}
\gamma_1 &:= 4^2 \cdot 5^2, \\
\gamma_2 &:= 14^2 \cdot 15^2, \\
\beta_1 &:= 2 \cdot 3^2 \cdot \gamma_1 \cdot 6^2 \cdot 7, \\
\beta_1' &:= 2 \cdot 9^2 \cdot \gamma_1 \cdot 10^2 \cdot 7, \\
\beta_2 &:= 12 \cdot 13^2 \cdot \gamma_2 \cdot 16^2 \cdot 17, \\
\beta_2' &:= 12 \cdot 19^2 \cdot \gamma_2 \cdot 20^2 \cdot 17, \\
\hat{\alpha}_1 &:= 1 \cdot \mathtt{a} \cdot \beta_1 \cdot \mathtt{b} \cdot 8 \cdot \mathtt{a} \cdot \beta_1' \cdot \mathtt{b} \cdot 11 \cdot \mathtt{c} \cdot \beta_2 \cdot \mathtt{d} \cdot 18 \cdot \mathtt{c} \cdot \beta_2' \cdot \mathtt{d} \cdot 21, \\
\hat{\alpha}_2 &:= (\gamma_2)^3 \cdot 22 \cdot (\gamma_1)^3.
\end{aligned}$$

Then $\alpha_{\mathtt{abc}} := 1 \cdot \mathtt{a} \cdot \beta_1 \cdot \mathtt{b} \cdot 8 \cdot \mathtt{a} \cdot \beta_1' \cdot \mathtt{b} \cdot 11$ *and* $\alpha_{\mathtt{abcd}} := \hat{\alpha}_1 \cdot \hat{\alpha}_2$.

As stated above, we wish to demonstrate that no substitution $\sigma : (\mathbb{N} \cup \{\mathtt{a}, \mathtt{b}, \mathtt{c}\})^* \longrightarrow \{\mathtt{a}, \mathtt{b}, \mathtt{c}\}^*$ is moderately ambiguous with respect to $\alpha^{\mathrm{qtf}}_{\mathtt{abc}}$, and no substitution $\sigma : (\mathbb{N} \cup \{\mathtt{a}, \mathtt{b}, \mathtt{c}, \mathtt{d}\})^* \longrightarrow \{\mathtt{a}, \mathtt{b}, \mathtt{c}, \mathtt{d}\}^*$ is moderately ambiguous with respect to $\alpha^{\mathrm{qtf}}_{\mathtt{abcd}}$. Because of our comprehensive result on the substantial ambiguity of every nonerasing morphism with respect to every morphically imprimitive pattern (cf. Theorem 6.1), this statement on $\alpha_{\mathtt{abc}}$ and $\alpha_{\mathtt{abcd}}$ of course is only noteworthy if the patterns are morphically primitive. Therefore we first state that our examples really show this property:

Proposition 6.6. *The patterns* $\alpha_{\mathtt{abc}}$ *and* $\alpha_{\mathtt{abcd}}$ *are morphically primitive.*

Proof. Just as our reasoning on Proposition 6.4, the proof utilises the fact that a pattern is morphically primitive if and only if it does not contain a redundant variable (cf. Proposition 2.1).

We focus on $\alpha_{\mathtt{abcd}}$ as introduced in the second version of Definition 6.7, and we shall demonstrate that no variable in $\alpha_{\mathtt{abcd}}$ is redundant. The variables $1, 8, 11, 18$ are not redundant as their only neighbours in $\alpha_{\mathtt{abcd}}$ are terminal symbols. Assume to the contrary that a variable i in β_1 or β_1' is redundant. Then, for the morphism ϕ with $\phi(i) := \varepsilon$ and $\phi(j) := j$, $j \in (\mathrm{var}(\beta_1) \cup \mathrm{var}(\beta_1')) \setminus \{i\}$, there exists a morphism ψ such that $\psi(\phi(\beta_1)) = \beta_1$ and $\psi(\phi(\beta_1')) = \beta_1'$. Thus, we may conclude from Lemma 5.1 that β_1 and β_1' contain an anchor variable with respect to ϕ and ψ. The variable 2 is not an anchor variable since $\psi(\phi(2))$ cannot equal both $2 \cdot 3 \cdot \ldots$ (which is necessary because of β_1) and $2 \cdot 9 \cdot \ldots$ (required due to the shape of β_1'). Similarly, as $\psi(\phi(7)) \neq \ldots \cdot 6 \cdot 7$ or $\psi(\phi(7)) \neq \ldots \cdot 10 \cdot 7$, the variable 7 cannot be an anchor variable. Thus, a variable in $\{3,4,5,6,9,10\} \subset \text{var}(\beta_1) \cup \text{var}(\beta_1')$ must be an anchor variable. However, since all of these variables occur as a square, this implies that $\psi(\phi(\beta_1)) = \ldots j' \ldots j'' \ldots j' \ldots j'' \ldots$ or $\psi(\phi(\beta_1')) = \ldots j' \ldots j'' \ldots j' \ldots j'' \ldots$ with $j' \neq j''$. As β_1 and β_1' do not have such a subpattern, there is no anchor variable in β_1 and no anchor variable in β_1', which contradicts Lemma 5.1. Consequently, in β_1 and β_1', there is no redundant variable.

If we replace in our argumentation on β_1 and β_1' each variable i by $i + 10$, then we immediately receive a proof showing that β_2 and β_2' do not contain a redundant variable either. Finally, with regard to the remaining variables 21 and 22, we can argue as follows: If a variable i among these variables is redundant then there must be the morphisms ϕ and ψ with $\phi(i) = \varepsilon$ and $\psi(\phi(21 \cdot \hat{\alpha}_2)) = 21 \cdot \hat{\alpha}_2$. However, since the only occurrences of the variables 21 and 22 are surrounded by terminal symbols or variables which additionally occur in β_1 or β_2, this also means that there exists a variable $i' \in (\text{var}(\beta_1) \cup \text{var}(\beta_2))$ with $\psi(\phi(i')) \neq i'$. Hence, we can again refer to our argumentation on β_1, which shows that corresponding morphisms ϕ, ψ do not exist. Therefore, the variables 21 and 22 are not redundant in $\alpha_{\mathtt{abcd}}$.

Consequently, $\alpha_{\mathtt{abcd}}$ does not contain any redundant variable, and therefore, by Proposition 2.1, it is morphically primitive. The analogous statement for $\alpha_{\mathtt{abc}}$ follows from our reasoning on β_1 and β_1' given above. □

Before we present the main lemmata of the present chapter and their rather technical proofs, we first wish to discuss the structure of $\alpha_{\mathtt{abc}}$ and $\alpha_{\mathtt{abcd}}$ in order to provide an intuitive understanding of why no nonerasing substitution is moderately ambiguous with respect to $\alpha_{\mathtt{abc}}$ and $\alpha_{\mathtt{abcd}}$. Additionally, our explanations are meant to give some hints about how patterns with such a property can be constructed (in this context note that, because of the same reasons as those mentioned in the footnote below Proposition 5.5, we do not know any promising possibility to use computers for searching for such examples). As $\alpha_{\mathtt{abc}}$ is a prefix of $\alpha_{\mathtt{abcd}}$, our explanations focus on the structure of the latter pattern. The properties of the former pattern can be easily derived therefrom or, alternatively, from the discussion of the pattern $\alpha_{\mathtt{abcd}}^{\mathrm{qtf}}$ presented in Chapter 6.1 (which in parts is equivalent to our subsequent remarks).

The components of $\alpha_{\mathtt{abcd}}$ are tailor-made for ensuring that every nonerasing substitution $\sigma : (\mathbb{N} \cup \{\mathtt{a},\mathtt{b},\mathtt{c},\mathtt{d}\})^* \longrightarrow \{\mathtt{a},\mathtt{b},\mathtt{c},\mathtt{d}\}^*$ is substantially ambiguous for the variables 4 or 14 in $\alpha_{\mathtt{abcd}}$. Hence, we need to explain that there exists a substitution τ satisfying $\tau(\alpha_{\mathtt{abcd}}) = \sigma(\alpha_{\mathtt{abcd}})$ and $\tau(4) = \varepsilon$ or $\tau(14) = \varepsilon$. We now examine those subpatterns of $\alpha_{\mathtt{abcd}}$ which guarantee the existence of such a substitution τ, and we begin with the kernels of the pattern, i.e. γ_1 and γ_2 (cf. the second version of Definition 6.7). These subpatterns imply the ambiguity of σ provided that $\sigma(4)$ or $\sigma(14)$ is a word over a unary alphabet: if $|\,\text{term}(\sigma(\gamma_1))| = 1$ or $|\,\text{term}(\sigma(\gamma_2))| = 1$ then the morphism τ given by $\tau(4) := \varepsilon$, $\tau(5) := \sigma(4 \cdot 5)$ or $\tau(14) := \varepsilon$, $\tau(15) := \sigma(14 \cdot 15)$ and, for all other variables j, $\tau(j) := \sigma(j)$ satisfies $\tau(\alpha) = \sigma(\alpha)$. In our formal argumentation, we utilise this fact only for $\sigma(\gamma_1) \in \{\mathtt{c}\}^* \cup \{\mathtt{d}\}^*$ or $\sigma(\gamma_2) \in \{\mathtt{a}\}^* \cup \{\mathtt{b}\}^*$. The other cases are covered by the terminal-comprising ambiguity of $\sigma(1 \cdot \mathtt{a} \cdot \beta_1 \cdot \mathtt{b} \cdot 8)$ (and, alternatively, $\sigma(11 \cdot \mathtt{c} \cdot \beta_2\ \mathtt{d} \cdot 18)$) whenever $\sigma(\gamma_1)$ (or $\sigma(\gamma_2)$, respectively) contains – possibly among others – the letters $\mathtt{a}$ or $\mathtt{b}$ (or $\mathtt{c}$ or $\mathtt{d}$, respectively), leading again to the substantial ambiguity of σ for 4 (or 14, respectively). Note that this special property of the subpattern $1 \cdot \mathtt{a} \cdot \beta_1 \cdot \mathtt{b} \cdot 8$ is sufficient for our proof on $\alpha_{\mathtt{abc}}$ given in Lemma 6.8 (see below). Returning to $\alpha_{\mathtt{abcd}}$, the last case to be considered is that $\sigma(\gamma_1)$ consists of the letters c and d and $\sigma(\gamma_2)$ of a and b. This is surely the most promising choice for avoiding terminal-comprising ambiguity and corresponds to our initial considerations given below Proposition 6.5. However, for that case, there exist some words w_i, $1 \leq i \leq 5$, such that $\sigma(\hat{\alpha}_2) = w_1 \, \mathtt{ab} \, w_2 \, \mathtt{ab} \, w_3 \, \mathtt{cd} \, w_4 \, \mathtt{cd} \, w_5$. Consequently, $\sigma(\alpha_{\mathtt{abcd}})$ can be generated by the scattered subpattern $1 \cdot \mathtt{a} \cdot \mathtt{b} \cdot 8 \cdot \mathtt{a} \cdot \mathtt{b} \cdot 11 \cdot \mathtt{c} \cdot \mathtt{d} \cdot 18 \cdot \mathtt{c} \cdot \mathtt{d} \cdot 21$ of $\alpha_{\mathtt{abcd}}$, and therefore the corresponding morphism τ can map the variables 4 and 14 onto the empty word again. The variables with single occurrences in $\alpha_{\mathtt{abcd}}$, such as 8 and 22, are generally used to compensate the side effects of any assignment $\tau(4) := \varepsilon$ or $\tau(14) := \varepsilon$; the modified repetitions of β_1 (as β_1') and β_2 (as β_2') and, particularly, those variables that distinguish β_1 from β_1' (e.g. 3) and β_2 from β_2' (e.g. 13) guarantee that $\alpha_{\mathtt{abcd}}$ is morphically primitive. Note that the latter point is demonstrated by Proposition 6.6 (see above).

Referring to those parts of these explanations that deal with the structure of $\alpha_{\mathtt{abc}}$, we now formally state and prove that, for every morphism $\sigma : (\mathbb{N} \cup \{\mathtt{a},\mathtt{b},\mathtt{c}\})^+ \longrightarrow \{\mathtt{a},\mathtt{b},\mathtt{c}\}^+$ (and, actually, even for every morphism $\sigma : (\mathbb{N} \cup \{\mathtt{a},\mathtt{b},\mathtt{c}\})^* \longrightarrow \{\mathtt{a},\mathtt{b},\mathtt{c}\}^*$ mapping the variables 4 or 14 onto a nonempty word), there exists a morphism τ satisfying $\tau(\alpha_{\mathtt{abc}}) = \sigma(\alpha_{\mathtt{abc}})$ and, for $j = 4$ or $j = 14$, $\tau(j) = \varepsilon$:

Lemma 6.8. *Let* $\Sigma := \{\mathtt{a},\mathtt{b},\mathtt{c}\}$ *be an alphabet. Then every nonerasing substitution* $\sigma : (\mathbb{N} \cup \Sigma)^* \longrightarrow \Sigma^*$ *is substantially ambiguous for the variable* 4 *in* $\alpha_{\mathtt{abc}}$.

Proof. Let $\sigma : (\mathbb{N} \cup \Sigma)^* \longrightarrow \Sigma^*$ be an arbitrary nonerasing substitution. We show that there exists a substitution $\tau : (\mathbb{N} \cup \Sigma)^* \longrightarrow \Sigma^*$ such that $\tau(\alpha_{\mathtt{abc}}) = \sigma(\alpha_{\mathtt{abc}})$ and $\tau(4) = \varepsilon$. To this end, we refer to $\alpha_{\mathtt{abc}}$ as declared in the second version of Definition 6.7 and regard the following cases:

Case 1 $\sigma(\gamma_1) \in \{\mathtt{a},\mathtt{b},\mathtt{c}\}^+ \setminus \{\mathtt{b},\mathtt{c}\}^+$:

Define
$$\begin{aligned}
\tau(1) &:= \sigma(1 \cdot \mathtt{a} \cdot 2 \cdot 3^2)\,[\sigma(\gamma_1)/\mathtt{a}\langle 1\rangle],\\
\tau(2) &:= [\mathtt{a}\langle 1\rangle \backslash \sigma(\gamma_1)],\\
\tau(8) &:= \sigma(8 \cdot \mathtt{a} \cdot 2 \cdot 9^2)\,[\sigma(\gamma_1)/\mathtt{a}\langle 1\rangle],\\
\tau(j) &:= \sigma(j),\ j \in \{6,7,10,11\},\\
\tau(j) &:= \varepsilon,\ j \in \mathrm{var}(\gamma_1) \cup \{3,9\}.
\end{aligned}$$

Case 2 $\sigma(\gamma_1) \in \{\mathtt{b},\mathtt{c}\}^+ \setminus \{\mathtt{c}\}^+$:

Define
$$\begin{aligned}
\tau(7) &:= [\sigma(\gamma_1)/\mathtt{b}\langle 1\rangle],\\
\tau(8) &:= [\mathtt{b}\langle 1\rangle \backslash \sigma(\gamma_1)]\,\sigma(6^2 \cdot 7 \cdot \mathtt{b} \cdot 8),\\
\tau(11) &:= [\mathtt{b}\langle 1\rangle \backslash \sigma(\gamma_1)]\,\sigma(10^2 \cdot 7 \cdot \mathtt{b} \cdot 11),\\
\tau(j) &:= \sigma(j),\ j \in \{1,2,3,9\},\\
\tau(j) &:= \varepsilon,\ j \in \mathrm{var}(\gamma_1) \cup \{6,10\}.
\end{aligned}$$

Case 3 $\sigma(\gamma_1) \in \{\mathtt{c}\}^+$:

Define
$$\begin{aligned}
\tau(4) &:= \varepsilon,\\
\tau(5) &:= \sigma(4 \cdot 5),\\
\tau(j) &:= \sigma(j),\ j \in \mathrm{var}(\alpha_{\mathtt{abc}}) \setminus \mathrm{var}(\gamma_1).
\end{aligned}$$

These three cases are exhaustive and, in each case, $\tau(4) = \varepsilon$. The verification of $\tau(\alpha_{\mathtt{abcd}}) = \sigma(\alpha_{\mathtt{abcd}})$ is straightforward. □

We proceed with the analogous result for $|\Sigma| = 4$ and $\alpha_{\mathtt{abcd}}$:

Lemma 6.9. *Let* $\Sigma := \{\mathtt{a}, \mathtt{b}, \mathtt{c}, \mathtt{d}\}$ *be an alphabet. Then every nonerasing substitution* $\sigma : (\mathbb{N} \cup \Sigma)^* \longrightarrow \Sigma^*$ *is substantially ambiguous for the variable* 4 *or the variable* 14 *in* $\alpha_{\mathtt{abcd}}$.

Proof. Since $\alpha_{\mathtt{abc}}$ is a prefix of $\alpha_{\mathtt{abcd}}$, the following proof largely extends our proof for Lemma 6.8. However, Case 7 (see below) has no counterpart in the reasoning on $\alpha_{\mathtt{abc}}$.

Let $\sigma : (\mathbb{N} \cup \Sigma)^* \longrightarrow \Sigma^*$ be an arbitrary nonerasing substitution. We show that there exists a substitution $\tau : (\mathbb{N} \cup \Sigma)^* \longrightarrow \Sigma^*$ such that $\tau(\alpha_{\mathtt{abcd}}) = \sigma(\alpha_{\mathtt{abcd}})$ and $\tau(4) = \varepsilon$ or $\tau(14) = \varepsilon$. To this end, we refer to $\alpha_{\mathtt{abcd}}$ as declared in the second version of Definition 6.7 and regard the following cases:

Case 1 $\sigma(\gamma_1) \in \{\mathtt{a}, \mathtt{b}, \mathtt{c}, \mathtt{d}\}^+ \setminus \{\mathtt{b}, \mathtt{c}, \mathtt{d}\}^+$:

Define
$$\begin{aligned}
\tau(1) &:= \sigma(1 \cdot \mathtt{a} \cdot 2 \cdot 3^2)\, [\sigma(\gamma_1)/\mathtt{a}\langle 1\rangle],\\
\tau(2) &:= [\mathtt{a}\langle 1\rangle \backslash \sigma(\gamma_1)],\\
\tau(8) &:= \sigma(8 \cdot \mathtt{a} \cdot 2 \cdot 9^2)\, [\sigma(\gamma_1)/\mathtt{a}\langle 1\rangle],\\
\tau(22) &:= \sigma(22 \cdot (\gamma_1)^3),\\
\tau(j) &:= \sigma(j),\ j \in \mathrm{var}(\beta_2 \cdot \beta_2') \cup \{6, 7, 10, 11, 18, 21\},\\
\tau(j) &:= \varepsilon,\ j \in \mathrm{var}(\gamma_1) \cup \{3, 9\}.
\end{aligned}$$

Case 2 $\sigma(\gamma_1) \in \{\mathtt{b}, \mathtt{c}, \mathtt{d}\}^+ \setminus \{\mathtt{c}, \mathtt{d}\}^+$:

Define
$$\begin{aligned}
\tau(7) &:= [\sigma(\gamma_1)/\mathtt{b}\langle 1\rangle],\\
\tau(8) &:= [\mathtt{b}\langle 1\rangle \backslash \sigma(\gamma_1)]\, \sigma(6^2 \cdot 7 \cdot \mathtt{b} \cdot 8),\\
\tau(11) &:= [\mathtt{b}\langle 1\rangle \backslash \sigma(\gamma_1)]\, \sigma(10^2 \cdot 7 \cdot \mathtt{b} \cdot 11),\\
\tau(22) &:= \sigma(22 \cdot (\gamma_1)^3),\\
\tau(j) &:= \sigma(j),\ j \in \mathrm{var}(\beta_2 \cdot \beta_2') \cup \{1, 2, 3, 9, 18, 21\},\\
\tau(j) &:= \varepsilon,\ j \in \mathrm{var}(\gamma_1) \cup \{6, 10\}.
\end{aligned}$$

Case 3 $\sigma(\gamma_1) \in \{\mathtt{c}\}^+ \cup \{\mathtt{d}\}^+$:

Define
$$\begin{aligned}
\tau(4) &:= \varepsilon,\\
\tau(5) &:= \sigma(4 \cdot 5),\\
\tau(j) &:= \sigma(j),\ j \in \mathrm{var}(\alpha_{\mathtt{abcd}}) \setminus \mathrm{var}(\gamma_1).
\end{aligned}$$

Case 4 $\sigma(\gamma_1) \in \{\mathtt{c}, \mathtt{d}\}^+ \setminus (\{\mathtt{c}\}^+ \cup \{\mathtt{d}\}^+)$ and $\sigma(\gamma_2) \in \{\mathtt{a}, \mathtt{b}, \mathtt{c}, \mathtt{d}\}^+ \setminus \{\mathtt{a}, \mathtt{b}, \mathtt{d}\}^+$:

Define
$$\begin{aligned}
\tau(11) &:= \sigma(11 \cdot \mathtt{c} \cdot 12 \cdot 13^2)\, [\sigma(\gamma_2)/\mathtt{c}\langle 1\rangle],\\
\tau(12) &:= [\mathtt{c}\langle 1\rangle \backslash \sigma(\gamma_2)],\\
\tau(18) &:= \sigma(18 \cdot \mathtt{c} \cdot 12 \cdot 19^2)\, [\sigma(\gamma_2)/\mathtt{c}\langle 1\rangle],\\
\tau(22) &:= \sigma((\gamma_2)^3 \cdot 22),\\
\tau(j) &:= \sigma(j),\ j \in \mathrm{var}(\beta_1 \cdot \beta_1') \cup \{1, 8, 16, 17, 20, 21\},\\
\tau(j) &:= \varepsilon,\ j \in \mathrm{var}(\gamma_2) \cup \{13, 19\}.
\end{aligned}$$

Case 5 $\sigma(\gamma_1) \in \{\mathtt{c}, \mathtt{d}\}^+ \setminus (\{\mathtt{c}\}^+ \cup \{\mathtt{d}\}^+)$ and $\sigma(\gamma_2) \in \{\mathtt{a}, \mathtt{b}, \mathtt{d}\}^+ \setminus \{\mathtt{a}, \mathtt{b}\}^+$:

$$\begin{aligned} \text{Define} \quad \tau(17) &:= [\sigma(\gamma_2)/\mathtt{d}\langle 1\rangle], \\ \tau(18) &:= [\mathtt{d}\langle 1\rangle \backslash \sigma(\gamma_2)]\, \sigma(16^2 \cdot 17 \cdot \mathtt{d} \cdot 18), \\ \tau(21) &:= [\mathtt{d}\langle 1\rangle \backslash \sigma(\gamma_2)]\, \sigma(20^2 \cdot 17 \cdot \mathtt{d} \cdot 21), \\ \tau(22) &:= \sigma((\gamma_2)^3 \cdot 22), \\ \tau(j) &:= \sigma(j),\ j \in \mathrm{var}(\beta_1 \cdot \beta_1') \cup \{1, 8, 11, 12, 13, 19\}, \\ \tau(j) &:= \varepsilon,\ j \in \mathrm{var}(\gamma_2) \cup \{16, 20\}. \end{aligned}$$

Case 6 $\sigma(\gamma_1) \in \{\mathtt{c}, \mathtt{d}\}^+ \setminus (\{\mathtt{c}\}^+ \cup \{\mathtt{d}\}^+)$ and $\sigma(\gamma_2) \in \{\mathtt{a}\}^+ \cup \{\mathtt{b}\}^+$:

$$\begin{aligned} \text{Define} \quad \tau(14) &:= \varepsilon, \\ \tau(15) &:= \sigma(14 \cdot 15), \\ \tau(j) &:= \sigma(j),\ j \in \mathrm{var}(\alpha_{\mathtt{abcd}}) \setminus \mathrm{var}(\gamma_2). \end{aligned}$$

Case 7 $\sigma(\gamma_1) \in \{\mathtt{c}, \mathtt{d}\}^+ \setminus (\{\mathtt{c}\}^+ \cup \{\mathtt{d}\}^+)$ and $\sigma(\gamma_2) \in \{\mathtt{a}, \mathtt{b}\}^+ \setminus (\{\mathtt{a}\}^+ \cup \{\mathtt{b}\}^+)$:

Consequently, $\sigma((\gamma_1)^3)$ contains at least two occurrences of the subword $\mathtt{c\,d}$ and $\sigma((\gamma_2)^3)$ contains at least two occurrences of the subword $\mathtt{a\,b}$. Furthermore, due to the shape of these subwords, their occurrences must be non-overlapping. Therefore τ can be given as follows:

$$\begin{aligned} \text{Define} \quad \tau(1) &:= \sigma(\hat{\alpha}_1)\, [\sigma(\hat{\alpha}_2)/\, \mathtt{a\,b}\langle 1\rangle], \\ \tau(8) &:= [\mathtt{a\,b}\langle 1\rangle \backslash \sigma(\hat{\alpha}_2)/\, \mathtt{a\,b}\langle 2\rangle], \\ \tau(11) &:= [\mathtt{a\,b}\langle 2\rangle \backslash \sigma(\hat{\alpha}_2)/\, \mathtt{c\,d}\langle 1\rangle], \\ \tau(18) &:= [\mathtt{c\,d}\langle 1\rangle \backslash \sigma(\hat{\alpha}_2)/\, \mathtt{c\,d}\langle 2\rangle], \\ \tau(21) &:= [\mathtt{c\,d}\langle 2\rangle \backslash \sigma(\hat{\alpha}_2)], \\ \tau(j) &:= \varepsilon,\ j \in \mathrm{var}(\alpha_{\mathtt{abcd}}) \setminus \{1, 8, 11, 18, 21\}. \end{aligned}$$

These seven cases are exhaustive and, in each case, $\tau(4) = \varepsilon$ or $\tau(14) = \varepsilon$. The verification of $\tau(\alpha_{\mathtt{abcd}}) = \sigma(\alpha_{\mathtt{abcd}})$ is straightforward. □

Obviously, there is a remarkable difference between Lemma 6.8 and Lemma 6.9 since the former proves the existence of a single variable j such that every nonerasing morphism $\sigma : (\mathbb{N} \cup \{\mathtt{a}, \mathtt{b}, \mathtt{c}\})^* \longrightarrow \{\mathtt{a}, \mathtt{b}, \mathtt{c}\}^*$ is substantially ambiguous for j in $\alpha_{\mathtt{abc}}$, whereas the latter can merely state that, for every nonerasing morphism $\sigma : (\mathbb{N} \cup \{\mathtt{a}, \mathtt{b}, \mathtt{c}, \mathtt{d}\})^* \longrightarrow \{\mathtt{a}, \mathtt{b}, \mathtt{c}, \mathtt{d}\}^*$, there exists a related variable j_σ (in a set of two variables) such that σ is substantially ambiguous for j_σ in $\alpha_{\mathtt{abcd}}$. While this phenomenon has a major impact on our considerations dealing with the equivalence problem for E-pattern languages (cf. Chapter 6.2.1), it does not affect the main result of the present chapter, which describes the nearly immediate consequences of the above lemmata on the question of whether or not, for every morphically primitive pattern over some alphabet, there exists a moderately ambiguous substitution:

Theorem 6.6. *Let Σ be an alphabet, $|\Sigma| \in \{3, 4\}$. Then there exists a morphically primitive pattern $\alpha \in \mathrm{Pat}_\Sigma$ such that no substitution $\sigma : (\mathbb{N} \cup \Sigma)^* \longrightarrow \Sigma^*$ is moderately ambiguous with respect to α.*

Proof. If σ is nonerasing then Theorem 6.6 follows from Proposition 6.6, Lemma 6.8 and Lemma 6.9. If σ is not nonerasing, then we do not need any extensive argumentation, but we can simply refer to Proposition 6.2. □

Since every unambiguous nonerasing morphism necessarily is moderately ambiguous (cf. Proposition 6.3), we can immediately conclude from Theorem 6.6 that there exist certain patterns (namely $\alpha_{\mathtt{abc}}$ and $\alpha_{\mathtt{abcd}}$ again), with respect to which there is no unambiguous nonerasing morphism, either:

Corollary 6.5. *Let* Σ *be an alphabet,* $|\Sigma| \in \{3, 4\}$. *Then there exists a morphically primitive pattern* $\alpha \in \mathrm{Pat}_\Sigma$ *such that no nonerasing substitution* $\sigma : (\mathbb{N} \cup \Sigma)^* \longrightarrow \Sigma^*$ *is unambiguous with respect to* α.

Proof. Directly from Theorem 6.6 and Proposition 6.3. □

Theorem 6.6 is the last of our main results on moderate ambiguity, and it substantiates our claim according to which the combinatorial properties of terminal-preserving morphisms significantly differ from those of common morphisms. In this regard, our first crucial Theorem 5.3 demonstrates that, for every morphically primitive *terminal-free* pattern, there exist numerous moderately ambiguous morphisms mapping any such pattern onto a word over a binary alphabet, and Lemma 5.9 shows that the same holds for telltale candidates as soon as we allow these morphisms to range over words over a ternary alphabet. If we proceed to *quasi-terminal-free* patterns then the situation is not as pleasant anymore: while Theorem 6.2 states that the abovementioned positive result on common morphisms in Theorem 5.3 can be extended to substitutions when applying them to morphically primitive quasi-terminal-free patterns, the corresponding result for telltale candidates turns out not to be extendable (cf. Theorem 6.3) (in case that the terminal alphabet consists of three or four distinct letters). Finally, as shown by Theorem 6.6, if we consider *arbitrary* (or *general*, as we prefer to call it) patterns over some alphabet with at most four letters then none of the above positive results is extendable, i. e. there exist morphically primitive patterns α such that no substitution is moderately ambiguous with respect to α. From this insight, we can immediately conclude that, for general patterns and contrary to terminal-free and quasi-terminal-free patterns, morphic primitivity is of minor importance since it is just a necessary (cf. Theorem 6.1), but not a characteristic condition for the existence of a moderately ambiguous morphism.

Additionally, a similar chain of results is presented with regard to the existence of unambiguous morphisms (though we do not discuss this topic for telltale candidates). Theorem 5.7 shows that, for every morphically primitive terminal-free pattern, there is a tailor-made unambiguous morphic image in $\{\mathtt{a}, \mathtt{b}\}^*$, and Chapter 6.1 claims that this result also holds for quasi-terminal-free patterns. The corresponding proof is considered to be cumbersome yet straightforward, and therefore it is omitted. Concerning general patterns, Corollary 6.5 states that the existence of general patterns (over alphabets of at most four letters) with respect to which there is no moderately ambiguous morphism, by definition, implies the analogous result for nonerasing unambiguous morphisms.

Very roughly speaking, we thus can summarise the main outcome of our studies as follows: The more terminal symbols are used in the pattern under consideration (relative to the number of distinct letters in the alphabet) the fewer moderately ambiguous and unambiguous morphisms exist. In this regard, however, it is important to notice that many of our results on terminal-preserving morphisms only hold for alphabets with up to four letters, and the difference between Lemma 6.8 and Lemma 6.9 suggests that perhaps the positive results on terminal-free and quasi-terminal-free patterns might be extendable to general patterns as soon as there are more than four letters in the alphabet. This is caused by the fact that the avoidability of terminal-comprising ambiguity seems easier for larger alphabets. Due to the same reason, we do not see any straightforward method to extend our way of composing example patterns with unpleasant properties (i. e. $\alpha_{\mathtt{abc}}$ and $\alpha_{\mathtt{abcd}}$, cf. Definition 6.7) to larger alphabets: Our argumentation on Lemma 6.9 is based on the fact that every substitution σ either matches the "easier" Cases 1 - 6, or it exactly reconstructs the terminal substring of the pattern, i. e. $\mathrm{res}(\alpha_{\mathtt{abcd}})$ occurs in $\sigma(\alpha_{\mathtt{abcd}})$ as a scattered subword (see Case 7). We expect that this problem can be avoided for all larger alphabets.

As mentioned above, the question partially answered in the present chapter – namely that for extensibility of the properties of common to terminal-preserving morphisms – again implicitly can be found in the literature on pattern languages. We extensively discuss the corresponding consequences of our results in the next chapter.

6.2.1 Application: The equivalence problem for E-pattern languages

In the present chapter, we describe the impact of the results presented in Chapter 6.2 on the equivalence problem for the full class of E-pattern languages as stated in Open Problem 3.3.

The equivalence problem is considered to be one of the most important open problems on the full class of E-patterns languages. Definitely, besides the question for its inferrability from positive data, it is among the problems most discussed in literature. It has first been tackled by Filè [19] and Jiang et al. [32] and later by Jiang et al. [33], Dányi and Fülöp [15], and Ohlebusch and Ukkonen [69]. In Chapter 3.2.1, a brief survey on the major results obtained by these examinations is given (cf. Corollary 3.3, Theorem 3.9, Theorem 3.10). Furthermore the decidability of the equivalence problem for each class of E-pattern languages generated by quasi-terminal-free patterns follows immediately from Theorem 6.4.

In addition to these preliminary insights, Ohlebusch and Ukkonen [69] introduce a procedure that stepwise removes all redundant variables from a pattern and, thus, computes a shortest *normal form*. They conjecture that, for alphabets with at least three letters, two patterns generate the same language if and only if the canonical forms of their normal forms are the same, and the authors paraphrase their conjecture as follows:

Conjecture 6.1 (Ohlebusch, Ukkonen [69]). *For an alphabet Σ, $|\Sigma| \geq 3$, and patterns $\alpha_1, \alpha_2 \in \mathrm{Pat}_\Sigma$, $L_{\mathrm{E},\Sigma}(\alpha_1) = L_{\mathrm{E},\Sigma}(\alpha_2)$ if and only if α_1 and α_2 are morphically coincident.*

This conjecture is based on the expectation that the inclusion of general similar patterns can be characterised in the same way as that of quasi-terminal-free similar patterns (see Theorem 6.4):

Question 6.1 (Ohlebusch, Ukkonen [69]). *Let Σ be an alphabet, $|\Sigma| \geq 3$, and let $\alpha_1, \alpha_2 \in$ Pat_Σ be similar patterns. Is $L_{\mathrm{E},\Sigma}(\alpha_1) \subseteq L_{\mathrm{E},\Sigma}(\alpha_2)$ if and only if there exists a residual preserving morphism $\phi : (\mathbb{N} \cup \Sigma)^* \longrightarrow (\mathbb{N} \cup \Sigma)^*$ with $\phi(\alpha_2) = \alpha_1$?*

In this context, note that, for $|\Sigma| \geq 3$, two patterns must be similar if they generate the same language (cf. Theorem 3.9).

Furthermore, Ohlebusch and Ukkonen ask for the way the relation of two languages generated by some patterns changes as soon as an additional terminal symbol is added to the alphabet Σ. More precisely, they investigate whether the equivalence for ePAT_Σ is *preserved under alphabet extension*, i. e. whether, for each pair of patterns $\alpha, \beta \in \mathrm{Pat}_\Sigma$ and a second alphabet $\Sigma' := \Sigma \cup \{\mathtt{A}\}$ (where $\mathtt{A} \notin \Sigma$), $L_{\mathrm{E},\Sigma}(\alpha) = L_{\mathrm{E},\Sigma}(\beta)$ if and only if $L_{\mathrm{E},\Sigma'}(\alpha) = L_{\mathrm{E},\Sigma'}(\beta)$:

Question 6.2 (Ohlebusch, Ukkonen [69]). *Let Σ be an alphabet, $|\Sigma| \geq 3$. Is the equivalence for ePAT_Σ preserved under alphabet extension?*

From Theorem 6.4 and Theorem 3.9 it follows immediately that an affirmative answer to this question is sufficient for the correctness of Conjecture 6.1 and, thus, the decidability of the equivalence problem for ePAT_Σ provided that this answer holds for a "double" extension, i. e. the equivalence is preserved under alphabet extension for both ePAT_Σ and $\mathrm{ePAT}_{\Sigma'}$ where the alphabet Σ' satisfies $\Sigma' \supset \Sigma$ and $|\Sigma'| = |\Sigma| + 1$. In this context note that the *if* part of the definition of preservedness under alphabet extension holds trivially: If two pattern $\alpha, \beta \in \mathrm{Pat}_\Sigma$ generate the same language over the enlarged alphabet Σ' then they necessarily also generate the same language over Σ.

With regard to the correctness of the abovementioned decision procedure proposed by Ohlebusch and Ukkonen [69], it is a necessary condition that the converse of Proposition 2.3 holds as well:

Question 6.3. *Let Σ be an alphabet, $|\Sigma| \geq 3$. For every pattern $\alpha \in \mathrm{Pat}_\Sigma$ and for every variable $j \in \mathrm{var}(\alpha)$, is j superfluous in α if and only if j is redundant in α?*

Referring to our explanatory remarks on the ambiguity of morphisms over unary and binary alphabets given in Chapter 6.2, it can be easily explained why these alphabet sizes are excluded from Conjecture 6.1. Our considerations are based on the following fact which describes an immediate relation between substantial ambiguity and variables being superfluous:

Lemma 6.10. *Let Σ be an alphabet and $\alpha \in \mathrm{Pat}_\Sigma$ a pattern. If there exists an $i \in \mathrm{var}(\alpha)$ such that every substitution $\sigma : (\mathbb{N} \cup \Sigma)^* \longrightarrow \Sigma^*$ with $\sigma(i) \neq \varepsilon$ is substantially ambiguous for i in α then i is superfluous in α with respect to Σ.*

Proof. If σ with $\sigma(i) \neq \varepsilon$ is substantially ambiguous for i in α then there exists a substitution $\tau : (\mathbb{N} \cup \Sigma)^* \longrightarrow \Sigma^*$ with $\tau(\alpha) = \sigma(\alpha)$ and $\tau(i) = \varepsilon$. Let the residual-preserving morphism $\phi : (\mathbb{N} \cup \Sigma)^* \longrightarrow (\mathbb{N} \cup \Sigma)^*$ be given by $\phi(i) := \varepsilon$ and $\phi(j) := j$, $j \in \mathbb{N} \setminus \{i\}$. Then, for every $\sigma : (\mathbb{N} \cup \Sigma)^* \longrightarrow \Sigma^*$, $\tau(\phi(\alpha)) = \sigma(\alpha)$. Consequently, $L_{\mathrm{E},\Sigma}(\phi(\alpha)) = L_{\mathrm{E},\Sigma}(\alpha)$, and therefore i is superfluous in α. □

While this phenomenon is by no means remarkable as long as i is a redundant variable (cf. Proposition 2.3), it deserves attention provided that i is not redundant:

Lemma 6.11. *Let Σ be an alphabet and $\alpha \in \text{Pat}_\Sigma$ a pattern. If α is morphically primitive and there exists an $i \in \text{var}(\alpha)$ such that i is superfluous in α with respect to Σ then there exists a pattern α' with*

- *α and α' are morphically semi-coincident and*
- *$L_{\text{E},\Sigma}(\alpha) = L_{\text{E},\Sigma}(\alpha')$.*

Proof. We define the residual-preserving morphism $\phi : (\mathbb{N} \cup \Sigma)^* \longrightarrow (\mathbb{N} \cup \Sigma)^*$ by $\phi(i) := \varepsilon$ and $\phi(j) := j$, $j \in \mathbb{N} \setminus \{i\}$. Since α is morphically primitive, there is no residual-preserving morphism $\psi : (\mathbb{N} \cup \Sigma)^* \longrightarrow (\mathbb{N} \cup \Sigma)^*$ such that $\psi(\phi(\alpha)) = \alpha$. Hence, α and $\alpha' := \phi(\alpha)$ are semi-coincident. The proof for $L_{\text{E},\Sigma}(\alpha') = L_{\text{E},\Sigma}(\alpha)$ is conducted in the proof for Lemma 6.10. □

Returning to our explanation on why Conjecture 6.1 does not consider unary and binary alphabets, we now can easily apply Lemma 6.11 to Proposition 6.5 (and to the patterns $\alpha_{\mathtt{a}}$ and $\alpha_{\mathtt{ab}}$ introduced in the proof thereof):

Proposition 6.7. *Let Σ be an alphabet, $|\Sigma| \in \{1, 2\}$. Then there exist semi-coincident patterns $\alpha, \alpha' \in \text{Pat}_\Sigma$ such that $L_{\text{E},\Sigma}(\alpha) = L_{\text{E},\Sigma}(\alpha')$.*

Proof. According to the proof for Proposition 6.5 there exist morphically primitive patterns in Pat_Σ – namely $\alpha_{\mathtt{a}}$ and $\alpha_{\mathtt{ab}}$, depending on the alphabet size – and a distinct variable i in these patterns such that each corresponding *nonerasing* substitution is substantially ambiguous for i. Additionally, it can be easily verified that the same holds for any other substitution σ with $\sigma(i) \neq \varepsilon$. Hence, the correctness of Proposition 6.7 follows immediately from Lemma 6.11. □

Consequently, as defined in the proof for Lemma 6.11, our simple semi-coincident example patterns substantiating Proposition 6.7 read

$$\begin{aligned} \alpha_{\mathtt{a}} &= 1 \cdot \mathtt{a} \cdot 2, \\ \alpha'_{\mathtt{a}} &= 1 \cdot \mathtt{a} \end{aligned}$$

for alphabet size 1 and

$$\begin{aligned} \alpha_{\mathtt{ab}} &= 1 \cdot \mathtt{a} \cdot 2 \cdot \mathtt{b} \cdot 3, \\ \alpha'_{\mathtt{ab}} &= 1 \cdot \mathtt{a} \cdot \mathtt{b} \cdot 3 \end{aligned}$$

for alphabet size 2.

Conjecture 6.1 now expects that suchlike examples do not exist for larger alphabets. The previous chapter, however, provides an example pattern for alphabet size 3 with analogous properties and therefore we can extend Proposition 6.7 to such alphabets:

Theorem 6.7. *Let Σ be an alphabet, $|\Sigma| = 3$. Then there exist semi-coincident patterns $\alpha, \alpha' \in \text{Pat}_\Sigma$ such that $L_{\text{E},\Sigma}(\alpha) = L_{\text{E},\Sigma}(\alpha')$.*

Proof. Let $\Sigma := \{\mathtt{a}, \mathtt{b}, \mathtt{c}\}$. As stated by Lemma 6.8, each substitution $\sigma : (\mathbb{N} \cup \Sigma)^+ \longrightarrow \Sigma^+$ is substantially ambiguous for the variable 4 in $\alpha_{\mathtt{abc}}$ (see Definition 6.7), and with regard to any other substitution $\sigma : (\mathbb{N} \cup \Sigma)^* \longrightarrow \Sigma^*$ satisfying $\sigma(4) \neq \varepsilon$, the proof for the existence of a substitution $\tau : (\mathbb{N} \cup \Sigma)^* \longrightarrow \Sigma^*$ with $\tau(4) = \varepsilon$ and $\tau(\alpha) = \sigma(\alpha)$ verbatim equals that of Lemma 6.8. Hence, $\alpha_{\mathtt{abc}}$ satisfies the conditions of Lemma 6.11, which thus proves the theorem. □

Consequently, since Lemma 6.8 identifies the variable 4 to be redundant in $\alpha_{\mathtt{abc}}$, the proof for Theorem 6.7 refers to the patterns

$$\begin{aligned} \alpha_{\mathtt{abc}} &= 1 \cdot \mathtt{a} \cdot 2 \cdot 3^2 \cdot 4^2 \cdot 5^2 \cdot 6^2 \cdot 7 \cdot \mathtt{b} \cdot 8 \cdot \mathtt{a} \cdot 2 \cdot 9^2 \cdot 4^2 \cdot 5^2 \cdot 10^2 \cdot 7 \cdot \mathtt{b} \cdot 11, \\ \alpha'_{\mathtt{abc}} &= 1 \cdot \mathtt{a} \cdot 2 \cdot 3^2 \cdot 5^2 \cdot 6^2 \cdot 7 \cdot \mathtt{b} \cdot 8 \cdot \mathtt{a} \cdot 2 \cdot 9^2 \cdot 5^2 \cdot 10^2 \cdot 7 \cdot \mathtt{b} \cdot 11. \end{aligned}$$

The second main lemma of the previous Chapter 6.2 presented in Lemma 6.9 does not allow for an application of Lemma 6.11 as we cannot guarantee that a particular single variable is superfluous in $\alpha_{\mathtt{abcd}}$, but we can merely prove that each nonerasing substitution is substantially ambiguous for one among two distinct variables. Nevertheless, we can use it for gaining (perhaps even deeper) insights into the correctness of Conjecture 6.1 and, in particular, into the relevance of Question 6.3 for the equivalence problem for E-pattern languages. Our considerations are based on the following observation which, within the present thesis, is the counterpart of Lemma 6.11 provided that the corresponding terminal alphabet consists of four distinct letters:

Lemma 6.12. *Let Σ be an alphabet and $\alpha \in \mathrm{Pat}_\Sigma$ a pattern. If α is morphically primitive and there exist two variables $i, j \in \mathrm{var}(\alpha)$, $i \neq j$, such that, for every substitution $\sigma : (\mathbb{N} \cup \Sigma)^* \longrightarrow \Sigma^*$ with $\sigma(i) \neq \varepsilon \neq \sigma(j)$, there is a substitution $\tau : (\mathbb{N} \cup \Sigma)^* \longrightarrow \Sigma^*$ with $\tau(\alpha) = \sigma(\alpha)$ and*

- *$\tau(i) = \varepsilon = \tau(j)$ or*
- *$\tau(i) = \varepsilon$ and $\tau(j) = \sigma(j)$ or*
- *$\tau(j) = \varepsilon$ and $\tau(i) = \sigma(i)$*

then there exist patterns $\hat{\alpha}, \hat{\alpha}' \in \mathrm{Pat}_\Sigma$ such that

- *$\hat{\alpha}$ and $\hat{\alpha}'$ are morphically incoincident and*
- *$L_{\mathrm{E},\Sigma}(\hat{\alpha}) = L_{\mathrm{E},\Sigma}(\hat{\alpha}')$.*

Proof. Initially, for an arbitrary $\mathtt{A} \in \Sigma$, we define

$$\begin{aligned} \hat{\alpha} &:= \alpha \cdot \mathtt{A} \cdot k \cdot i \cdot j \cdot k', \\ \hat{\alpha}' &:= \alpha \cdot \mathtt{A} \cdot k \cdot j \cdot i \cdot k' \end{aligned}$$

with $k, k' \in \mathbb{N}$, $k \neq k'$ and $k, k' \notin \mathrm{var}(\alpha)$.

We first prove that these patterns are morphically incoincident. Assume to the contrary that there is a residual-preserving morphism $\phi : (\mathbb{N} \cup \Sigma)^* \longrightarrow (\mathbb{N} \cup \Sigma)^*$ with $\phi(\hat{\alpha}) = \hat{\alpha}'$. Then – due to the last occurrence of the terminal symbol $\mathtt{A}$ in the patterns and the fact that ϕ is residual preserving and $\hat{\alpha}$ and $\hat{\alpha}'$ are similar – necessarily $\phi(k \cdot i \cdot j \cdot k') = k \cdot j \cdot i \cdot k'$. In particular, this implies that $\phi(i) \neq i$ or $\phi(j) \neq j$. On the other hand, because of the said property of ϕ and the similarity of the patterns, $\phi(\alpha) = \alpha$. As $k, k' \notin \mathrm{var}(\alpha)$ and $i, j \in \mathrm{var}(\alpha)$ this leads to $k, k' \notin \mathrm{var}(\phi(i)) \cup \mathrm{var}(\phi(j))$. Consequently, $\phi(i) = \varepsilon$ or $\phi(j) = \varepsilon$. W.l.o.g. we assume that $\phi(i) = \varepsilon$, and we define a residual-preserving morphism $\psi : (\mathbb{N} \cup \Sigma)^* \longrightarrow (\mathbb{N} \cup \Sigma)^*$ by $\psi(i) := \varepsilon$ and $\psi(i') = \phi(i')$, $i' \in \mathrm{var}(\hat{\alpha}) \setminus \{i\}$. Then $\phi(\psi(\alpha)) = \alpha$. In other words, α and $\psi(\alpha)$ are morphically coincident, and, additionally, $|\psi(\alpha)| < |\alpha|$. Thus, α is morphically imprimitive, which is a contradiction. Hence, there is no residual-preserving morphism mapping $\hat{\alpha}$ onto $\hat{\alpha}'$. Analogously, we can prove the non-existence of a morphism mapping $\hat{\alpha}'$ onto $\hat{\alpha}$ and therefore the patterns are incoincident.

We now prove that $\hat{\alpha}$ and $\hat{\alpha}'$ generate the same E-pattern language. To this end, we first show that for every substitution $\sigma : (\mathbb{N} \cup \Sigma)^* \longrightarrow \Sigma^*$ there exists a substitution $\sigma' : (\mathbb{N} \cup \Sigma)^* \longrightarrow \Sigma^*$ with $\sigma(\hat{\alpha}) = \sigma'(\hat{\alpha}')$ – a statement which implies $L_{\mathrm{E},\Sigma}(\hat{\alpha}) \subseteq L_{\mathrm{E},\Sigma}(\hat{\alpha}')$. If $\sigma(i) = \varepsilon$ or $\sigma(j) = \varepsilon$ then we define $\sigma'(i') := \sigma(i')$, $i' \in \mathbb{N}$; due to the shape of both patterns under consideration, this immediately leads to $\sigma(\hat{\alpha}) = \sigma'(\hat{\alpha}')$. Hence, let $\sigma(i) \neq \varepsilon \neq \sigma(j)$. Then, according to the condition of the theorem, there exists a substitution $\tau : (\mathbb{N} \cup \Sigma)^* \longrightarrow \Sigma^*$ such that one of the following cases is satisfied:

<u>Case 1:</u> $\tau(i) = \varepsilon = \tau(j)$.

Then we simply define

$$\sigma'(i') := \begin{cases} \tau(i') & , \quad i' \in \mathrm{var}(\alpha)\,, \\ \sigma(k \cdot i) & , \quad i' = k\,, \\ \sigma(j \cdot k') & , \quad i' = k'\,. \end{cases}$$

<u>Case 2:</u> $\tau(i) = \varepsilon$ and $\tau(j) = \sigma(j)$.

In the present case, σ' is given by

$$\sigma'(i') := \begin{cases} \tau(i') & , \quad i' \in \mathrm{var}(\alpha)\,, \\ \sigma(k \cdot i) & , \quad i' = k\,, \\ \sigma(k') & , \quad i' = k'\,. \end{cases}$$

<u>Case 3:</u> $\tau(j) = \varepsilon$ and $\tau(i) = \sigma(i)$.

Then σ' looks as follows:

$$\sigma'(i') := \begin{cases} \tau(i') & , \quad i' \in \mathrm{var}(\alpha)\,, \\ \sigma(k) & , \quad i' = k\,, \\ \sigma(j \cdot k') & , \quad i' = k'\,. \end{cases}$$

In each of these three cases, the verification of $\sigma(\hat{\alpha}) = \sigma'(\hat{\alpha}')$ is straightforward from the respective definitions. Consequently, $L_{\mathrm{E},\Sigma}(\hat{\alpha}) \subseteq L_{\mathrm{E},\Sigma}(\hat{\alpha}')$. The proof for $L_{\mathrm{E},\Sigma}(\hat{\alpha}') \subseteq L_{\mathrm{E},\Sigma}(\hat{\alpha})$ is analogous, and therefore $L_{\mathrm{E},\Sigma}(\hat{\alpha}') = L_{\mathrm{E},\Sigma}(\hat{\alpha})$. This proves the lemma. □

Thus, the proof for Lemma 6.12 does not construct a pattern α' to a given pattern α (as done by the proof for Lemma 6.10) such that these two patterns generate the same language, but it introduces two new patterns by adding two different particular suffixes to α. These suffixes contain occurrences of two non-redundant variables, the order of which is not reflected by their E-pattern languages (so that these languages are the same).

Referring to the particular properties of the pattern α_{abcd} (cf. Definition 6.7 and Lemma 6.9), Lemma 6.12 allows for an easy application of our considerations on the substantial ambiguity of substitutions over alphabets with four letters to the equivalence problem for E-pattern languages:

Theorem 6.8. *Let Σ be an alphabet, $|\Sigma| = 4$. Then there exist incoincident patterns $\alpha, \alpha' \in \mathrm{Pat}_\Sigma$ such that $L_{\mathrm{E},\Sigma}(\alpha) = L_{\mathrm{E},\Sigma}(\alpha')$.*

Proof. From the seven cases in the proof for Lemma 6.9 (or, more precisely, the definitions of $\tau(4)$ and $\tau(14)$ in these cases) it can be immediately derived that, for each nonerasing substitution, the pattern $\alpha_{\mathtt{abcd}}$ satisfies the conditions of Lemma 6.12. Additionally, with respect to any other substitution, the complete argumentation in the proof for Lemma 6.9 holds as well. Hence, defining $i := 4$ and $j := 14$, we can apply Lemma 6.12, which yields two patterns $\hat{\alpha}_{\mathtt{abcd}}$ and $\hat{\alpha}'_{\mathtt{abcd}}$ proving the theorem. □

If we define, e. g., $\mathtt{A} := \mathtt{a}$ then the patterns $\hat{\alpha}_{\mathtt{abcd}}$ and $\hat{\alpha}'_{\mathtt{abcd}}$ alleged in the above proof are given by

$$\begin{aligned}
\hat{\alpha}_{\mathtt{abcd}} &= 1 \cdot \mathtt{a} \cdot 2 \cdot 3^2 \cdot 4^2 \cdot 5^2 \cdot 6^2 \cdot 7 \cdot \mathtt{b} \cdot 8 \cdot \mathtt{a} \cdot 2 \cdot 9^2 \cdot 4^2 \cdot 5^2 \cdot 10^2 \cdot 7 \cdot \mathtt{b} \cdot 11 \\
&\quad \cdot \mathtt{c} \cdot 12 \cdot 13^2 \cdot 14^2 \cdot 15^2 \cdot 16^2 \cdot 17 \cdot \mathtt{d} \cdot 18 \cdot \mathtt{c} \cdot 12 \cdot 19^2 \cdot 14^2 \cdot 15^2 \cdot 20^2 \cdot 17 \cdot \mathtt{d} \cdot 21 \\
&\quad \cdot 14^2 \cdot 15^2 \cdot 14^2 \cdot 15^2 \cdot 14^2 \cdot 15^2 \cdot 22 \cdot 4^2 \cdot 5^2 \cdot 4^2 \cdot 5^2 \cdot 4^2 \cdot 5^2 \cdot \underline{\mathtt{a} \cdot 23 \cdot 4 \cdot 14 \cdot 24}\,, \\
\hat{\alpha}'_{\mathtt{abcd}} &= 1 \cdot \mathtt{a} \cdot 2 \cdot 3^2 \cdot 4^2 \cdot 5^2 \cdot 6^2 \cdot 7 \cdot \mathtt{b} \cdot 8 \cdot \mathtt{a} \cdot 2 \cdot 9^2 \cdot 4^2 \cdot 5^2 \cdot 10^2 \cdot 7 \cdot \mathtt{b} \cdot 11 \\
&\quad \cdot \mathtt{c} \cdot 12 \cdot 13^2 \cdot 14^2 \cdot 15^2 \cdot 16^2 \cdot 17 \cdot \mathtt{d} \cdot 18 \cdot \mathtt{c} \cdot 12 \cdot 19^2 \cdot 14^2 \cdot 15^2 \cdot 20^2 \cdot 17 \cdot \mathtt{d} \cdot 21 \\
&\quad \cdot 14^2 \cdot 15^2 \cdot 14^2 \cdot 15^2 \cdot 14^2 \cdot 15^2 \cdot 22 \cdot 4^2 \cdot 5^2 \cdot 4^2 \cdot 5^2 \cdot 4^2 \cdot 5^2 \cdot \underline{\mathtt{a} \cdot 23 \cdot 14 \cdot 4 \cdot 24}\,,
\end{aligned}$$

where the underlined parts describe the part differing from $\alpha_{\mathtt{abcd}}$. Note that, in the concrete case considered, the rightmost occurrence of the terminal symbol a in $\hat{\alpha}_{\mathtt{abcd}}$ and $\hat{\alpha}'_{\mathtt{abcd}}$ is not required for ensuring morphical incoincidence of the patterns. In our general proof for Lemma 6.12, however, it is necessary.

As explained above, Conjecture 6.1 expects that, for alphabets with at least three distinct letters, two patterns generate the same language if and only if they are morphically coincident. Contrary to this, the properties of our patterns $\alpha_{\mathtt{abc}}$, $\alpha'_{\mathtt{abc}}$ and $\hat{\alpha}_{\mathtt{abcd}}$, $\hat{\alpha}'_{\mathtt{abcd}}$ demonstrate that the equivalence of E-pattern languages is more sophisticated. Thus, we have to state that Ohlebusch and Ukkonen's conjecture is not completely correct:

Corollary 6.6. *Conjecture 6.1 is incorrect.*

Proof. Directly from Theorem 6.7 (for $|\Sigma| = \{3\}$) and Theorem 6.8 (for $|\Sigma| = \{4\}$). □

As our results on the ambiguity of terminal-preserving morphisms are restricted to alphabets with at most four distinct letters, we cannot provide any insights on the correctness of Conjecture 6.1 for larger alphabets. In fact, according to our notes concluding Chapter 6.2 we do not see any straightforward method to extend our method of constructing examples to $|\Sigma| \geq 5$. Therefore, from our current state of knowledge, it might be possible that Conjecture 6.1 is correct in that case.

Although there are significant differences (to be further discussed below) between the proof of Theorem 6.7 and that of Theorem 6.8, the equivalence of E-pattern languages over alphabets with at most four letters has some common properties which nicely contrast with the expectations (potentially) involved in Theorem 3.9, Theorem 3.10, and Conjecture 6.1. This can be easily derived from a minor paraphrase of Corollary 6.6:

Corollary 6.7. *Let Σ be an alphabet, $|\Sigma| \leq 4$. Then there exist morphically incoincident or semi-coincident patterns $\alpha_1, \alpha_2 \in \mathrm{Pat}_\Sigma$ such that $L_\Sigma(\alpha_1) = L_\Sigma(\alpha_2)$.*

Proof. Directly from Proposition 6.7, Theorem 6.7 and Theorem 6.8. □

Furthermore, from Corollary 6.6 it directly follows that, for some alphabet sizes, Question 6.2 must have an answer in the negative, too. Still, we need to state for which alphabet sizes exactly the equivalence is not preserved under alphabet extension. In this regard, we have the same alphabet sizes as in Corollary 6.7 – which, as to be discussed in the proof, is not completely evident:

Theorem 6.9. *Let* Σ *be an alphabet,* $|\Sigma| \leq 4$*. Then the equivalence for* ePAT_Σ *is not preserved under alphabet extension.*

Proof. With respect to alphabet size 1, the patterns $\alpha_{\mathtt{a}} = 1 \cdot \mathtt{a} \cdot 2$ and $\alpha'_{\mathtt{a}} = 1 \cdot \mathtt{a}$ immediately prove Theorem 6.9. As explained by the proof of Proposition 6.7, these patterns generate the same language over $\Sigma := \{\mathtt{a}\}$, but concerning $\Sigma' := \{\mathtt{a}, \mathtt{b}\}$ we can refer to, e. g. the word $w_{1,2} := \mathtt{a}\,\mathtt{b}$ which satisfies $w_{1,2} \in L_{\mathrm{E},\Sigma'}(\alpha_{\mathtt{a}}) \setminus L_{\mathrm{E},\Sigma'}(\alpha'_{\mathtt{a}})$.

If we consider $|\Sigma| = 2$ then we can refer to the patterns $\alpha_{\mathtt{ab}} = 1 \cdot \mathtt{a} \cdot 2 \cdot \mathtt{b} \cdot 3$ and $\alpha'_{\mathtt{ab}} = 1 \cdot \mathtt{a} \cdot \mathtt{b} \cdot 3$ introduced above. The proof for Proposition 6.7 states that $L_{\mathrm{E},\Sigma}(\alpha_{\mathtt{ab}}) = L_{\mathrm{E},\Sigma}(\alpha'_{\mathtt{ab}})$, and for $\Sigma' := \{\mathtt{a}, \mathtt{b}, \mathtt{c}\}$ the word $w_{2,3} := \mathtt{a}\,\mathtt{c}\,\mathtt{b}$ yields $L_{\mathrm{E},\Sigma'}(\alpha_{\mathtt{ab}}) \neq L_{\mathrm{E},\Sigma'}(\alpha'_{\mathtt{ab}})$.

While these insights on unary and binary alphabets do not require any significant effort (and, though not explicitly stated in literature, are well-known), our argumentation on $\Sigma := \{\mathtt{a}, \mathtt{b}, \mathtt{c}\}$ is based on the much more complex example patterns $\alpha_{\mathtt{abc}}$ and $\alpha'_{\mathtt{abc}}$ as introduced below the proof for Theorem 6.7. Nevertheless, due to the tools provided so far, our reasoning is simple: According to the proof of Theorem 6.7, $L_{\mathrm{E},\Sigma}(\alpha_{\mathtt{abc}}) = L_{\mathrm{E},\Sigma}(\alpha'_{\mathtt{abc}})$. Contrary to this, if we examine $\Sigma' := \{\mathtt{a}, \mathtt{b}, \mathtt{c}, \mathtt{d}\}$ then both $\alpha_{\mathtt{abc}}$ and $\alpha'_{\mathtt{abc}}$ are quasi-terminal free on Σ'. Hence, by Theorem 6.4, the fact that the patterns are semi-coincident (cf. the relevant parts of Proposition 6.6 and Lemma 6.11) immediately implies that they generate different languages over Σ'.

Finally, concerning $\Sigma := \{\mathtt{a}, \mathtt{b}, \mathtt{c}, \mathtt{d}\}$, we again utilise the corresponding patterns introduced above (more precisely, below the proof for Theorem 6.8), namely $\hat{\alpha}_{\mathtt{abcd}}$ and $\hat{\alpha}'_{\mathtt{abcd}}$. Since $|\operatorname{term}(\hat{\alpha}_{\mathtt{abcd}})| = |\operatorname{term}(\hat{\alpha}'_{\mathtt{abcd}})| = |\Sigma| = 4$, an alphabet extension from Σ to $\Sigma' := \{\mathtt{a}, \mathtt{b}, \mathtt{c}, \mathtt{d}, \mathtt{e}\}$, however, does not suffice for rendering $\hat{\alpha}_{\mathtt{abcd}}$ and $\hat{\alpha}'_{\mathtt{abcd}}$ quasi-terminal-free on Σ'. Consequently, an application of Theorem 6.7 – which states that the patterns generate the same language over Σ – and Proposition 6.6 and Lemma 6.12 – which, by Theorem 6.4, implies that their languages differ over alphabets of at least six distinct letters – only allows for the conclusion that ePAT_Σ *or* $\mathrm{ePAT}_{\Sigma'}$ is not preserved under alphabet extension. Hence, we now have to show that this indeed holds for ePAT_Σ.

To this end we construct a word in $L_{\mathrm{E},\Sigma'}(\hat{\alpha}_{\mathtt{abcd}}) \setminus L_{\mathrm{E},\Sigma'}(\hat{\alpha}'_{\mathtt{abcd}})$. Note that $\hat{\alpha}_{\mathtt{abcd}} = \alpha_{\mathtt{abcd}} \cdot \mathtt{a} \cdot 23 \cdot 4 \cdot 14 \cdot 24$ and $\hat{\alpha}'_{\mathtt{abcd}} = \alpha_{\mathtt{abcd}} \cdot \mathtt{a} \cdot 23 \cdot 14 \cdot 4 \cdot 24$; hence, we can refer to major parts of the patterns by the terms introduced in the second version of Definition 6.7, which describes $\alpha_{\mathtt{abcd}}$.

Using these notations we first can conclude from Theorem 5.2 that the patterns $\delta_1 := \beta_1 \cdot \beta'_1 \in \mathrm{Pat_{tf}}$ and $\delta_2 := \beta_2 \cdot \beta'_2 \in \mathrm{Pat_{tf}}$ are morphically primitive. Consequently, according to Theorem 5.7, the morphism $\sigma_{\mathrm{un},\delta_1}$ is unambiguous with respect to δ_1 and the morphism $\sigma_{\mathrm{un},\delta_2}$ is unambiguous with respect to δ_2. We now introduce a morphism $\phi_1 : \{\mathtt{a}, \mathtt{b}\}^+ \longrightarrow \{\mathtt{c}, \mathtt{e}\}^+$ by $\phi_1(\mathtt{a}) := \mathtt{c}$ and $\phi_1(\mathtt{b}) := \mathtt{e}$; analogously, we consider the morphism $\phi_2 : \{\mathtt{a}, \mathtt{b}\}^+ \longrightarrow \{\mathtt{a}, \mathtt{e}\}^+$ given by $\phi_2(\mathtt{a}) := \mathtt{a}$ and $\phi_2(\mathtt{b}) := \mathtt{e}$. Obviously, these morphisms ϕ_1, ϕ_2 simply rename letters, and therefore – and due to the fact that we so far only deal with terminal-free patterns – we may state that

(i) $\phi_1(\sigma_{\mathrm{un},\delta_1}(\delta_1))$ is unambiguous with respect to δ_1 and $\phi_2(\sigma_{\mathrm{un},\delta_2}(\delta_2))$ is unambiguous with respect to δ_2 and

(ii) $\phi_1(\sigma_{\mathrm{un},\delta_1}(4)) \in \{\mathtt{c},\mathtt{e}\}^+ \setminus \{\mathtt{e}\}^+$ and $\phi_2(\sigma_{\mathrm{un},\delta_2}(14)) \in \{\mathtt{a},\mathtt{e}\}^+ \setminus \{\mathtt{e}\}^+$, which implies that $\phi_1(\sigma_{\mathrm{un},\delta_1}(4))$ is not a suffix of $\phi_2(\sigma_{\mathrm{un},\delta_2}(14))$ and $\phi_2(\sigma_{\mathrm{un},\delta_2}(14))$ is not a suffix of $\phi_1(\sigma_{\mathrm{un},\delta_1}(4))$.

Since $\mathrm{var}(\delta_1) \cap \mathrm{var}(\delta_2) = \emptyset$ we can construct a substitution $\sigma : (\mathbb{N} \cup \Sigma')^* \longrightarrow \Sigma'^*$ by, for every $j \in \mathbb{N}$,

$$\sigma(j) := \begin{cases} \phi_1(\sigma_{\mathrm{un},\delta_1}(j)) & , \quad j \in \mathrm{var}(\delta_1), \\ \phi_2(\sigma_{\mathrm{un},\delta_2}(j)) & , \quad j \in \mathrm{var}(\delta_2), \\ \varepsilon & , \quad \text{else.} \end{cases}$$

We now wish to show that $\sigma(\hat{\alpha}_{\mathtt{abcd}}) \in L_{\mathrm{E},\Sigma'}(\hat{\alpha}_{\mathtt{abcd}}) \setminus L_{\mathrm{E},\Sigma'}(\hat{\alpha}'_{\mathtt{abcd}})$. Assume to the contrary that there is a substitution σ' with $\sigma'(\hat{\alpha}'_{\mathtt{abcd}}) = \sigma(\hat{\alpha}_{\mathtt{abcd}})$. Evidently, the letters $\mathtt{b}$ and $\mathtt{d}$ each occur exactly twice in $\sigma(\hat{\alpha}_{\mathtt{abcd}}) = \sigma'(\hat{\alpha}'_{\mathtt{abcd}})$ since, for every $j \in \mathbb{N}$, $\sigma(j) \in \{\mathtt{a},\mathtt{c},\mathtt{e}\}^*$. Therefore, we may conclude that $\sigma'(\beta) = \sigma(\beta)$ for $\beta \in \{\beta_1, \beta'_1, \beta_2, \beta'_2\}$, or, in other words, that σ' is not terminal-comprisingly ambiguous with respect to $\hat{\alpha}'_{\mathtt{abcd}}$. Hence (and due to (i)), for every $j \in \mathrm{var}(\delta_1) \cup \mathrm{var}(\delta_2)$, $\sigma'(j) = \sigma(j)$. Additionally, according to the assumption, we have $|\sigma'(\hat{\alpha}'_{\mathtt{abcd}})| = |\sigma(\hat{\alpha}_{\mathtt{abcd}})|$ and therefore, for every $j \notin \mathrm{var}(\delta_1) \cup \mathrm{var}(\delta_2)$, necessarily $\sigma'(j) = \sigma(j) = \varepsilon$. Thus, $\sigma(14)$ is a suffix of $\sigma(\hat{\alpha}_{\mathtt{abcd}})$ and $\sigma'(4) = \sigma(4)$ is a suffix of $\sigma'(\hat{\alpha}'_{\mathtt{abcd}})$. Due to our assumption $\sigma'(\hat{\alpha}'_{\mathtt{abcd}}) = \sigma(\hat{\alpha}_{\mathtt{abcd}})$, this implies that $\sigma(4)$ is a suffix of $\sigma(14)$ or $\sigma(14)$ is a suffix of $\sigma(4)$. Since $\sigma(4) = \phi_1(\sigma_{\mathrm{un},\delta_1}(4))$ and $\sigma(14) = \phi_2(\sigma_{\mathrm{un},\delta_2}(14))$, however, this statement contradicts (ii). Consequently, there is no substitution σ' satisfying $\sigma'(\hat{\alpha}'_{\mathtt{abcd}}) = \sigma(\hat{\alpha}_{\mathtt{abcd}})$ and therefore $L_{\mathrm{E},\Sigma'}(\hat{\alpha}_{\mathtt{abcd}}) \neq L_{\mathrm{E},\Sigma'}(\hat{\alpha}'_{\mathtt{abcd}})$. Thus, the equivalence for ePAT_Σ is not preserved under alphabet extension. □

Additionally, our example patterns $\alpha_{\mathtt{a}}$, $\alpha_{\mathtt{ab}}$ and $\alpha_{\mathtt{abc}}$ demonstrate that, for terminal alphabets Σ with up to three distinct letters, the set of prolix patterns in Pat_Σ does not equal the set of fixed points of nontrivial residual-preserving morphisms (see Schneider [93] for a charactarisation of these patterns, which is based on a decomposition similar to that in Theorem 5.2). This also contrasts with the corresponding results for terminal-free patterns (cf. Chapter 5.1.1) and quasi-terminal-free patterns (cf. Chapter 6.1.1).

While our above results mainly emphasise the common properties of "small" alphabets with respect to the equivalence problem (apart from the above remark on fixed points, which does not cover $|\Sigma| = 4$), we wish to conclude our explanations on Ohlebusch and Ukkonen's Conjecture by a closer look at the *differences* between the single alphabet sizes. In this regard, we do not see any newly discovered properties worth mentioning which differ for alphabets with up to three letters. Contrary to this, alphabet size 4 seems to have some particular characteristics: evidently, for $|\Sigma| = 3$, our proof on Corollary 6.6 is based on the semi-coincident patterns $\alpha_{\mathtt{abc}}$ and $\alpha'_{\mathtt{abc}}$, whereas, with respect to $|\Sigma| = 4$, we use the incoincident patterns $\hat{\alpha}_{\mathtt{abcd}}$ and $\hat{\alpha}'_{\mathtt{abcd}}$ – a fact which, by the way, implies that our argumentation on alphabet size 4 is much more sophisticated than that on size 3. The reason for our choice is fairly simple, as we do not know any semi-coincident examples generating the same languages over an alphabet of size 4, and we even suppose that such patterns do not exist:

Conjecture 6.2. *Let Σ be an alphabet, $|\Sigma| \geq 4$. Then there do not exist any morphically semi-coincident patterns α, α' such that $L_{\mathrm{E},\Sigma}(\alpha) = L_{\mathrm{E},\Sigma}(\alpha')$.*

In other words, we conjecture that there do not exist any semi-coincident counter-examples to Conjecture 6.1 in case of $|\Sigma| \geq 4$. Moreover, it seems as if neither $\hat{\alpha}_{\mathtt{abcd}}$ nor $\hat{\alpha}'_{\mathtt{abcd}}$ contain any superfluous variables such that these patterns disprove Conjecture 6.1 for alphabet size 4 although they seem to be compatible with a positive answer to Question 6.3. Consequently, and roughly speaking, we consider $L_{\mathrm{E},\Sigma}(\hat{\alpha}_{\mathtt{abcd}})$ not to be able to precisely reflect the order of the variables in any of its succinct generating patterns. Thus, we conjecture $L_{\mathrm{E},\Sigma}(\hat{\alpha}_{\mathtt{abcd}})$, $|\Sigma| = 4$, to be a pattern language that has two different succinct generating patterns in canonical form!

If this is correct then $\hat{\alpha}_{\mathtt{abcd}}$ and $\hat{\alpha}'_{\mathtt{abcd}}$ show that a modification of the definition of redundant variables leading to a correct "syntactic" characterisation of superfluous variables – as suggested by the examples on alphabets with exactly three letters – surprisingly has no impact on the decidability of the equivalence problem for E-pattern languages over four distinct letters. Therefore, if our conjecture holds and if, nevertheless, there exists a decision procedure for the equivalence of E-pattern languages over the said alphabet then it must strongly differ from the procedure proposed by Ohlebusch and Ukkonen [69] initially mentioned in this chapter.

6.3 A counter-intuitive observation

From our current state of knowledge, the considerations in the previous chapter – which, for alphabets with up to four distinct letters, provides deep insights into the non-extensibility of our positive findings for common morphisms and terminal-free patterns to terminal-preserving morphisms and general patterns – do not give a reliable hint on the properties of substitutions and general patterns over larger alphabets. As stated above and substantiated by the growing intricacy of the example patterns and by the difference between Lemma 6.8 and Lemma 6.9, this is caused by the fact that it seems easier to construct morphisms which are not terminal-comprisingly ambiguous (with respect to a fixed pattern) if a larger number of terminal symbols is available in the alphabet. In the present chapter we therefore can merely present a small observation on alphabets with at least five letters (which by the way also holds for smaller alphabets). Nevertheless, we consider it rather surprising and, thus, worth mentioning; furthermore it leads to unexpected insights into E-pattern languages which are described in the subsequent Chapter 6.3.1.

One of the most fundamental results presented so far states that neither common nonerasing morphisms nor their terminal-preserving counterparts can be unambiguous provided that the pattern under considerations is morphically imprimitive – this is formally explained by Theorem 5.1 and Theorem 6.1, and particularly the proof for the latter demonstrates the reason for this phenomenon. Consequently, one of the main problems discussed in the previous chapters is the question of whether at least for each morphically primitive pattern there exists an unambiguous nonerasing morphism. As shown by Theorem 5.7, the question has an answer in the affirmative for terminal-free patterns, Chapter 6.1 discusses why this also holds for quasi-terminal-free patterns and, finally, Theorem 6.5 demonstrates that there are particular general patterns such that the question has an answer in the negative if the corresponding alphabet does not contain more than four letters.

While the most important open question on unambiguity thus asks for the existence of analogous morphically primitive patterns for larger alphabets, we now take a closer look at morphically imprimitive patterns. Since these patterns are characterised by the fact that they contain redundant variables (cf. Propostion 2.1), it is by no means surprising that such variables cause substantial ambiguity. Hence, we can give a more precise version of Theorem 6.1:

Corollary 6.8. *Let Σ be an alphabet. Let $\alpha \in (\mathbb{N} \cup \Sigma)^+$ be morphically imprimitive. Then every nonerasing substitution $\sigma : (\mathbb{N} \cup \Sigma)^* \longrightarrow \Sigma^*$ is substantially ambiguous for every redundant variable i in α.*

Proof. The corollary follows from a straightforward extension of the proof of Theorem 6.1. □

Thus, roughly speaking, for redundant variables substantial ambiguity is inevitable.

Contrary to this, as explained above, our previous results show that if a morphism σ – chosen among the types of "useful" nonerasing morphisms examined in this thesis – is substantially ambiguous for a non-redundant variable in a pattern α then σ (cf. Lemma 6.3) or α (cf. Lemmata 6.8 and 6.9) are rather special. Although the corresponding results in Chapter 6.1 and, in particular, Chapter 6.2 merely consider morphically primitive patterns which, thus, do not contain any redundant variables, one might conjecture that well-chosen morphisms also can only be substantially ambiguous for non-redundant variables provided that the pattern under considerations is fairly sophisticated. Furthermore, it seems reasonable to expect that we again cannot give examples using more than four distinct letters.

We now shall demonstrate these two hypotheses to be incorrect. To this end, we consider the following patterns:

Definition 6.8 (Pattern $\alpha_{\pi,n}$). *Let $\Sigma := \{\mathtt{a}_1, \mathtt{a}_2, \dots \mathtt{a}_n\}$ be an alphabet, $n \geq 2$. Then the pattern $\alpha_{\pi,n} \in (\mathbb{N} \cup \Sigma)^+$ is given by*

$$\begin{aligned} \alpha_{\pi,n} \;:=\; & 1 \cdot \mathtt{a}_1 \cdot (2n+3) \cdot 2 \cdot \mathtt{a}_1 \cdot 3 \cdot \mathtt{a}_2 \cdot (2n+3) \cdot 4 \cdot \mathtt{a}_2 \cdot 5 \cdot \mathtt{a}_3 \cdot (2n+3) \cdot 6 \cdot \mathtt{a}_3 \cdot 7 \\ & [\dots] \cdot (2n-1) \cdot \mathtt{a}_n \cdot (2n+3) \cdot 2n \cdot \mathtt{a}_n \cdot (2n+1)\,. \end{aligned}$$

These patterns are given in a non-canonical form for the sake of a more convient presentation of the proof of Lemma 6.14 presented below.

Since the patterns $\alpha_{\pi,n}$ as introduced in Definition 6.8 at first glance might look more sophisticated than claimed above, we proceed with some examples:

Example 6.1. Let $\Sigma \subseteq \{\mathtt{a}, \mathtt{b}, \mathtt{c}, \mathtt{d}, \mathtt{e}\}$. Then the patterns $\alpha_{\pi,2}$, $\alpha_{\pi,3}$, $\alpha_{\pi,4}$, $\alpha_{\pi,5}$ read as follows:

$$\begin{aligned} \alpha_{\pi,2} &= 1 \cdot \underbrace{\mathtt{a} \cdot 7 \cdot 2 \cdot \mathtt{a}} \cdot 3 \cdot \underbrace{\mathtt{b} \cdot 7 \cdot 4 \cdot \mathtt{b}} \cdot 5, \\ \alpha_{\pi,3} &= 1 \cdot \underbrace{\mathtt{a} \cdot 9 \cdot 2 \cdot \mathtt{a}} \cdot 3 \cdot \underbrace{\mathtt{b} \cdot 9 \cdot 4 \cdot \mathtt{b}} \cdot 5 \cdot \underbrace{\mathtt{c} \cdot 9 \cdot 6 \cdot \mathtt{c}} \cdot 7, \\ \alpha_{\pi,4} &= 1 \cdot \underbrace{\mathtt{a} \cdot 11 \cdot 2 \cdot \mathtt{a}} \cdot 3 \cdot \underbrace{\mathtt{b} \cdot 11 \cdot 4 \cdot \mathtt{b}} \cdot 5 \cdot \underbrace{\mathtt{c} \cdot 11 \cdot 6 \cdot \mathtt{c}} \cdot 7 \cdot \underbrace{\mathtt{d} \cdot 11 \cdot 8 \cdot \mathtt{d}} \cdot 9, \\ \alpha_{\pi,5} &= 1 \cdot \underbrace{\mathtt{a} \cdot 13 \cdot 2 \cdot \mathtt{a}} \cdot 3 \cdot \underbrace{\mathtt{b} \cdot 13 \cdot 4 \cdot \mathtt{b}} \cdot 5 \cdot \underbrace{\mathtt{c} \cdot 13 \cdot 6 \cdot \mathtt{c}} \cdot 7 \cdot \underbrace{\mathtt{d} \cdot 13 \cdot 8 \cdot \mathtt{d}} \cdot 9 \cdot \\ & \quad \underbrace{\mathtt{e} \cdot 13 \cdot 10 \cdot \mathtt{e}} \cdot 11, \end{aligned}$$

Note that the braces are solely meant to emphasise the structure of the patterns. ◇

According to our overall goal, we first identify some of the non-redundant variables in $\alpha_{\pi,n}$ that we wish to deal with in Lemma 6.14 and Theorem 6.10:

Lemma 6.13. *Let Σ be an alphabet, $|\Sigma| \geq 2$, and let $i \in \{2, 4, 6, \ldots, 2n\}$. Then i is not redundant in $\alpha_{\pi,n}$.*

Proof. The lemma follows by straightforward combinatorial considerations from the structure of $\alpha_{\pi,n}$ and the definition of redundant variables. □

It can be easily seen that, actually, there are more non-redundant variables in each $\alpha_{\pi,n}$; in fact, only $(2n + 3) \in \text{var}(\alpha_{\pi,n})$ is redundant.

Now we can present the main lemma of the present chapter which specifies the substantial ambiguity of nonerasing morphisms with respect to $\alpha_{\pi,n}$:

Lemma 6.14. *Let $\Sigma := \{\mathtt{a}_1, \mathtt{a}_2, \ldots, \mathtt{a}_n\}$ be an alphabet, $n \geq 2$. Then, for every nonerasing substitution $\sigma : (\mathbb{N} \cup \Sigma)^* \longrightarrow \Sigma^*$, there exists a variable $i \in \{2, 4, 6, \ldots, 2n\}$ such that σ is substantially ambiguous for i in $\alpha_{\pi,n}$.*

Proof. Since σ is nonerasing, there exists some $\mathtt{a}_k \in \Sigma$, such that $\sigma(2n + 3) = \mathtt{a}_k \ldots$. Then the substitution $\tau : (\mathbb{N} \cup \Sigma)^* \longrightarrow \Sigma^*$ given by

$$\tau(j) := \begin{cases} \varepsilon & , \quad j \in \{2k, 2k+3\}, \\ [\mathtt{a}_k \setminus \sigma((2n+3) \cdot (j-1) \cdot \mathtt{a}_k \cdot j)] & , \quad j = 2k+1, \\ \sigma((2n+3) \cdot j) & , \quad j \text{ is even and } j \neq 2k, \\ \sigma(j) & , \quad j \text{ is odd and } j \notin \{2k+1, 2k+3\}, \end{cases}$$

satisfies $\tau(\alpha_{\pi,n}) = \sigma(\alpha_{\pi,n})$. Furthermore, $\tau(2k) = \varepsilon$, and therefore σ is substantially ambiguous for $2k$ in $\alpha_{\pi,n}$. □

Summarising the lemmata, we now can describe the main result of the present chapter:

Theorem 6.10. *Let Σ be an alphabet, $|\Sigma| \geq 2$. Then there exists an $\alpha \in (\mathbb{N} \cup \Sigma)^+$ with the following property: for every nonerasing substitution $\sigma : (\mathbb{N} \cup \Sigma)^* \longrightarrow \Sigma^*$, there exists a non-redundant $i \in \text{var}(\alpha)$ such that σ is substantially ambiguous for i in α.*

Proof. Directly from Lemma 6.13 and Lemma 6.14. □

Consequently, Theorem 6.10 provides first insights into the existence of *morphically imprimitive* patterns with respect to which each nonerasing substitution is substantially ambiguous for a *non-redundant* variable. In the light of our above explanations, the facts that these patterns are simply structured, exist for each alphabet with at least two distinct letters and are not derived from corresponding morphically primitive patterns strongly conflict with our intuition. In addition to this, as to be explained by the subsequent Chapter 6.3.1, even their mere existence does not meet the expectations involved in the literature on pattern languages (more precisely, the equivalence problem for E-pattern languages). Nevertheless, from a combinatorial point of view, we do not consider Theorem 6.10 a major insight which could match with the main results of the previous chapters since we already know from Theorem 6.1 that each nonerasing substitution is substantially ambiguous for some variable (which, according to the proof for Theorem 6.1, however is a redundant one) in $\alpha_{\pi,n}$.

6.3.1 Application: The Pattern Coding Paradox

In the present chapter, we further examine the pattern $\alpha_{\pi,n}$ (cf. Definition 6.8) and its property described in Theorem 6.10. Our considerations lead to the stunning insight, that injectivity of morphisms – in the previous parts of this thesis largely used as a convenient, but not always necessary feature of moderate ambiguity (or even unambiguity) with respect to morphically primitive patterns – can even be an *undesirable* property if terminal-preserving morphisms are sought that preserve the features of a pattern as precisely as possible. Hence, in a combinatorial context, if we have to choose *either* an injective *or* an unambiguous morphism for mapping a pattern onto a meaningful word (according to some task to be specified below) then it can be mandatory to pick the unambiguous one. We shall discuss this phenomenon with regard to the standard proof technique that in literature is normally used for showing the decidability of the equivalence problem for classes of E-pattern languages (see Chapter 3.2.1, Chapter 6.2.1 and Open Problem 3.3). Our analysis implies that this approach has to be modified when tackling the full class of E-pattern languages – provided that it is applicable at all in that case.

We first extensively discuss the said proof technique: If Conjecture 6.1 is correct for $|\Sigma| \geq 5$ – i. e. if the *inclusion* of general similar patterns (cf. Question 6.1) has the same properties as, e. g., that of quasi-terminal-free similar patterns (cf. Theorem 6.4) – then the method introduced for terminal-free patterns by Filè [19] and Jiang et al. [33] and pursued by Dányi and Fülöp [15] and Ohlebusch and Ukkonen [69] seems to be promising. Subject to any fixed pair of terminal-free or similar quasi-terminal-free patterns α, β, it is based on a particular substitution τ such that the following statements are equivalent:

- $\tau(\alpha) \in L(\beta)$,
- there is a terminal-preserving morphisms ϕ such that $\phi(\beta) = \alpha$ and
- $L_{\mathrm{E}}(\alpha) \subseteq L_{\mathrm{E}}(\beta)$.

With regard to this substitution τ, two properties are crucial:

(i) τ depends on the length of β and,

(ii) in a sense, τ is injective.

Property (i) mainly guarantees that $\tau(\alpha) \in L_{\mathrm{E}}(\beta)$ if *and only if* $L_{\mathrm{E}}(\alpha) \subseteq L_{\mathrm{E}}(\beta)$ – or, in other words, that neither $L_{\mathrm{E}}(\alpha) \# L_{\mathrm{E}}(\beta)$ nor $L_{\mathrm{E}}(\alpha) \supset L_{\mathrm{E}}(\beta)$ provided that $\tau(\alpha)$ can be generated by β. Preferably, for such a task, one might wish to choose a substitution τ' such that, for *every* pattern γ with $L_{\mathrm{E}}(\gamma) \not\supseteq L_{\mathrm{E}}(\alpha)$, the word $\tau'(\alpha)$ is not contained in $L_{\mathrm{E}}(\gamma)$. However, there are many patterns α such that for *every* word $w \in L_{\mathrm{E}}(\alpha)$ a pattern γ can be constructed with $w \in L_{\mathrm{E}}(\gamma)$ and $L_{\mathrm{E}}(\gamma) \subset L_{\mathrm{E}}(\alpha)$.[†] Thus, in general, for a given pattern α there is no *single* substitution which generates a significant word separating $L_{\mathrm{E}}(\alpha)$ from all E-pattern languages that are incomparable or strictly included. This is demonstrated by the following example (which is well-known from the negative learnability result for the class of terminal-free E-pattern languages over binary alphabets, cf. Theorem 3.15):

[†]Note that, due to our considerations in Chapter 5.2.1 and Chapter 6.1.1 on the closely related topic of telltales for E-pattern languages (cf. Theorem 3.12), this is by no means surprising.

Example 6.2. Let $\Sigma = \{\mathtt{a}, \mathtt{b}\}$ and $\alpha := \alpha^{\text{tf}}_{\mathtt{ab}} = 1 \cdot 1 \cdot 2 \cdot 2 \cdot 3 \cdot 3$. Let the inverse substitution $\bar{\sigma} : \Sigma^* \longrightarrow \mathbb{N}^*$ be given by $\bar{\sigma}(\mathtt{a}) := 1$ and $\bar{\sigma}(\mathtt{b}) := 2$. Then, for every morphism $\sigma : \mathbb{N}^* \longrightarrow \Sigma^*$ that assigns a nonempty word to α, we define $\gamma := \bar{\sigma}(\sigma(\alpha))$; thus, $|\sigma(\alpha)| = |\gamma|$, $\sigma(\alpha) \in L_{\text{E},\Sigma}(\gamma)$ and $L_{\text{E},\Sigma}(\gamma) \subset L_{\text{E},\Sigma}(\alpha)$ (due to Theorem 3.6). ◇

Consequently, *for every* $k \in \mathbb{N}$, Jiang et al. [33] define a particular substitution τ_k (which is based on a number of *segments* as described in Chapter 5.2) such that, for every pair of patterns α, β with $|\beta| = k$, $|\tau_k(\alpha)| \gg |\beta|$. Thereby they can conclude that the generated word $\tau_k(\alpha)$ because of its length and shape is not contained in the language of any pattern γ with $|\gamma| = |\beta|$ and $L(\gamma) \not\supseteq L(\alpha)$.

Due to technical reasons and with regard to our subsequent needs, we combine α with a finite number of arbitrary suchlike γ and, thus, introduce the following concept:

Definition 6.9 (Separable tuple). *Let Σ be an alphabet, and let $\alpha, \gamma_1, \gamma_2, \ldots, \gamma_n \in \text{Pat}_\Sigma$, $n \geq 1$. The $(n+1)$-tuple $(\alpha, \gamma_1, \gamma_2, \ldots, \gamma_n)$ is called* separable *if and only if $L_{\text{E},\Sigma}(\alpha) \not\subseteq (L_{\text{E},\Sigma}(\gamma_1) \cup L_{\text{E},\Sigma}(\gamma_2) \cup \ldots \cup L_{\text{E},\Sigma}(\gamma_n))$.*

Hence, for each separable tuple there must be a word $w \in L_{\text{E},\Sigma}(\alpha)$ which is not contained in any of the $L_{\text{E},\Sigma}(\gamma_i)$, $1 \leq i \leq n$. We name such a word as follows:

Definition 6.10 (Coding of a language). *Let Σ be an alphabet, and $(\alpha, \gamma_1, \gamma_2, \ldots, \gamma_n) \in \text{Pat}_\Sigma^{n+1}$ a separable tuple. Then a word w is called a* coding of $L_{\text{E},\Sigma}(\alpha)$ *(*with respect to $L_{\text{E},\Sigma}(\gamma_1), \ldots, L_{\text{E},\Sigma}(\gamma_n)$*) if and only if $w \in L_{\text{E},\Sigma}(\alpha) \setminus (L_{\text{E},\Sigma}(\gamma_1) \cup L_{\text{E},\Sigma}(\gamma_2) \cup \ldots \cup L_{\text{E},\Sigma}(\gamma_n))$.*

Before we proceed with the actual results of this chapter we now briefly discuss property (ii) of τ. In this context, it is stated in the end of the introduction of Chapter 6 that no substitution is injective on $(\mathbb{N} \cup \Sigma)^+$, and therefore we call a substitution injective if it is injective on $\mathbb{N}^+$. We wish to emphasise that, within the present chapter, this decision does not lead to any restrictions: As we shall exclusively deal with similar patterns we can utilise that, for every set Pat_{sim} of pairwise similar patterns, there exist various substitutions $\sigma_{\text{inj,sim}}$ that are injective on Pat_{sim}. The τ_k as introduced by Dányi and Fülöp [15] constitute an example for substitutions with such a property.† If we deal with a set Pat_{sim} which contains *all* patterns that are similar to some given pattern α then each $\sigma_{\text{inj,sim}}$ additionally is injective on $\mathbb{N}^+$, and therefore the set of all substitutions that are injective on $\mathbb{N}^+$ comprises all injective substitutions that we are interested in.

Property (ii) of τ mainly addresses the fact that $\tau(\alpha) \in L_{\text{E}}(\beta)$ if and only if there exists a morphism ϕ mapping β onto α. The existence of such a morphism ϕ with $\phi(\beta) = \alpha$ is caused by the special shape of any τ_k, which allows for the definition of a particular morphism $\bar{\tau}_k$ that is inverse to τ_k, i.e. $\bar{\tau}_k(\tau_k(\alpha)) = \alpha$. In other words, if a pattern β can generate $\tau_k(\alpha)$ then it can also morphically generate α and, thus, all words in $L(\alpha)$ – more precisely, if there is a substitution τ' with $\tau'(\beta) = \tau_k(\alpha)$ then $\phi = \bar{\tau}_k \circ \tau'$, and consequently, for any word $w = \sigma'(\alpha)$, $w = \sigma'(\bar{\tau}_k(\tau'(\beta)))$. Thus, as the existence of $\bar{\tau}_k$ apparently is closely connected to the injectivity of τ_k, it seems mandatory to search for a coding of an E-pattern language L among the set of all words in L generated by injective substitutions. For the sake of convenience and in accordance with literature, we henceforth designate a word generated by an injective substitution as follows:

†Note that the substitutions τ_k in [15] generally do not, when applied to some α, lead to a coding of $L(\alpha)$ since these τ_k do not sufficiently account for the impact of terminal-comprising ambiguity. This is pointed out by Ohlebusch and Ukkonen [69] and strengthened by our results in Chapter 6.2.1.

Definition 6.11 (Coding of a pattern). *Let α be a pattern. Then a word $w \in L(\alpha)$ is called a* coding *(of α) if and only if there is an injective substitution σ such that $\sigma(\alpha) = w$.*

We now show that the natural conjunction between codings of patterns and codings of pattern languages – which among others is backed by the abovementioned positive results on terminal-free and similar quasi-terminal-free E-pattern languages in [19, 33, 69] – does not hold any longer if the full class of E-pattern languages is considered. For any alphabet size n, this is demonstrated by $\alpha_{\pi,n}$ (see Definition 6.8) and the following patterns:

Definition 6.12 (Patterns $\breve{\gamma}_i$). *Let $\Sigma := \{\mathtt{a}_1, \mathtt{a}_2, \dots, \mathtt{a}_n\}$ be an alphabet, $n \geq 2$. Then, for every i, $1 \leq i \leq n$, let the pattern $\breve{\gamma}_i \in \mathrm{Pat}_\Sigma$ be given by*

$$\breve{\gamma}_i := 1 \cdot \gamma_{i,1} \cdot 3 \cdot \gamma_{i,2} \cdot 5 \cdot \gamma_{i,3} \cdot 7 \cdot [\dots] \cdot 2n-1 \cdot \gamma_{i,n} \cdot 2n+1$$

with, for every i', $1 \leq i' \leq n$,

$$\gamma_{i,i'} := \begin{cases} \mathtt{a}_{i'} \cdot 2i' \cdot 2i' \cdot \mathtt{a}_{i'} & , \quad i = i', \\ \mathtt{a}_{i'} \cdot 2i' \cdot \mathtt{a}_{i'} & , \quad i \neq i'. \end{cases}$$

Obviously, the patterns $\alpha_{\pi,n}, \breve{\gamma}_1, \breve{\gamma}_2, \dots, \breve{\gamma}_n$ are pairwise similar and, apart from $\alpha_{\pi,n}$, have the same length.

We first prove that the given patterns can constitute a separable tuple:

Lemma 6.15. *Let Σ be an alphabet, and $n := |\Sigma| \geq 2$. Then the tuple $(\alpha_{\pi,n}, \breve{\gamma}_1, \dots, \breve{\gamma}_n)$ is separable.*

Proof. Let $\Sigma := \{\mathtt{a}_1, \mathtt{a}_2, \dots \mathtt{a}_n\}$. For every $j \in \mathrm{var}(\alpha_{\pi,n})$, let the substitution $\sigma : (\mathbb{N} \cup \Sigma)^* \longrightarrow \Sigma^*$ be given by

$$\sigma(j) := \begin{cases} \varepsilon & , \quad j \text{ is odd}, \\ \mathtt{a}_n & , \quad j = 2, \\ \mathtt{a}_{\frac{j}{2}-1} & , \quad j \text{ is even and } j \neq 2. \end{cases}$$

Then $\sigma(\alpha_{\pi,n}) = \mathtt{a}_1\, \mathtt{a}_n\, \mathtt{a}_1\;\; \mathtt{a}_2\, \mathtt{a}_1\, \mathtt{a}_2\;\; \mathtt{a}_3\, \mathtt{a}_2\, \mathtt{a}_3\, [\dots]\; \mathtt{a}_n\, \mathtt{a}_{n-1}\, \mathtt{a}_n =: w_0$. Thus, $|w_0| = 3n$ and, for any i, $1 \leq i \leq n$, w_0 does not contain the subword $\mathtt{a}_i^2$. Now assume to the contrary that there is a substitution σ' and a k, $1 \leq k \leq n$, such that $\sigma'(\breve{\gamma}_k) = w_0$. Then, since $\sigma'(\breve{\gamma}_k)$ must not contain any subword $\mathtt{a}_i^2$, σ' has to assign nonempty words to all even variables in $\breve{\gamma}_k$. Consequently, $|\sigma'(\breve{\gamma}_k)| \geq 3n+1$, which contradicts the length of w_0. Thus, $(\alpha_{\pi,n}, \breve{\gamma}_1, \dots, \breve{\gamma}_n)$ is separable. □

Note that the proof for Lemma 6.15 indirectly demonstrates that the even variables in $\alpha_{\pi,n}$ are not only non-redundant (as extensively discussed in Chapter 6.3), but also non-superfluous in $\alpha_{\pi,n}$. This is essential for the statements in the present chapter.

In the proof of Lemma 6.15, w_0 evidently is generated by an unambiguous, non-injective morphism. However – since the patterns in the separable tuple $(\alpha_{\pi,n}, \breve{\gamma}_1, \dots, \breve{\gamma}_n)$ are pairwise similar and all $\breve{\gamma}_i$, $1 \leq i \leq n$, have the same length – the proof technique discussed in this chapter can only be applicable to general E-pattern languages provided that, additionally, there exists an *injective* morphism τ such that $\tau(\alpha_{\pi,n})$ is not contained in the union of all $L_{\mathrm{E}}(\breve{\gamma}_i)$, $1 \leq i \leq n$. But this is disproven by the following lemma:

Lemma 6.16. *Let Σ be an alphabet, and $n := |\Sigma| \geq 2$. Then, for every injective substitution $\sigma : (\mathbb{N} \cup \Sigma)^* \longrightarrow \Sigma^*$, there exists a substitution σ' and an index k, $1 \leq k \leq n$, such that $\sigma'(\breve{\gamma}_k) = \sigma(\alpha_{\pi,n})$.*

Proof. From Lemma 6.14 we know that each nonerasing (and, hence, each injective) substitution is substantially ambiguous for a variable $i \in \{2, 4, \dots, 2n\}$ in $\alpha_{\pi,n}$. Hence there exists a substitution $\tau : (\mathbb{N} \cup \Sigma)^* \longrightarrow \Sigma^*$ such that $\tau(\alpha_{\pi,n}) = \sigma(\alpha_{\pi,n})$ and $\tau(i) = \varepsilon$. Additionally, we can immediately derive from the proof of Lemma 6.14 that $\tau(2n+3) = \varepsilon$, too. Thus, if we define $\sigma' := \tau$ and $k := \frac{i}{2}$ then, due to the structure of $\breve{\gamma}_k$, this yields $\sigma'(\breve{\gamma}_k) = \sigma(\alpha_{\pi,n})$. □

Consequently, sometimes an inaccurate morphic image of a pattern α represents the set of all morphic images of α more accurately than the most accurate ones:

Theorem 6.11. *Let Σ be an alphabet, $|\Sigma| \geq 2$. Then there exists a separable tuple $(\alpha, \gamma_1, \gamma_2, \dots, \gamma_{|\Sigma|})$ of patterns such that no word in $L_\Sigma(\alpha)$ is both a coding of α and a coding of $L_\Sigma(\alpha)$.*

Proof. Directly from Lemma 6.15 and Lemma 6.16. □

With reference to the common understanding of codes (i.e., in our sense, injective morphisms) we consider this a paradoxical phenomenon.

Additionally, with respect to Theorem 6.11, it seems worth mentioning that each example pattern $\breve{\gamma}_i$, $1 \leq i \leq n$, generates a proper sublanguage of $L(\alpha_{\pi,n})$. Thus, Theorem 6.11 actually implies an even stronger result. Moreover, the given patterns can slightly be shortened (e.g. by removing $2n + 1$ and the last occurrence of $\mathtt{a}_n$) without losing the stated property, but this leads to a less transparent structure of the examples.

We conclude this chapter with some remarks describing the consequences of Theorem 6.11 on the equivalence problem for E-pattern languages. As discussed in Chapter 6.3, the ambiguity of $\tau(\alpha_{\pi,n})$ for injective substitutions τ is caused by the particular arrangement of the redundant variable $2n+3$. Consequently, $\alpha_{\pi,n}$ is a prolix pattern, the succinct counterpart of which can be evaluated easily and does not show a comparable problem. Hence, we consider it one of the key questions on the equivalence problem whether or not the substantial ambiguity of injective substitutions for non-superfluous variables can only be brought about by existence of some other, redundant variables in the pattern under consideration. If this can be answered in the affirmative then there is a simple solution preventing the problem explained in Theorem 6.11: In that case, the search for an appropriate injective substitution can simply be restricted to morphically primitive patterns. Nevertheless, this implies that the current perception of redundant variables – as symbols that have no impact on a language and, thus, *may* be mapped onto the empty word – has to be redetermined since Theorem 6.11 provides the insight that sometimes they *must* be mapped onto the empty word. Additionally note that this proposed modification is not sufficient for correcting the proof of Lemma 1 in [15]. On the other hand, if an analogous result to Theorem 6.11 can also be obtained for morphically primitive patterns then, for two patterns α, β, it might still be possible to extend the examined proof technique by introducing a set $\{\tau_{k,1}, \tau_{k,2}, \dots, \tau_{k,m}\}$, $m \geq 2$, of non-injective substitutions – each of which is "partially injective", i.e. it assigns the empty word to a nonempty set $X_\varepsilon \subset \mathrm{var}(\alpha)$ and is injective on $(\mathrm{var}(\alpha) \setminus X_\varepsilon)^+$ – such that $\{\tau_{k,1}(\alpha), \tau_{k,2}(\alpha), \ldots, \tau_{k,m}(\alpha)\} \subseteq L(\beta)$ if and only if $L(\alpha) \subseteq L(\beta)$. We expect Conjecture 6.1 to be incorrect for each alphabet size that does not allow any of these two proposed proof methods.

Chapter 7

Conclusion and open questions

In the present chapter, we summarise the main results of this thesis, and we discuss some further research directions which can be derived from our studies. For the sake of an easier access to the main lines of reasoning in the previous chapters, we describe the crucial statements on the ambiguity of morphisms and their impact on properties of pattern languages separately.

7.1 Ambiguity of morphisms

Our terminology on the ambiguity of morphisms has been introduced in Chapter 4. Therein, we have described the concepts of unambiguity, weak unambiguity, moderate ambiguity, substantial ambiguity and terminal-comprising ambiguity. Additionally, we have posed our main questions on the subject, which ask for

- the existence of an unambiguous nonerasing morphism with respect to an arbitrarily chosen pattern,
- the existence of a moderately ambiguous morphism with respect to an arbitrarily chosen pattern,
- the existence of a (preferably injective) morphism which is weakly unambiguous with respect to every pattern,
- the avoidability of terminal-comprising ambiguity, and, as a consequence thereof,
- the extensibility of the results on common morphisms to terminal-comprising morphisms, i. e., more precisely, to substitutions.

In Chapter 5, for arbitrary alphabets σ satisfying $|\Sigma| \geq 2$, we have studied the ambiguity of common morphisms $\sigma : \mathbb{N}^* \longrightarrow \Sigma^*$; hence, we have restricted ourselves to terminal-free patterns. We first have effortlessly proven a fundamental negative result which states that every nonerasing morphism is substantially ambiguous with respect to every *morphically imprimitive* pattern (cf. Theorem 5.1). Hence, as an immediate consequence thereof, we cannot unambiguously or moderately ambiguously map any morphically imprimitive pattern by a nonerasing morphism onto a string in an arbitrarily chosen free monoid. Subsequent to this, in order to gain a deeper understanding of this phenomenon and to have a tool for our subsequent considerations, we have characterised the morphically impritive patterns by means of a particular decomposition which these patterns have to show (cf. Theorem 5.2).

This characterisation of morphic imprimitivity – and, hence, of morphic primitivity as well – has significantly contributed to our proof for Lemma 5.8 (which requires the perhaps most difficult argumentation in the present thesis) that, in turn, is crucial for the proof of our first main statement on *moderate ambiguity* given in Theorem 5.3. Therein, we have described that every morphism $\sigma_{\text{3-seg}}$, which maps each variable in the pattern under consideration onto three distinct segments $\mathtt{A}\,\mathtt{B}^n\,\mathtt{A}$, $\mathtt{A}, \mathtt{B} \in \{\mathtt{a}, \mathtt{b}\}$, $\mathtt{A} \neq \mathtt{B}$, $n \in \mathbb{N}$ (cf. Definition 5.1), is moderately ambiguous with respect to every *morphically primitive* pattern. Consequently, and referring to Theorem 5.1, the existence of a morphism which is moderately ambiguous with respect to a pattern α essentially depends on the question of whether α is morphically primitive (cf. Corollary 5.6). As our result on moderate ambiguity is based on a concrete (set of) morphism(s), we have been able to briefly study whether the particular structure of the morphisms $\sigma_{\text{3-seg}}$ is necessary for guaranteeing moderate ambiguity. Our corresponding result has suggested an answer in the affirmative, as it introduces a pattern α and a morphism $\sigma_{\text{2-seg}}$ mapping each variable in α onto a word that consists of two distinct segments such that $\sigma_{\text{2-seg}}$ is not moderately ambiguous with respect to α (cf. Proposition 5.1).

We have started our considerations on *un*ambiguity by some remarks on *weak unambiguity*. Our main result on this topic has demonstrated that – while there are nonerasing morphisms that are weakly unambiguous with respect to every (morphically primitive and morphically imprimitive) pattern – there is no injective morphisms which is weakly unambiguous with respect to every pattern in $\mathbb{N}^+$ (cf. Theorem 5.6).

Turning our attention to *unambiguity* in free monoids, we have immediately concluded from Theorem 5.1 that, contrary to the outcome for weak unambiguity, there is no unambiguous nonerasing morphism with respect to any morphically imprimitive pattern. Hence, we have found a fundamental analogy between moderate ambiguity and unambiguity and a major difference between weak unambiguity and unambiguity. In Proposition 5.4, however, we have shown that, unlike our considerations on moderate ambiguity, there is no single nonerasing morphism that is unambiguous with respect to every morphically primitive pattern. Thus, we have studied whether or not, for every morphically primitive pattern $\alpha \in \mathbb{N}^+$, there is at least a tailor-made nonerasing morphism. Referring to the morphism $\sigma_{\text{un},\alpha}$ (cf. Definition 5.6), which incorporates the structure of α in a sophisticated manner, we have answered this question in the affirmative (cf. Theorem 5.7). The corresponding proof essentially depends on the fact that $\sigma_{\text{un},\alpha}$ again is a morphism which maps each variable in the pattern onto a word that consists of three distinct segments, so that $\sigma_{\text{un},\alpha}$ definitely is moderately ambiguous with respect to α. Similarly to our argumentation on $\sigma_{\text{2-seg}}$, we have demonstrated that these three segments of $\sigma_{\text{un},\alpha}$ are not only convenient for our proof of Theorem 5.7, but also necessary for the unambiguity of $\sigma_{\text{un},\alpha}$ since there exists a pattern α such that the morphism $\sigma_{\neg\text{un},\alpha}$ – which merely maps the variables j in α onto words that consists of the left and the right segments of $\sigma_{\text{un},\alpha}(j)$ – is not unambiguous (cf. Proposition 5.5). Remarkably, the corresponding pattern α is significantly more intricate than that proving Proposition 5.1. Finally, we have once more addressed the problem that there exists a nonerasing morphism which is moderately ambiguous with respect to every morphically primitive pattern, whereas there is no such morphism guaranteeing unambiguity. In order to partly compensate for this insurmountable problem, we have proposed to seek for major and natural sets Π of morphically primitive patterns such that there exists a single (and preferably simply structured) morphism σ_Π which is unambiguous with respect to each pattern in Π. We have illustrated this approach by giving a corrected proof for a theorem incompletely proven by Reidenbach [76], according to which the morphism σ_c given by $\sigma_c(j) := \mathtt{a}\,\mathtt{b}^j$, $j \in \mathbb{N}$, is unambiguous with respect to every terminal-free non-cross pattern (cf. Theorem 5.8).

Consequently, by Theorems 5.1, 5.3, 5.6, 5.7 and the related corollaries, we have provided comprehensive insights into what we consider the most fundamental problems on the ambiguity of common morphisms $\sigma : \mathbb{N}^* \longrightarrow \Sigma^*$, $|\Sigma| \geq 2$. In particular, we have exhaustively answered Questions 4.1, 4.2 and 4.4, and have given a first nontrivial result related to Question 4.3. Our argumentation on the positive statements among these results has been widely constructive, and therefore we have not only discussed the *(non-)existence* of moderately ambiguous, weakly unambiguous and unambiguous morphisms, but, at the same time, we have also analysed concrete important example morphisms.

In Chapter 6, we have examined the ambiguity of terminal-preserving morphisms; more precisely, we have largely focussed on substitutions $\sigma : (\mathbb{N} \cup \Sigma)^* \longrightarrow \Sigma^*$ for various alphabets Σ. Our considerations have started with the simple yet momentous observation that every nonerasing substitution is substantially ambiguous with respect to every morphically imprimitive pattern (cf. Theorem 6.1). Thus, in this regard, the behaviour of common morphisms (cf. Theorem 5.1) and of substitutions is identical. Consequently, and since our other negative results on common morphisms can be immediately extended to substitutions, too, we have concentrated on the question of whether or not the positive results on the existence of moderately ambiguous and unambiguous morphisms for morphically primitive patterns provided by Chapter 5 also hold for substitutions. From a rather technical point of view, we have explained that any answer to this question strongly depends on the *avoidability of terminal-comprising ambiguity*, i. e. the question of whether, for any given pattern $\alpha \in (\mathbb{N} \cup \Sigma)^+$, there exists a substitution σ such that σ is not terminal-comprisingly ambiguous with respect to α. With respect to any fixed alphabet Σ, $|\Sigma| \geq 3$, we have first tackled this problem for the *quasi-terminal-free* patterns $\alpha \in (\mathbb{N} \cup \Sigma)^+$, which are characterised by the fact that Σ contains at least two letters $\mathtt{A}, \mathtt{B}$ not occurring in α (cf. Definition 6.2). Since for these patterns terminal-comprising ambiguity can be avoided easily – simply by choosing substitutions which map the variables onto words over these two letters $\mathtt{A}, \mathtt{B}$ (cf. Lemma 6.2) – we have been able to extend our insights in the ambiguity of common morphisms with little effort. Formally, we have merely stated in Theorem 6.2 that, for every morphically primitive quasi-terminal-free pattern α, there exists a substitution $\sigma_{3\text{-seg},\mathtt{A},\mathtt{B}}$ such that $\sigma_{3\text{-seg},\mathtt{A},\mathtt{B}}$ is moderately ambiguous with respect to α. We have not explicitly examined the analogous question for unambiguity, as we consider it straightforward and of little interest. Instead of this, we have turned our attention to a particular type of substitutions which are crucial for problems related to inductive inference, namely the *telltale candidates*. Many telltale candidates have to map the variables in a pattern onto a word over three distinct letters, so that the abovementioned simple method of avoiding terminal-comprising ambiguity as formally described by Lemma 6.2 in general is not applicable for these substitutions (when considering quasi-terminal-free patterns). Referring to this observation and utilising the particular patterns $\alpha^{\mathrm{qtf}}_{\mathtt{abc}}$ (which is quasi-terminal-free on alphabets with three distinct letters) and $\alpha^{\mathrm{qtf}}_{\mathtt{abcd}}$ (which is quasi-terminal-free as soon as the alphabet contains four letters), we have demonstrated in Theorem 6.3 that there exist alphabets Σ, morphically primitive patterns α (that are quasi-terminal-free on Σ) and a variable i in these patterns such that no telltale candidate for i is moderately ambiguous with respect to α. Thus, as this result does not hold for terminal-free patterns (cf. Lemma 5.9), we have found the first (and, concerning quasi-terminal-free patterns, our only) significant difference between the ambiguity of common and terminal-preserving morphisms. The impact of this insight into the inferrability of E-pattern languages is summarised in Chapter 7.2.

Subsequent to this, we have suspended all restrictions on the shape of the patterns, and, thus, we have examined whether the positive insights on morphically primitive terminal-free and quasi-terminal-free patterns also hold for general patterns in $(\mathbb{N} \cup \Sigma)^+$. With respect to unary and binary alphabets Σ, it is a well-known fact that there exist simple morphically primitive patterns in $(\mathbb{N} \cup \Sigma)^+$ such that terminal-comprising ambiguity cannot be avoided and, furthermore, every nonerasing substitution is substantially ambiguous with respect to these patterns. Consequently, for binary alphabets, we have immediately seen that the ambiguity of terminal-preserving morphisms differs from that of common morphisms. With regard to larger alphabets, we have noted that the situation is less evident since, in this case, we can always choose a substitution which maps each variable j in the pattern α under consideration onto a word over a letter $\mathtt{C}$ such that, for at least one occurrence j_k of j in α, $\mathtt{C}$ does not belong to the (at most) two distinct terminal symbols which are closest to j_k in α, i.e. if $\alpha = \ldots \cdot \mathtt{A} \cdot \beta_1 \cdot j_k \cdot \beta_2 \cdot \mathtt{B} \cdot \ldots$ with $\mathtt{A} \neq \mathtt{C} \neq \mathtt{B}$ and $\beta_1, \beta_2 \in \mathbb{N}^*$ then we can choose a morphism σ with $\sigma(j) \in \{\mathtt{C}\}^+$, which is a major obstacle to terminal-comprising ambiguity. Note that, for alphabets with four or more letters and for every variable in every pattern, we even have at least two such letters $\mathtt{C}, \mathtt{D}$, $\mathtt{C} \neq \mathtt{D}$, so that we can conveniently map every variable onto a word over $\{\mathtt{C}, \mathtt{D}\}^+$, which, in the case of terminal-free patterns, allows for the construction of an unambiguous nonerasing morphism. Nevertheless, our examination of general patterns over alphabets of size 3 or 4 has yielded two morphically primitive examples, namely $\alpha_{\mathtt{abc}}$ and $\alpha_{\mathtt{abcd}}$ (cf. Definition 6.7), for which no relevant substitution σ can avoid terminal-comprising ambiguity, and this fact even results in the substantial ambiguity of σ. Hence, we have concluded in Theorem 6.6 that there exist morphically primitive patterns over alphabets Σ with three or four distinct letters with respect to which no substitution is moderately ambiguous. Consequently, the fundamental difference between common and terminal-preserving morphisms stated for binary alphabets also holds for $|\Sigma| \in \{3, 4\}$.

With regard to other alphabet sizes we have merely been able to note a minor yet counter-intuitive observation. It describes that, for every alphabet Σ and for $n := |\Sigma|$, we can give a particular *morphically imprimitive* pattern $\alpha_{\pi,n}$ such that every nonerasing substitution is substantially ambiguous for a *non-redundant* variable in $\alpha_{\pi,n}$ (cf. Theorem 6.10). This phenomenon is by no means trivial since, normally, the substantial ambiguity of nonerasing morphisms when applied to morphically imprimitive patterns is caused by the redundant variables (cf., e. g., the proof for Theorem 6.1).

Summarising Chapter 6, we have to state that our main results, namely Theorems 6.1, 6.2, 6.6 and 6.10 have provided some first substantial results on the ambiguity of terminal-preserving morphisms, but some of these insights are not comprehensive yet. We have demonstrated that, based on our results on the ambiguity of common morphisms, the (non-)avoidability of terminal-comprising ambiguity as introduced by Question 4.6 can be considered a key problem to the (non-)existence of moderately ambiguous and unambiguous substitutions for morphically primitive patterns. Therefore, our results on quasi-terminal-free patterns (apart from the substantial ambiguity of certain telltale candidates) largely correspond to those on terminal-free patterns, whereas, for $|\Sigma| \leq 4$, we have been able to construct morphically primitive general patterns in $(\mathbb{N} \cup \Sigma)^+$ with respect to which every nonerasing substitution is substantially ambiguous. Hence, in particular, we have proven that, at least with regard to these alphabets Σ, the morphic primitivity of a pattern $\alpha \in (\mathbb{N} \cup \Sigma)^+$ is not a characteristic condition for the existence of a moderately ambiguous or unambiguous substitution. Some final minor yet surprising observations on morphically imprimitive patterns have substantiated our expectations that any remarkable progress on patterns over larger alphabets and the ambiguity of the related substitutions requires a deep understanding of very complex phenomena.

In addition to their intrinsic interest, our combinatorial considerations have yielded several (partial) solutions to prominent and long-standing open problems for E-pattern languages. We summarise our respective results in the subsequent chapter.

7.2 Pattern languages

Our first result on E-pattern language has been rather technical. In Chapter 5.1.1, we have stated that the set of *succinct* terminal-free patterns exactly corresponds to the set of morphically primitive patterns (cf. Corollary 5.3). While this statement has been an immediate consequence of Theorem 3.6 and, hence, not an original insight of our thesis, it has turned our structural characterisation of morphically primitive terminal-free patterns (cf. Theorem 5.2) into an easily applicable characterisation of succinct terminal-free patterns.

In Chapter 5.2.1, we have dealt with inductive inference of terminal-free E-pattern languages. By Theorem 5.5, we have demonstrated that the full class of *terminal-free E-pattern languages* $\mathrm{ePAT}_{\mathrm{tf},\Sigma}$ is inferrable from positive data provided that the corresponding terminal alphabet Σ consists of at least three distinct letters. Thus, we have provided a comprehensive answer to Question 3.1. With reference to the negative result for binary alphabets presented by Reidenbach [76, 81], we have completed the characterisation of the learnability of $\mathrm{ePAT}_{\mathrm{tf},\Sigma}$ subject to the alphabet size, and we have found the first natural class of pattern languages to be known, the inferrability of which varies subject to the size of the terminal alphabet. Since the learnability of any indexed family of nonempty recursive languages depends on the existence of telltales (cf. Theorem 3.12), this result additionally demonstrates that, within the scope of pattern languages, the expressive power of words over a ternary alphabet exceeds that of words over binary alphabets. While this phenomenon from a coding theoretical point of view is rather counter-intuitive, it is not overly surprising when considering the impact of the existence of moderately ambiguous morphisms for morphically primitive patterns on the inferrability of classes of E-pattern languages as described by our main tool for the proof of Theorem 5.5, namely Theorem 5.4.

Due to the affirmative answer to Question 3.1, it has been necessary to discuss the inferrability of the full class of E-pattern languages separately, and we have done so in Chapter 6.1.1. Although our abovementioned results on the inferrability of $\mathrm{ePAT}_{\mathrm{tf},\Sigma}$ have strongly suggested a positive result on this problem, our combinatorial insights into the ambiguity of certain telltale candidates with respect to particular *quasi-terminal-free* patterns have led to the conclusion that, concerning $|\Sigma| \in \{3, 4\}$, $\mathrm{ePAT}_{\text{q-tf},\Sigma}$ and, hence, ePAT_{Σ} are not inferrable from positive data (cf. Theorem 6.5 and Corollary 6.3). Thus, we have provided a definite answer to the well-known and widely-discussed Question 3.2, but this answer is restricted to two selected alphabet sizes. Furthermore, as illustrated by Lemma 6.6, our corresponding argumentation has strengthened the insight that the existence of particular moderately ambiguous morphisms does not only provably influence the inferrability of the class of terminal-free E-pattern languages (cf. Theorem 5.4), but is also crucial for inductive inference of other classes of E-pattern languages.

Incorporating the results presented in this thesis, the subsequent table summarises the new state of knowledge on the learnability of selected classes of E-pattern languages (in this regard, $\mathrm{ePAT}_{\mathrm{nc,tf},\Sigma}$ stands for the class of terminal-free non-cross E-pattern languages over some alphabet Σ):

	Size of alphabet Σ					
Class	1	2	3	4	$5 \leq n < \infty$	∞
$\mathrm{ePAT}_{\mathrm{tf},\Sigma}$	Yes [1]	No [2]	Yes [3]			Yes [1]
$\mathrm{ePAT}_{\mathrm{nc,tf},\Sigma}$	Yes [1]	Yes [4]				Yes [1]
$\mathrm{ePAT}_{\text{q-tf},\Sigma}$	Undefined		No [5]		Open	Yes [1]
ePAT_{Σ}	Yes [1]	No [2]	No [5]		Open	Yes [1]

In addition to questions related to inductive inference, we have tackled a second algorithmic problem for E-pattern languages by means of insights related to the ambiguity of morphisms, namely the *equivalence problem* (see Chapter 6.2.1). In this context, we have demonstrated that if there exists a morphically primitive pattern α and a variable $j \in \mathrm{var}(\alpha)$ such that every substitution σ with $\sigma(j) \neq \varepsilon$ is substantially ambiguous for j then a pattern α' can be constructed such that $L_{\mathrm{E}}(\alpha') = L_{\mathrm{E}}(\alpha)$ and α' and α are semi-coincident (cf. Lemmata 6.10 and 6.11). Referring to this insight and considering $|\Sigma| = 3$, we have been effortlessly able to turn one of our main results on the ambiguity of terminal-preserving morpisms – namely the fact that every nonerasing substitution $\sigma : (\mathbb{N} \cup \Sigma)^* \longrightarrow \Sigma^*$ is substantially ambiguous with respect to α_{abc} – into the statement that there exist *semi-coincident* patterns over a ternary alphabet which generate the same language (cf. Theorem 6.7). We have explained that, in the light of previous literature, this is an unexpected result as it disproves the well-known conjecture by Ohlebusch and Ukkonen [69] on the equivalence problem for E-pattern languages (cf. Conjecture 6.1); furthermore, it spoils the hope that there might be a simple decision procedure solving this problem (provided that such a procedure exists). In a similar (yet more sophisticated) manner, we have intensively studied our second combinatorial main result of Chapter 6.2, according to which there are two variables i, j in α_{abcd} such

[1] Mitchell [59]
[2] Reidenbach [76, 81]
[3] present thesis, Chapter 5.2.1
[4] Reidenbach [76, 81] and present thesis, Chapter 5.4
[5] present thesis, Chapter 6.1.1

that, with respect to $|\Sigma| = 4$, every nonerasing substitution is substantially ambiguous for i or for j, and we have shown that this implies the existence of *incoincident* patterns generating the same language over alphabets of size 4 (cf. Lemma 6.12 and Theorem 6.8). Thus, we have disproven Conjecture 6.1 for E-pattern languages over three and over four letters, but we have left its correctness open for finite alphabets with at least five letters. Furthermore, we have not been able to use our results for proving or disproving the decidability of the equivalence problem for ePAT_Σ in case of $|\Sigma| \in \{3, 4\}$.

In addition to a number of supplementary conclusions on the equivalence problem for E-pattern languages presented in Chapter 6.2.1, we have demonstrated in Chapter 6.3.1 that, for every finite alphabet Σ, $n := |\Sigma|$, *all* injective substitutions $(\mathbb{N} \cup \Sigma)^* \longrightarrow \Sigma^*$ fail in mapping $\alpha_{\pi,n}$ (as introduced in Chapter 6.3) onto a word which can sufficiently precisely describe $L_{\text{E},\Sigma}(\alpha)$, whereas there exist non-injective substitutions succeeding in this task. We have not only explained why we consider this a paradoxical insight, but we have also demonstrated that, as a consequence thereof, a prominent proof technique considered several times in literature is not applicable to the equivalence problem for E-pattern languages.

Consequently, our studies on the ambiguity of morphisms have yielded major contributions to the research on inductive inference of E-pattern languages initiated in 1982, and they have provided new directions for the research on the equivalence problem for E-pattern languages, which has first been explicitly discussed in 1988. We expect, that any further significant progress on these two topics will again show strong connections to several questions related to the ambiguity of morphisms. Therefore we now shall briefly describe those questions on this topic which we expect to be crucial for future examinations.

7.3 Further research directions

For every alphabet Σ satisfying $|\Sigma| \geq 2$, Chapter 5 exhaustively answers the most fundamental questions on the *existence* of moderately ambiguous, weakly unambiguous and unambiguous *common nonerasing morphisms* $\sigma : \mathbb{N}^* \longrightarrow \Sigma^*$ with respect to arbitrary patterns. Therefore we consider it particularly interesting to seek for criteria describing a set of patterns, with respect to which an arbitrarily chosen *fixed* nonerasing morphism is moderately ambiguous or even unambiguous. Our examination of the morphism σ_c as conducted by Theorem 5.8 provides a first minor example for such an analysis. Consequently, we feel that Question 4.3 deserves further attention. In this regard, we think that it is crucial to find morphisms which map each pattern α in some preferably large set Π onto an unambiguous or moderately ambiguous word that is shorter than $\sigma_{\text{un},\alpha}(\alpha)$, so as to receive telltales for terminal-free E-pattern languages that are less complex than those introduced by Definition 5.3. As a side effect, we expect that a profound analysis of such a topic should yield a deeper understanding of the impact of morphic heterogeneity of the equivalence classes introduced in Definition 5.5 on the unambiguity of morphisms (note that we have merely used this property of $\sigma_{\text{un},\alpha}$ as a convenient tool for the proof of Lemma 5.10, but we have not studied it in detail).

Of course, one might alternatively wish to give a general procedure constructing, for *every* morphically primitive pattern α, an unambiguous or moderately ambiguous morphism σ_α satisfying $\sigma_\alpha(\alpha) \ll \sigma_{\text{un},\alpha}(\alpha)$. We feel, however, that such an examination might be extremly difficult and, due to Propositions 5.1 and 5.5, perhaps even necessarily unsuccessful.

In addition to this, it seems a worthwile idea to study the ambiguity of morphisms without our implicit goal of finding structure-preserving morphisms, which has made us investigating injective or at least nonerasing morphisms mapping a string over an infinite alphabet onto a string over a finite (and normally even binary) alphabet. Hence, one could focus on terminal-free patterns over finite sets of variables or, alternatively, on non-injective morphisms or even those morphisms mapping a variable onto the empty word. In this context, we wish to emphasise that the latter type is definitely necessary whenever morphisms are sought for that are unambiguous or moderately ambiguous with respect to morphically imprimitive patterns. Thus, such considerations could lead to a characterisation of those morphically imprimitive patterns with respect to which there is an unambiguous morphisms – a subject which would overcome the restrictions involved in our approach that cause the comprehensive negative result for morphically imprimitive patterns (cf. Theorem 5.1). Moreover, the studies on the ambiguity of non-injective morphisms could strive for an explanation, to which extent injectivity is a necessary property of a morphism which is moderately ambiguous or unambiguous with respect to each pattern in a given set.

Finally, we consider the weak unambiguity a wortwhile topic for further considerations since our corresponding notes are not as deep as our statements on moderate ambiguity and unambiguity. For instance, we feel that a characterisation of those patterns for which there exists an injective weakly unambiguous morphism is an interesting problem, as an answer to this question necessarily yields a partition of $\mathbb{N}^+$ that differs from the partition in morphically primitive and morphically imprimitive patterns (and, hence, in non-fixed points and fixed points) induced by the analogous problem for unambiguity.

With regard to *terminal-preserving morphisms*, our results have not been as comprehensive as those on common morphisms. Hence, though it is of course reasonable to adapt the questions introduced above to substitutions, we consider it more useful to tackle the problems explicitly left open in Chapter 6 first. Doubtlessly, among these topics, there is a truly outstanding question:

Open Problem 7.1. *Let Σ be an alphabet, $|\Sigma| \geq 5$. For every morphically primitive pattern $\alpha \in (\mathbb{N} \cup \Sigma)^*$, is there a nonerasing or even injective substitution $\sigma : (\mathbb{N} \cup \Sigma)^* \longrightarrow \Sigma^*$ such that σ is moderately ambiguous or unambiguous with respect to α?*

Hence, we ask whether or not, for larger alphabets, there exist patterns showing the same properties as $\alpha_{\mathtt{a}}$, $\alpha_{\mathtt{ab}}$, $\alpha_{\mathtt{abc}}$ and $\alpha_{\mathtt{abcd}}$ (cf. Proposition 6.5 and Theorem 6.6).

From a more technical point of view, we can give a question which might turn out to be almost equivalent to Open Problem 7.1, namely that for the avoidability of terminal-comprising ambiguity for larger alphabets:

Open Problem 7.2. *Let Σ be an alphabet, $|\Sigma| \geq 5$. For every morphically primitive pattern $\alpha \in (\mathbb{N} \cup \Sigma)^*$, is there an injective substitution $\sigma : (\mathbb{N} \cup \Sigma)^* \longrightarrow \Sigma^*$ such that σ is not terminal-comprisingly ambiguous with respect to α?*

Note that, for $\alpha_{\mathtt{abc}}$ and $\alpha_{\mathtt{abcd}}$, there exist nonerasing morphisms which are not terminal-comprisingly ambiguous. These morphisms, however, are not injective as they map at least two variables in the patterns onto words over the same unary alphabet.

If, for any alphabet Σ, Open Problem 7.2 has an answer in the affirmative then we can straightforward extend most of our insights into the ambiguity of common morphisms to substitutions – just as illustrated by our argumentation on quasi-terminal-free patterns, where, due to Lemma 6.2, the avoidability of terminal-comprising ambiguity is a trivial task.

Though Open Problem 7.1 and, in particular, Open Problem 7.2 do not look overly hard, we anticipate that it might be extraordinarily complicated to gain deeper insights into these questions. This feeling can be substantiated by the fact that our corresponding negative answer for alphabet sizes 3 and 4 is based on the very sophisticated example patterns $\alpha_{\mathtt{abc}}$ and $\alpha_{\mathtt{abcd}}$. Furthermore, we expect that any solution to these problems is closely related to the probably most prominent open questions on E-pattern languages: For any alphabet Σ, if both questions have an answer in the negative then this should imply that, first, ePAT_Σ is not inferrable from positive data and, second, Conjecture 6.1 is incorrect. On the other hand, if we can answer Open Problem 7.1 or Open Problem 7.2 in the affirmative then it seems that, among the resulting moderately ambiguous substitutions, there should be, first, suitable telltale candidates (cf. Definition 6.4) and, second, substitutions which (just as the morphisms τ introduced by Jiang et al. [33] and extensively studied in Chapter 6.3.1 of the present thesis) generate appropriate codings of the pattern language under consideration (cf. Definition 6.10). Consequently, a positive answer to the above open problems should yield both the inferrability of ePAT_Σ from positive data and the decidability of equivalence problem for that class. Unfortunately – as our corresponding considerations, which we have not documented in the present thesis, have led to conflicting suggestions – we feel unable to give any well-founded conjecture on the problems.

Concerning smaller alphabet sizes, such as $|\Sigma| = 4$, it is a canonical problem to seek for a characterisation of those patterns with respect to which there exist moderately ambiguous or unambiguous morphisms; in this regard, our results merely imply that the set of these patterns is a *proper* subset of that of the morphically primitive ones. While this problem from a combinatorial point of view is surely fascinating, it seems to be extremely difficult, and a corresponding answer might not be extendable to other alphabets. Therefore we consider it rather advisable to study the above open problems dealing with larger alphabets.

Turning our attention to those topics discussed in the present thesis that are concerned with *E-pattern languages*, we consider it an interesting problem to determine the time complexity of the test for succinctness of terminal-free patterns. Since Corollary 3.3 demonstrates that this problem is related to the NP-complete membership problem for E-pattern languages, one might expect the same complexity for the succinctness test. Contrary to this, we feel that an application of Theorem 5.2 – which, by Corollary 5.3, does not only characterise the morphically primitive patterns, but also the succinct ones – perhaps has a flavour of a polynomial time algorithm. Hence, we are uncertain about which result to expect on this question (that surely does not match with the main subjects of the present thesis).

Our studies on inductive inference of *terminal-free* E-pattern languages have led to a characterisation of those alphabets Σ for which $\text{ePAT}_{\text{tf},\Sigma}$ is inferrable from positive data (namely all but the binary ones, cf. Corollary 5.8). In spite of this seemingly exhaustive result, numerous challenging questions on this topic are still open, which is a consequence of both our non-algorithmic argumentation and the unsatisfactory complexity of the telltales introduced in the present thesis (cf. Definition 5.3). As the former problem has been discussed in Chapter 5.2.1 and the latter is closely related to the abovementioned question on the existence of preferably short moderately ambiguous morphic images for morphically primitive patterns, we do not consider it necessary to give our corresponding remarks in this chapter once more.

Within the present thesis, we have regarded the *quasi-terminal-free* patterns as a mere vehicle for describing the consequences of avoidability of terminal-comprising ambiguity. In addition, we have utilised that inclusion is decidable for every class of E-pattern languages generated by similar quasi-terminal-free patterns (cf. Theorem 6.4) – a fact which, due to Theorem 3.12, is helpful for learning-theoretical considerations. Consequently, we have studied the quasi-terminal-free pattern languages mainly because of technical reasons and not so much on account of any intrinsic interest. Nevertheless, as the inferrability of $\mathrm{ePAT}_{\text{q-tf},\Sigma}$ is open for $|\Sigma| \geq 5$, we consider it a worthwhile task to further investigate this class. We conjecture, however, that our negative result on alphabet sizes 3 and 4 (cf. Theorem 6.5) is not extendable to larger alphabets:

Conjecture 7.1. *Let Σ be an alphabet, $|\Sigma| \geq 5$. Then $\mathrm{ePAT}_{\text{q-tf},\Sigma}$ is inferrable from positive data.*

Thus, we expect that the quasi-terminal-free E-pattern languages cannot be used for solving the crucial question on inductive inference of E-pattern languages over larger alphabets determined by Open Problem 7.3 to be given below.

In fact, our result on inductive inference of the *full class* of E-pattern languages (cf. Corollary 6.3) has been an immediate consequence of the abovementioned insight into the inferrability of quasi-terminal-free E-pattern languages, and we have not presented any additional considerations solely dealing with the learnability of ePAT. Therefore, for every finite alphabet Σ satisfying $|\Sigma| \geq 5$, Question 3.2 is still unresolved:

Open Problem 7.3. *Let Σ be a finite alphabet, $|\Sigma| \geq 5$. Is ePAT_Σ inferrable from positive data?*

With regard to a potential approach solving Open Problem 7.3, we refer to the method briefly proposed below Open Problem 7.2.

Finally, Corollary 6.7 demonstrates that we can adapt the conjecture given by Ohlebusch and Ukkonen on the equivalence of E-pattern languages (cf. Conjecture 6.1), but, due to our considerations in Chapter 6.2.1, we prefer to consider the modified version an open problem rather than a conjecture:

Open Problem 7.4. *Let Σ be an alphabet, $|\Sigma| \geq 5$, and $\alpha_1, \alpha_2 \in \mathrm{Pat}_\Sigma$. Does $L_{\mathrm{E},\Sigma}(\alpha_1)$ equal $L_{\mathrm{E},\Sigma}(\alpha_2)$ if and only if α_1 and α_2 are morphically coincident?*

We expect that any significant progress on Open Problem 7.4 requires deep insights into the (non-)avoidability of terminal-comprising ambiguity for the alphabet sizes under consideration. Thus, our notes below Open Problem 7.2 can also be applied to Open Problem 7.4.

By our current state of knowledge, we unfortunately are unable to use Corollary 6.6 for drawing any conclusions on the (un-)decidability of the equivalence problem for ePAT_Σ, $|\Sigma| \geq 2$; hence, we cannot give any refinement of Open Problem 3.3 whatsoever.

This unpleasant fact of course suggests to further study the equivalence of E-pattern languages over alphabets with at most four distinct letters (where, at least, morphic coincidence now is known not to characterise patterns generating the same language), and we have given some corresponding notes and a minor conjecture in Chapter 6.2.1 (cf. Conjecture 6.2). In accordance with our above remarks on a potential characterisation of those morphically primitive patterns (such as $\alpha_{\mathtt{abcd}}$) with respect to which, for an appropriate alphabet, there is no moderately ambiguous morphism, we however fear that such an examination might raise extraordinarily intricate technical questions.

Bibliography

[1] A distinguished mathematician. Personal communication at DLT'04, 2004.

[2] J.-P. Allouche and J. Shallit. *Automatic Sequences.* Cambridge University Press, Cambridge, New York, 2003.

[3] D. Angluin. Finding patterns common to a set of strings. *Journal of Computer and System Sciences*, 21:46–62, 1980.

[4] D. Angluin. Inductive inference of formal languages from positive data. *Information and Control*, 45:117–135, 1980.

[5] D. Angluin and C. Smith. Inductive inference: Theory and methods. *Computing Surveys*, 15:237–269, 1983.

[6] H. Arimura, T. Shinohara, and S. Otsuki. Finding minimal generalizations for unions of pattern languages and its application to inductive inference from positive data. In *Proc. 11th Annual Symposium on Theoretical Aspects of Computer Science, STACS 1994*, volume 775 of *Lecture Notes in Computer Science*, pages 649–660, 1994.

[7] G.R. Baliga, J. Case, and S. Jain. The synthesis of language learners. *Information and Computation*, 152:16–43, 1999.

[8] Ja.M. Barzdin and R.V. Freivald. On the prediction of general recursive functions. *Soviet Mathematics Doklady*, 13:1224–1228, 1972.

[9] D.R. Bean, A. Ehrenfeucht, and G.F. McNulty. Avoidable patterns in strings of symbols. *Pacific Journal of Mathematics*, 85:261–294, 1979.

[10] J. Berstel and D. Perrin. *Theory of Codes.* Academic Press, Orlando, 1985.

[11] A. Brazma, I. Jonassen, I. Eidhammer, and D. Gilbert. Approaches to the automatic discovery of patterns in biosequences. *Journal of Computational Biology*, 5:279–305, 1998.

[12] C. Choffrut and J. Karhumäki. Combinatorics of words. In G. Rozenberg and A. Salomaa, editors, *Handbook of Formal Languages*, volume 1, chapter 6, pages 329–438. Springer, 1997.

[13] N. Chomsky. Three models for the description of a language. *IRE Transactions on Information Theory*, 2:113–124, 1956.

[14] K. Culik II. A purely homomorphic characterization of recursively enumerable sets. *Journal of the ACM*, 26:345–350, 1979.

[15] G. Dányi and Z. Fülöp. A note on the equivalence problem of E-patterns. *Information Processing Letters*, 57:125–128, 1996.

[16] J. Dassow, G. Păun, and A. Salomaa. Grammars based on patterns. *International Journal of Foundations of Computer Science*, 4:1–14, 1993.

[17] A. Ehrenfeucht and G. Rozenberg. Finding a homomorphism between two words is NP-complete. *Information Processing Letters*, 9:86–88, 1979.

[18] T. Erlebach, P. Rossmanith, H. Stadtherr, A. Steger, and T. Zeugmann. Learning one-variable pattern languages very efficiently on average, in parallel, and by asking queries. *Theoretical Computer Science*, 261:119–156, 2001.

[19] G. Filè. The relation of two patterns with comparable language. In *Proc. 5th Annual Symposium on Theoretical Aspects of Computer Science, STACS 1988*, volume 294 of *Lecture Notes in Computer Science*, pages 184–192, 1988.

[20] D.D. Freydenberger. Wortkombinatorische Überlegungen zu Patternsprachen. Diplomarbeit, Fachbereich Informatik, Technische Universität Kaiserslautern, 2006. In German.

[21] D.D. Freydenberger, D. Reidenbach, and J.C. Schneider. Unambiguous morphic images of strings. In *Proc. 9th Conference on Developments in Language Theory, DLT 2005*, volume 3572 of *Lecture Notes in Computer Science*, pages 248–259, 2005.

[22] D.D. Freydenberger, D. Reidenbach, and J.C. Schneider. Unambiguous morphic images of strings. *International Journal of Foundations of Computer Science*, 17:601–628, 2006.

[23] G. Georgescu. Infinite hierarchies of pattern languages. *Revue Roumaine de Mathématiques Pures et Appliquées*, 41:341–355, 1996.

[24] E.M. Gold. Language identification in the limit. *Information and Control*, 10:447–474, 1967.

[25] C. Gutiérrez. Solving equations in strings: On Makanin's algorithm. In *Proc. 3rd Latin American Theoretical Informatics Symposium, LATIN 1998*, volume 1380 of *Lecture Notes in Computer Science*, pages 358–373, 1998.

[26] D. Hamm and J. Shallit. Characterization of finite and one-sided infinite fixed points of morphisms on free monoids. Technical Report CS-99-17, Dep. of Computer Science, University of Waterloo, 1999. `http://www.cs.uwaterloo.ca/~shallit/papers.html`.

[27] T. Harju and J. Karhumäki. Morphisms. In G. Rozenberg and A. Salomaa, editors, *Handbook of Formal Languages*, volume 1, chapter 7, pages 439–510. Springer, 1997.

[28] T. Head. Fixed languages and the adult languages of 0L schemes. *International Journal of Computer Mathematics*, 10:103–107, 1981.

[29] F. Hechler. Induktive Inferenz von erweiterten Patternsprachen über binären Alphabeten. Staatsexamensarbeit, Fachbereich Informatik, Technische Universität Kaiserslautern, 2005. In German.

[30] J.E. Hopcroft and J.D. Ullman. *Introduction to Automata Theory, Languages, and Computation.* Addison-Wesley, Reading, MA, 1979.

[31] S. Jain, D. Osherson, J.S. Royer, and A. Sharma. *Systems That Learn.* MIT Press, Cambridge, MA, second edition, 1999.

[32] T. Jiang, E. Kinber, A. Salomaa, K. Salomaa, and S. Yu. Pattern languages with and without erasing. *International Journal of Computer Mathematics*, 50:147–163, 1994.

[33] T. Jiang, A. Salomaa, K. Salomaa, and S. Yu. Decision problems for patterns. *Journal of Computer and System Sciences*, 50:53–63, 1995.

[34] H. Jürgensen and S. Konstantinidis. Codes. In G. Rozenberg and A. Salomaa, editors, *Handbook of Formal Languages*, volume 1, chapter 8, pages 511–607. Springer, 1997.

[35] L. Kari, A. Mateescu, G. Păun, and A. Salomaa. Multi-pattern languages. *Theoretical Computer Science*, 141:253–268, 1995.

[36] L. Kari, G. Rozenberg, and A. Salomaa. L systems. In G. Rozenberg and A. Salomaa, editors, *Handbook of Formal Languages*, volume 1, chapter 5, pages 253–328. Springer, 1997.

[37] S.C. Kleene. Representation of events in nerve nets and finite automata. In C.E. Shannon and J. McCarthy, editors, *Automata Studies*, volume 34 of *Annals of Mathematics Studies*, pages 3–41. Princeton University Press, 1956.

[38] R. Klette and R. Wiehagen. Research in the theory of inductive inference by GDR mathematicians – a survey. *Information Sciences*, 22:149–169, 1980.

[39] D.E. Knuth, J.H. Morris, and V.R. Pratt. Fast pattern matching in strings. *SIAM Journal on Computing*, 6:323–350, 1977.

[40] S. Lange. *Algorithmic Learning of Recursive Languages.* Mensch & Buch, Berlin, 2000.

[41] S. Lange. Personal communication, 2004.

[42] S. Lange, J. Nessel, and R. Wiehagen. Learning recursive languages from good examples. *Annals of Mathematics and Artificial Intelligence*, 23:27–52, 1998.

[43] S. Lange and R. Wiehagen. Polynomial-time inference of arbitrary pattern languages. *New Generation Computing*, 8:361–370, 1991.

[44] S. Lange and S. Zilles. On the learnability of erasing pattern languages in the query model. In *Proc. 14th International Conference on Algorithmic Learning Theory, ALT 2003*, volume 2842 of *Lecture Notes in Artificial Intelligence*, pages 129–143, 2003.

[45] S. Lange and S. Zilles. Relations between Gold-style learning and query learning. *Information and Computation*, 203:211–237, 2005.

[46] F. Levé and G. Richomme. On a conjecture about finite fixed points of morphisms. *Theoretical Computer Science*, 339:103–128, 2005.

[47] M. Lipponen and G. Păun. Strongly prime PCP words. *Discrete Applied Mathematics*, 63:193–197, 1995.

[48] M. Lothaire. *Combinatorics on Words*. Addison-Wesley, Reading, MA, 1983.

[49] M. Lothaire. *Algebraic Combinatorics on Words*. Cambridge University Press, Cambridge, New York, 2002.

[50] G.S. Makanin. The problem of solvability of equations in a free semi-group. *Soviet Mathematics Doklady*, 18:330–334, 1977.

[51] A.A. Markov. On the impossibility of certain algorithms in the theory of associative systems. *Comptes Rendus de l'Académie des Sciences de l'URSS*, 55:583–586, 1947.

[52] A. Mateescu and A. Salomaa. On simplest possible solutions for post correspondence problems. *Acta Informatica*, 30:441–457, 1993.

[53] A. Mateescu and A. Salomaa. PCP-prime words and primality types. *RAIRO Informatique théoretique et Applications*, 27:57–70, 1993.

[54] A. Mateescu and A. Salomaa. Finite degrees of ambiguity in pattern languages. *RAIRO Informatique théoretique et Applications*, 28:233–253, 1994.

[55] A. Mateescu and A. Salomaa. Patterns. In G. Rozenberg and A. Salomaa, editors, *Handbook of Formal Languages*, volume 1, chapter 4.6, pages 230–242. Springer, 1997.

[56] A. Mateescu, A. Salomaa, K. Salomaa, and S. Yu. P, NP, and the Post Correspondence Problem. *Information and Computation*, 121:135–142, 1995.

[57] Y. Matiyasevich. Enumerable sets are Diophantine. *Soviet Mathematics Doklady*, 11:354–357, 1970.

[58] A. Mitchell, T. Scheffer, A. Sharma, and F. Stephan. The VC-dimension of subclasses of pattern languages. In *Proc. 10th International Conference on Algorithmic Learning Theory, ALT 1999*, volume 1720 of *Lecture Notes in Artificial Intelligence*, pages 93–105, 1999.

[59] A.R. Mitchell. Learnability of a subclass of extended pattern languages. In *Proc. 11th Annual Conference on Computational Learning Theory, COLT 1998*, pages 64–71, 1998.

[60] V. Mitrana, G. Păun, G. Rozenberg, and A. Salomaa. Pattern systems. *Theoretical Computer Science*, 154:183–201, 1996.

[61] M. Morse. Recurrent geodesics on a surface of negative curvature. *Transactions of the American Mathematical Society*, 22:84–100, 1921.

[62] M. Morse and G. Hedlund. Unending chess, symbolic dynamics and a problem in semigroups. *Duke Mathematical Journal*, 11:1–7, 1944.

[63] J. Myhill. Finite automata and the representation of events. Technical Report WADD TR-57-624, Wright Patterson AFB, Ohio, 1957.

[64] A. Nerode. Linear automata transformation. *Proceedings of the American Mathematical Society*, 9:541–544, 1958.

[65] J. Nessel and S. Lange. Learning erasing pattern languages with queries. *Theoretical Computer Science*, 348:41–57, 2005.

[66] A. Nijenhuis and H.S. Wilf. *Combinatorial Algorithms for Computers and Calculators.* Academic Press, New York, second edition, 1978.

[67] P.G. Odifreddi. *Classical Recursion Theory.* Elsevier, Amsterdam, 1989.

[68] P.G. Odifreddi. *Classical Recursion Theory*, volume 2. Elsevier, Amsterdam, 1999.

[69] E. Ohlebusch and E. Ukkonen. On the equivalence problem for E-pattern languages. *Theoretical Computer Science*, 186:231–248, 1997.

[70] H.C. Papadimitriou. *Computational Complexity.* Addison-Wesley, Reading, MA, 1995.

[71] J.-E. Pin. Syntactic semigroups. In G. Rozenberg and A. Salomaa, editors, *Handbook of Formal Languages*, volume 1, chapter 10, pages 679–746. Springer, 1997.

[72] E.L. Post. A variant of a recursively unsolvable problem. *Bulletin of the American Mathematical Society*, 52:264–268, 1946.

[73] E.L. Post. Recursive unsolvability of a problem of Thue. *Journal of Symbolic Logic*, 12:1–11, 1947.

[74] D. Reidenbach. Discontinuities in pattern inference. *Theoretical Computer Science.* To appear.

[75] D. Reidenbach. An examination of Ohlebusch and Ukkonen's conjecture on the equivalence problem for E-pattern languages. *Journal of Automata, Languages and Combinatorics.* To appear.

[76] D. Reidenbach. Induktive Inferenz von erweiterten Patternsprachen. Diplomarbeit, Fachbereich Informatik, Universität Kaiserslautern, 2002. In German.

[77] D. Reidenbach. A negative result on inductive inference of extended pattern languages. In *Proc. 13th International Conference on Algorithmic Learning Theory, ALT 2002*, volume 2533 of *Lecture Notes in Artificial Intelligence*, pages 308–320, 2002.

[78] D. Reidenbach. A discontinuity in pattern inference. In *Proc. 21st Annual Symposium on Theoretical Aspects of Computer Science, STACS 2004*, volume 2996 of *Lecture Notes in Computer Science*, pages 129–140, 2004.

[79] D. Reidenbach. On the equivalence problem for E-pattern languages over small alphabets. In *Proc. 8th International Conference on Developments in Language Theory, DLT 2004*, volume 3340 of *Lecture Notes in Computer Science*, pages 368–380, 2004.

[80] D. Reidenbach. On the learnability of E-pattern languages over small alphabets. In *Proc. 17th Annual Conference on Learning Theory, COLT 2004*, volume 3120 of *Lecture Notes in Artificial Intelligence*, pages 140–154, 2004.

[81] D. Reidenbach. A non-learnable class of E-pattern languages. *Theoretical Computer Science*, 350:91–102, 2006.

[82] R. Reischuk and T. Zeugmann. An average-case optimal one-variable pattern language learner. *Journal of Computer and System Sciences*, 60:302–335, 2000.

[83] H. Rogers. *Theory of Recursive Functions and Effective Computability.* MIT Press, Cambridge, MA, 1992. 3rd print.

[84] P. Rossmanith and T. Zeugmann. Stochastic finite learning of the pattern languages. *Machine Learning*, 44:67–91, 2001.

[85] G.-C. Rota. The number of partitions of a set. *American Mathematical Monthly*, 71:498–504, 1964.

[86] G. Rozenberg and A. Salomaa. *Handbook of Formal Languages*, volume 1. Springer, Berlin, 1997.

[87] A. Salomaa. *Formal Languages.* Academic Press, New York, London, 1973.

[88] A. Salomaa. Patterns. *Bulletin of the EATCS*, 54:194–206, 1994.

[89] A. Salomaa. Return to patterns. *Bulletin of the EATCS*, 55:144–157, 1995.

[90] K. Salomaa. Patterns. In C. Martin-Vide, V. Mitrana, and G. Păun, editors, *Formal Languages and Applications*, number 148 in Studies in Fuzziness and Soft Computing, pages 367–379. Springer, 2004.

[91] K. Salomaa. Personal communication, 2006.

[92] R.E. Schapire. Pattern languages are not learnable. In *Proc. 3rd Annual Workshop on Computational Learning Theory, COLT 1990*, pages 122–129, 1990.

[93] J.C. Schneider. Kombinatorische Eigenschaften von prägnanten und prolixen Pattern. Projektarbeit, Fachbereich Informatik, Technische Universität Kaiserslautern, 2006. In German.

[94] K.U. Schulz. Word unification and transformation of generalized equations. *Journal of Automated Reasoning*, 11:149–184, 1995.

[95] M.P. Schützenberger. On an application of semi groups methods to some problems in coding. *IRE Transactions on Information Theory*, 2:47–60, 1956.

[96] C.E. Shannon. A mathematical theory of communication. *The Bell Systems Technical Journal*, 27:379–423, 623–656, 1948.

[97] T. Shinohara. Polynomial time inference of extended regular pattern languages. In *Proc. RIMS Symposia on Software Science and Engineering, Kyoto*, volume 147 of *Lecture Notes in Computer Science*, pages 115–127, 1982.

[98] T. Shinohara. Polynomial time inference of pattern languages and its application. In *Proc. 7th IBM Symposium on Mathematical Foundations of Computer Science*, pages 191–209, 1982.

[99] T. Shinohara and S. Arikawa. Pattern inference. In K.P. Jantke and S. Lange, editors, *Algorithmic Learning for Knowledge-Based Systems, GOSLER Final Report*, volume 961 of *Lecture Notes in Artificial Intelligence*, pages 259–291. Springer, Berlin, 1995.

[100] T. Shinohara and H. Arimura. Inductive inference of unbounded unions of pattern languages from positive data. *Theoretical Computer Science*, 241:191–209, 2000.

[101] R. Solomonoff. A formal theory of inductive inference. *Information and Control*, 7:1–22,234–254, 1964.

[102] A. Thue. Über unendliche Zeichenreihen. *Kra. Vidensk. Selsk. Skrifter. I Mat. Nat. Kl.*, 7, 1906.

[103] A. Thue. Über die gegenseitige Lage gleicher Teile gewisser Zeichenreihen. *Kra. Vidensk. Selsk. Skrifter. I Mat. Nat. Kl.*, 1, 1912.

[104] A. Thue. Probleme über die Veränderungen von Zeichenreihen nach gegebenen Regeln. *Kra. Vidensk. Selsk. Skrifter. I Mat. Nat. Kl.*, 10, 1914.

[105] R. Wiehagen and T. Zeugmann. Ignoring data may be the only way to learn efficiently. *Journal of Experimental and Theoretical Artificial Intelligence*, 6:131–144, 1994.

[106] K. Wright. Identification of unions of languages drawn from an identifiable class. In *Proc. 2nd Annual Workshop on Computational Learning Theory, COLT 1989*, pages 328–333, 1989.

[107] S. Yu. Regular languages. In G. Rozenberg and A. Salomaa, editors, *Handbook of Formal Languages*, volume 1, chapter 2, pages 41–110. Springer, 1997.

[108] T. Zeugmann and S. Lange. A guided tour across the boundaries of learning recursive languages. In K.P. Jantke and S. Lange, editors, *Algorithmic Learning for Knowledge-Based Systems, GOSLER Final Report*, volume 961 of *Lecture Notes in Artificial Intelligence*, pages 190–258. Springer, Berlin, 1995.

Index

Curriculum Vitae

Name:	Daniel Reidenbach
Address:	Pfaffenbergstr. 61 67663 Kaiserslautern Germany
Date of Birth:	November 7, 1973
Place of Birth:	Trier, Germany
Marital Status:	married

1980-1984	Grundschule Farschweiler, Germany
1984-1993	Friedrich-Wilhelm-Gymnasium Trier, Germany
1993-1994	Civilian service
1994-2003	Studies at the University of Kaiserslautern, Germany, Diploma in Computer Science, 2003
since June 2003	Research associate at the Department of Computer Science, University of Kaiserslautern, Germany